MICROLENSING 2000
A NEW ERA OF MICROLENSING ASTROPHYSICS

COVER ILLUSTRATION:

The cover illustration is taken from a poster that was produced for the "Friends with the Universe" project which formed part of South Africa's first year of Science and Technology, YEAST, in 1998. It was created by Braam Botha, and the copyright rests with SAAO. The various scenes depict legends of southern Africa that relate to the heavens. Further details can be obtained at http://www.saao.ac.za.

A SERIES OF BOOKS ON RECENT DEVELOPMENTS IN ASTRONOMY AND ASTROPHYSICS

Publisher

THE ASTRONOMICAL SOCIETY OF THE PACIFIC
390 Ashton Avenue, San Francisco, California, USA 94112-1722
Phone: (415) 337-1100   E-Mail: catalog@astrosociety.org
Fax: (415) 337-5205   Web Site: www.astrosociety.org

ASP CONFERENCE SERIES - EDITORIAL STAFF
Managing Editor: D. H. McNamara   LaTeX-Computer Consultant: T. J. Mahoney
Associate Managing Editor: J. W. Moody   Production Manager: Enid L. Livingston

PO Box 24453, Room 211 - KMB, Brigham Young University, Provo, Utah, 84602-4463
Phone: (801) 378-2111   Fax: (801) 378-4049   E-Mail: pasp@byu.edu

ASP CONFERENCE SERIES PUBLICATION COMMITTEE:
Alexei V. Filippenko   Geoffrey Marcy
Ray Norris   Donald Terndrup
Frank X. Timmes   C. Megan Urry

A listing of all other ASP Conference Series Volumes and IAU Volumes
published by the ASP is cited at the back of this volume

ASTRONOMICAL SOCIETY OF THE PACIFIC
CONFERENCE SERIES

Volume 239

MICROLENSING 2000
A NEW ERA IN MICROLENSING ASTROPHYSICS

Proceedings of a Meeting held at
Cape Town, South Africa
21-25 February 2000

Edited by

John Menzies
South African Astronomical Observatory (SAAO), South Africa

and

Penny D. Sackett
*Kapteyn Astronomical Institute, University of Groningen
Groningen, The Netherlands*

© 2001 by Astronomical Society of the Pacific. All Rights Reserved

No part of the material protected by this copyright notice may be reproduced or utilized in any form or by any means – graphic, electronic, or mechanical including photocopying, taping, recording or by any information storage and retrieval system, without written permission from the publisher.

Library of Congress Cataloging in Publication Data
Main entry under title

Card Number: 2001093952
ISBN: 1-58381-076-5

ASP Conference Series - First Edition

Printed in United States of America by Sheridan Books, Chelsea, Michigan

Table of Contents

Preface ix

Conference participants xii

Part 1. Introduction to Microlensing

Theory of Microlensing 3
    A. Gould
Additional Information from Astrometric Gravitational Microlensing
    Observations 18
    C. Han
Microlensing Observations 27
    K. Cook
The EROS Microlensing Alert System 28
    J.-F. Glicenstein
Difference Imaging Analysis of the MOA Image Data Base 33
    I. Bond
Baryonic Dark Matter in Galaxies 37
    B. J. Carr
Are there enough MACHOs to fill the Galactic Halo? 54
    T. Lasserre
MACHO Results from 5.7 Years of LMC Observations 63
    K. Cook and the MACHO Collaboration
LMC Self-Lensing Constraints 64
    G. Gyuk, N. Dalal and K. Griest
What are MACHOs? Interpreting LMC Microlensing 73
    D. S. Graff
Old White Dwarfs as a Microlensing Population 82
    B. M. S. Hansen

Part 2. Extra-Solar Planets

From Low-Mass Star Binaries down to Planetary Systems 91
    S. Udry, M. Mayor, J. -L. Halbwachs and F. Arenou

PLANET Observations of Anomalous Microlensing Events 109
    J. Menzies, M. D. Albrow, J. -P. Beaulieu, J. A. R. Caldwell,
    D. L. DePoy, B. S. Gaudi, A. Gould, J. Greenhill, K. Hill, S. Kane,
    R. Martin, M. Dominik, R. M. Naber, R. W. Pogge, K. R. Pollard,
    P. D. Sackett, K. C. Sahu, P. Vermaak, R. Watson and A. Williams

Doppler Search for Extrasolar Planets 116
    D. A. Fischer

Photometric Characterization of Stars with Planets 130
    A. Giménez

Microlensing Constraints on the Frequency of Jupiter-Mass Planets . . . 135
    B. S. Gaudi, M. D. Albrow, J. H. An, J. -P. Beaulieu, J. A. R. Caldwell,
    D. L. DePoy, M. Dominik, A. Gould, J. Greenhill, K. Hill, S. Kane,
    R. Martin, J. Menzies, R. W. Pogge, K. Pollard, P. D. Sackett,
    K. C. Sahu, P. Vermaak, R. Watson and A. Williams

Planet Detection via Microlensing: Consequences of Resolving the Source 144
    P. Vermaak

Planetary Microlensing Signatures in the High Magnification Events
MACHO 98-BLG-35 and MACHO 99-LMC-2 153
    I. Bond

Abundance of Terrestrial Planets by Microlensing 160
    P. Yock

Discussion Session I: Mass and Orbital Characteristics of Binaries and
Planets 164
    J. Menzies and P. D. Sackett

Part 3. Stellar Astrophysics

Stellar Atmospheres 175
    P. H. Hauschildt, F. Allard, J. Aufdenberg, T. Barman, A. Schweitzer
    and E. Baron

Microlensing Extended Stellar Sources 195
    H. M. Bryce, M. A. Hendry and D. Valls-Gabaud

Source Reconstruction as an Inverse Problem 204
    N. Gray and I. J. Coleman

Microlensing and the Physics of Stellar Atmospheres 213
    P. D. Sackett

A Free-Floating Planet Population in the Galaxy? 223
 H. Zinnecker

Part 4. Galactic Structure and Constituents

Microlensing and Galactic Structure 231
 J. Binney

Galactic Bulge Microlensing Events with Clump Giants as Sources . . . 244
 P. Popowski, C. Alcock, R. A. Allsman, D. R. Alves, T. S. Axelrod,
 A. C. Becker, D. P. Bennett, K. H. Cook, A. J. Drake, K. C. Freeman,
 M. Geha, K. Griest, M. J. Lehner, S. L. Marshall, D. Minniti,
 C. A. Nelson, B. A. Peterson, M. R. Pratt, P. J. Quinn, C. W. Stubbs,
 W. Sutherland, A. B. Tomaney, T. Vandehei and D. Welch

A Galactic Bar to Beyond the Solar Circle and its Relevance for
Microlensing 254
 M. Feast and P. Whitelock

New EROS2 Results towards the Galactic Disk 261
 J.-F. Glicenstein and the EROS Collaboration

Evidence for Isolated Black Hole Stellar Remnants from Microlensing
Parallax Events 270
 D. Bennett

A New Component of the Galaxy as the Origin of the LMC Microlensing
Events 271
 E. Gates and G. Gyuk

The Local Group 280
 E. K. Grebel

The Haloes of the Milky Way and Andromeda Galaxies 299
 N. W. Evans and M. I. Wilkinson

Pixel Lensing towards M31 in Principle and in Practice 309
 E. Kerins for the POINT-AGAPE Collaboration

Microlensing in M31 - The MEGA Survey's Prospects and Initial Results 318
 A. Crotts, R. Uglesich, A. Gould, G. Gyuk, P. Sackett, K. Kuijken,
 W. Sutherland and L. Widrow

Things That Go Blip in the Night: Microlensing and the Stellar/Substellar
Mass Function 327
 I. N. Reid

Discussion Session II: Mass Functions/Budgets of Dark and Luminous
Objects 341
 J. Menzies and P. D. Sackett

Part 5. And Beyond ...

Cosmological Microlensing 351
    J. Wambsganss

A Radio-microlensing Caustic Crossing in B1600+434? 363
    L. V. E. Koopmans, A. G. de Bruyn, J. Wambsganss and C. D. Fassnacht

Telescopes of the Future 372
    R. Gilmozzi

Microlensing observations with the 4-m International Liquid Mirror
    Telescope 373
    J. C. Claeskens, C. Jean and J. Surdej

Monitoring Light Variations from Space with the OMC 378
    A. Giménez

Telescope Design, Instrumentation and Status of SALT 382
    R. S. Stobie, D. O'Donoghue, D. A. H. Buckley and K. Meiring

Galactic Exoplanet Survey Telescope (GEST): A Proposed Space-Based
    Microlensing Survey for Terrestrial Extra-Solar Planets 393
    D. Bennett and S. H. Rhie

Some Closing Comments 394
    M. Feast

Author index 396

Subject index 398

Preface

In the short span of one decade, microlensing has grown from a theoretical concept to an observational science that is now being used as a tool to study a wide range of astronomical problems. The usefulness and diversity of microlensing as a tool is due to the simplicity of the physics that describes it and to the physical size scales that it allows the astronomer to probe. Microlensing can contribute to fields as diverse as Galactic structure and dynamics, the stellar mass function, the structure and dynamics of the Local Group, abundance of stellar binaries, abundance of baryonic dark matter in our and other galaxies, stellar atmospheres, and extra-solar planets. Other meetings have been held to bring together the most active researchers in microlensing, with an emphasis on the data collected by the highly sucessful surveys that were begun in the 1990s to identify and characterize the phenomenon in our own Galaxy. This meeting centred on the symbiotic interface between microlensing and other areas of astronomy, focusing on the new information that microlensing astrophysics can provide.

The stage was set by invited talks that reviewed the most important aspects of microlensing theory and observations for non-experts in the field. Other invited reviews from (non-microlensing) experts working in Galactic structure, the Local group, stellar atmospheres, binary fractions, baryonic dark matter, stellar mass function, and extra-solar planets served to educate microlensing scientists about the current status and remaining problems in the areas of astronomy that microlensing is most likely to touch. Contributed talks then focussed on making the bridge between microlensing and a broader astrophysical perspective. Two lengthy discussion sessions were included.

This volume contains most of the papers presented at the meeting together with the questions and answers that followed them. In transcribing the contents of the tape recordings of the two long discussion sessions, we have attempted to be as faithful to the originals as possible but have allowed ourselves some editorial licence. A few authors were unable to meet the deadline for submission of papers, and their contributions are represented here by the abstracts they submitted prior to the conference. Authors were free to write from within their own orthographic traditions. We have edited the papers on that basis. The fact that the editors themselves were raised in different hemispheres, on opposite sides of the Pacific, led to inevitable disagreements over spelling and punctuation, but all disputes were settled amicably. There is consequently a certain inhomogeneity in this collection, but we hope this will not deter readers, nor detract from the usefulness of the contributions.

The meeting was held at the Arthur's Seat Hotel in Cape Town, South Africa from 21 to 25 February 2000. There were 50 delegates from 12 countries. Although this was a smaller number than originally expected, it contributed to making discussions more lively. Opening remarks were made by Professor Sibusiso Sibisi, Vice Chancellor for Research at the University of Cape Town.

Several social events served to introduce the participants to various aspects of life in South Africa. At a reception in the Two Oceans Aquarium guests dined as sharks and various varieties of fish circulated on the other side of a thick window and Professor Brian Warner spoke eloquently of early astronomy at the Cape. Excursions were organised to Robben Island and Cape Point, while the conference dinner took the form of a traditional 'braai' in the grounds of the SAAO, with a troupe of local African dancers providing entertainment.

The meeting was sponsored by the South African Astronomical Observatory, and we wish to thank the Director, Dr Bob Stobie, for financial support and for allowing us to use the Observatory grounds for the braai. A grant was also received from the National Research Foundation, the parent body of the SAAO, to help defray the costs of producing these Proceedings. The smooth running of the meeting was largely due to the efforts of the local organizers, in particular Isobel Bassett.

John Menzies (SAAO) and Penny D. Sackett (Kapteyn Institute)
May 14, 2001

SCIENTIFIC ORGANIZING COMMITTEE

James Binney	Oxford University	United Kingdom
Nathalie Palanque-Delabrouille	CEA, DAPNIA, Saclay	France
Rosanne DiStefano	Center for Astrophysics, Boston	United States
Andrew Gould	Ohio State University	United States
Kenneth Freeman	MSSSO	Australia
Shude Mao	Max Planck Institute, Garching	Germany
John Menzies	SAAO	South Africa
Bohdan Paczyński	Princeton University	United States
Penny Sackett (Chair)	Kapteyn Institute	Netherlands
Andrzej Udalski	University of Warsaw	Poland

LOCAL ORGANIZING COMMITTEE

Isobel Bassett, John Caldwell, John Menzies (Chair), Glenda Snowball, Barbara Bohle

Participant List

D Bennett, Physics Department, University of Notre Dame, 225 Nieuwland Science Hall, Notre Dame, IN 46556 ⟨bennett@nd.edu⟩

J Binney, Theoretical Physics, Keble Road, Oxford OX1 3NP, England ⟨binney@thphys.ox.ac.uk⟩

R Bradbury, Aeiveos Corporation, P.O. Box 31877, Seattle, WA 98103, USA ⟨bradbury@aeiveos.com⟩

I Bond, Mount John Observatory, P.O. Box 56, Lake Tekapo 8770, New Zealand ⟨bondi@scitec.auckland.ac.nz⟩

H Bryce, Glasgow University, University Avenue, Glasgow, G12 8QQ, Scotland ⟨helen@astro.gla.ac.uk⟩

J Caldwell, SAAO, P.O.Box 9, Observatory, 7935, South Africa ⟨jac@saao.ac.za⟩

B Carr, Astronomy Unit, Queen Mary & Westfield College, Mile End Road, London E1 4NS, England ⟨B.J.Carr@qmw.ac.uk⟩

J.-F. Claeskens, Avenue de Cointe, 5 4000 Liège, Belgium ⟨claesken@astro.ulg.ac.be⟩

K Cook, Lawrence Livermore National Laboratory, MS L-413, P.O.Box 808, Livermore, CA 94550, USA ⟨kcook@llnl.gov⟩

A Crotts, Columbia Astrophysics Lab, 550 W. 120th Street, Mail Code 5240, New York, NY 10027 ⟨arlin@astro.columbia.edu⟩

W Evans, Theoretical Physics, 1 Keble Rd, University of Oxford, Oxford, OX1 3NP, England ⟨nwe@thphys.ox.ac.uk⟩

M Feast, Astronomy Department, University of Cape Town, Rondebosch 7701, South Africa ⟨mwf@artemisia.ast.uct.ac.za⟩

R Ferlet, Institut d'Astrophysique de Paris, CNRS, 98 bis Bd. Arago, 75014 Paris, France ⟨ferlet@iap.fr⟩

D Fischer, 601 Campbell Hall, Dept. of Astronomy, UC Berkeley, Berkeley, CA 94720, USA ⟨fischer@serpens.berkeley.edu⟩

E Gates, Adler Planetarium & Astronomy Museum, 1300 S. Lake Shore Drive, Chicago, IL 60605, USA ⟨gates@oddjob.uchicago.edu⟩

B Gaudi, 140 W. 18th Avenue, Columbus, OH 43210, USA ⟨gaudi@astronomy.ohio-state.edu⟩

R Gilmozzi, Paranal Observatory ESO, Alonso de Cordova 3107, Vitacura, Casilla 19001, Santiago ⟨rgilmozz@eso.org⟩

A Gimenez, LAEFF, INTA-CSIC, Apartado 50.727, 28080 Madrid, Spain ⟨ag@laeff.esa.es⟩

J.-F. Glicenstein, DAPNIA/SPP, CEA-Saclay, 91191 Gif-sur-Yvette, France ⟨glicens@hep.saclay.cea.fr⟩

A Gould, 140 W 18th Avenue, Columbus, OH 43210-1173, USA ⟨gould@astronomy.ohio-state.edu⟩

D Graff, Department of Physics, 174 W. 18th Avenue, Columbus, OH 43210, USA ⟨graff.25@osu.edu⟩

N Gray, Glasgow University, Glasgow, G12 8QQ, Scotland ⟨norman@astro.gla.ac.uk⟩

E Grebel, University of Washington, Department of Astronomy, Box 351580, Seattle, WA 98195-1580, USA ⟨grebel@astro.washington.edu⟩

G Gyuk, Department of Physics, University of California, San Diego, 9500 Gilman Drive, La Jolla, CA 92093 ⟨gyuk@mizar.ucsd.edu⟩

C Han, Department of Astronomy and Space Science, Chungbuk National University, Chongju, Korea 361-763 ⟨cheongho@ast.chungbuk.ac.kr⟩

B Hansen, Department of Astrophysical Sciences, Peyton Hall, Princeton University, Princeton. NJ 08544, USA ⟨hansen@astro.princeton.edu⟩

P Hauschildt, Department of Physics and Astronomy, University of Georgia, Athens, GA 30602-2451, USA ⟨yeti@hobbes.physast.uga.edu⟩

M Hendry, Department of Physics and Astronomy, University of Glasgow, Glasgow G12 8QQ, Scotland ⟨martin@astro.gla.ac.uk⟩

M Hoffman, Physics Dept., University of the Free State, Bloemfontein 9300, South Africa ⟨Hoffmanm@fsk.nw.uovs.ac.za⟩

E Kerins, Theoretical Physics, Oxford University, 1 Keble Road, Oxford OX1 3NP, England ⟨e.kerins1@physics.ox.ac.uk⟩

D Kilkenny, SAAO, P.O.Box 9 Observatory, Observatory, 7935, South Africa ⟨dmk@saao.ac.za⟩

L Koopmans, Kapteyn Astronomical Institute, P.O.Box 800, NL-9700 AV Groningen, The Netherlands ⟨leon@astro.rug.nl⟩

T Lloyd Evans, SAAO, P.O.Box 9, Observatory, 7935, South Africa ⟨tle@saao.ac.za⟩

T Lasserre, CEA/DSM/DAPNIA/SPP, Centre d'Etude de Saclay, 91191 Gif-sur-Yvette Cedex, France ⟨lasserre@hep.saclay.cea.fr⟩

P Martinez, SAAO, P.O. Box 9, Observatory, 7935, South Africa ⟨peter@saao.ac.za⟩

M Mayor, Ch.des Maillettes 51, CH-1290 SAUVERNY, Switzerland ⟨michel.mayor@obs.unige.ch⟩

P Meintjies, Physics Dept., University of the Free State, Bloemfontein 9300, South Africa ⟨MeintP@fsk.nw.uovs.ac.za⟩

J Menzies, SAAO, P.O. Box 9, Observatory, 7935, South Africa ⟨jwm@saao.ac.za⟩

P Popowski, Lawrence Livermore National Lab, L-413, P.O. Box 808, Livermore, CA 94550, USA ⟨popowski@igpp.llnl.gov⟩

I Reid, Department of Physics and Astronomy, 209 South 33rd Street, Philadelphia, PA 19104, USA ⟨inr@morales.physics.upenn.edu⟩

P D Sackett, Kapteyn Institute, 9700 AV Groningen, Postbus 800, Groningen, The Netherlands ⟨psackett@astro.rug.nl⟩

J Skuljan, Department of Physics and Astronomy, University of Canterbury, Private Bag 4800, Christchurch 8020, New Zealand ⟨j.skuljan@phys.canterbury.ac.nz⟩

R Stobie, SAAO, P.O. Box 9, Observatory, 7935, South Africa ⟨rss@saao.ac.za⟩

M Takeuti, Astronomical Institute, Tohoku University, Aoba, Sendai 980-8578, Japan ⟨takeuti@astr.tohoku.ac.jp⟩

P Vermaak, SAAO, P.O. Box 9, Observatory, 7935, South Africa ⟨pierre@saao.ac.za⟩

A Vidal-Madjar, Institut d'Astrophysique de Paris, 98bis Boulevard Arago, 75014 Paris, France ⟨alfred@iap.fr⟩

J Wambsganss, Institut fuer Physik, Am Neuen Palais 10, 14463 Potsdam, Germany ⟨jkw@astro.physik.uni-potsdam.de⟩

B Warner, Astronomy Department, University of Cape Town, Rondebosch 7701, South Africa ⟨warner@physci.uct.ac.za⟩

P Whitelock, SAAO, P.O.Box 9, Observatory, 7935, South Africa ⟨paw@saao.ac.za⟩

P Yock, Tamaki Campus, University of Auckland, Private Bag 92019, Auckland, New Zealand ⟨p.yock@auckland.ac.nz⟩

H Zinnecker, An der Sternwarte 16, D-14482, Potsdam, Germany ⟨hzinnecker@aip.de⟩

Part 1: Introduction to Microlensing

Theory of Microlensing

Andrew Gould

Dept. of Astronomy, Ohio State University, 140 W. 18th Ave., Columbus, OH, 43210-1173

Abstract. I present a somewhat selective review of microlensing theory, covering five major areas: (1) the derivation of the basic formulae, (2) the relation between the observables and the fundamental physical parameters, (3) binaries, (4) astrometric microlensing, and (5) femtolensing.

1. Introduction

All of gravitational microlensing is reducible to a single equation, the Einstein formula for α, the deflection of light from a distant source passing by a lens of mass M at an impact parameter b,

$$\alpha = \frac{4GM}{bc^2}. \qquad (1)$$

This equation has been verified experimentally by Hipparcos to an accuracy of 0.3% (Froeschle, Mignard, & Arenou 1997). Despite its apparent simplicity, equation (1) generates an incredibly rich phenomenology. The aim of this review is to present the reader with a concise introduction to microlensing phenomena from the standpoint of theory. It is impossible to cover all aspects of microlensing in the space permitted. Several good reviews of microlensing can be consulted to obtain a deeper appreciation for various aspects of the subject (Paczyński 1996; Gould 1996; Roulet & Mollerach 1997; Mao 1999b).

By definition, microlensing is gravitational lensing where the images are too close to be separately resolved. The main microlensing effect that has been discussed in the literature (and the only one that has actually been observed) is *photometric microlensing*: the magnification of the source due to the convex nature of the lens (Einstein 1936; Refsdal 1964; Paczyński 1986; Griest 1991; Nemiroff 1991). However, there are two other effects that deserve attention from the standpoint of theory: *astrometric microlensing*, the motion of the centroid of the images relative to the source, and *femtolensing*, which refers to interference effects in microlensing.

2. Observables and Physical Parameters

2.1. The Lens Diagram

Consider a lens at d_l and a more distant source at d_s that are perfectly coaligned with the observer (Fig. 1). By axial symmetry, the source is imaged into a ring

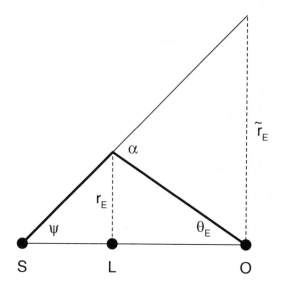

Figure 1. Basic geometry of lensing for the case when the source (S), lens (L), and observer (O), are aligned. The light is deflected by an angle α into a ring of radius θ_E. The resulting Einstein ring, r_E, projected onto the observer plane is $\tilde{r}_E$. Simple geometry relates the observables, θ_E and $\tilde{r}_E$, to the lens mass M and lens-source relative parallax π_{rel}. See eqs. (2) and (3).

(the "Einstein ring") whose angular radius is denoted θ_E. The impact parameter, r_E, is called the "physical Einstein ring", and its projection back onto the plane of the observer is called the "projected Einstein ring", $\tilde{r}_E$. As I will show below, θ_E and $\tilde{r}_E$ are two of the seven observables of the system. Using the small-angle approximation, they can be related to the physical parameters, M and the source-lens relative parallax, π_{rel} (Gould 2000b). First, $\alpha/\tilde{r}_E = \theta_E/r_E$, so using equation (1), one finds

$$\tilde{r}_E \theta_E = \alpha r_E = \frac{4GM}{c^2}. \qquad (2)$$

Secondly, using the exterior angle theorem, $\theta_E = \alpha - \psi = \tilde{r}_E/d_l - \tilde{r}_E/d_s$, so

$$\frac{\theta_E}{\tilde{r}_E} = \frac{\pi_{rel}}{AU}. \qquad (3)$$

These equations can easily be combined to yield,

$$\theta_E = \sqrt{\frac{4GM}{c^2}\frac{\pi_{rel}}{AU}}, \qquad \tilde{r}_E = \sqrt{\frac{4GM}{c^2}\frac{AU}{\pi_{rel}}}. \qquad (4)$$

Note that if θ_E and π_{rel} are measured in mas, $\tilde{r}_E$ is measured in AU, and M is measured in $M_\odot/8$, then all numerical factors and physical constants in these last three equations can be ignored.

2.2. The Lensing Event

If the lens is not perfectly aligned with the source, then the axial symmetry is broken and there are only two images. A similar use of the exterior-angle theorem then yields $\theta_I^2 - \theta_I \theta_s = \theta_E^2$, for the relation between the image positions, $\theta_{I,\pm}$ and the source position θ_s, relative to the lens. It is conventional to normalize the (vector) source position to θ_E, i.e., $\mathbf{u} \equiv \vec{\theta}_s/\theta_E$, so that the image positions are at

$$\vec{\theta}_{I,\pm} = \pm u_\pm \hat{\mathbf{u}}, \qquad u_\pm \equiv \frac{\sqrt{u^2+4} \pm u}{2}. \tag{5}$$

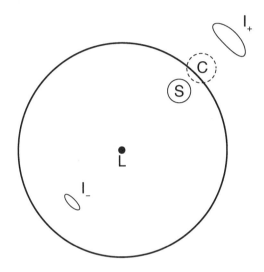

Figure 2. Source and images for a point lens. The bold line shows the Einstein ring centered on the lens (L). The two images (I_+ and I_-) of the source (S) are shown with their correct relative size and shape. The centroid of light (C) is shown at its correct position and with a size proportional to the magnification but, since the centroid is by definition unresolved, its shape is displayed arbitrarily as a circle.

By Liouville's theorem, surface brightness is conserved, so for a uniformly bright source, the magnification of each of the two images $A_\pm$ is given by the ratio of the area of the image to the area of the source. See Figure 2. In the limit of a point source, this ratio reduces to the Jacobian of the transformation,

$$A_\pm = \left| \frac{\partial \vec{\theta}_{I,\pm}}{\partial \vec{\theta}_s} \right| = \frac{u_\pm^2}{u_+^2 - u_-^2}, \qquad A = A_+ + A_- = \frac{u^2+2}{u\sqrt{u^2+4}} \tag{6}$$

Since A is a monotonic function of u, the signature of a *microlensing event* is that the source becomes brighter and then fainter as the line of sight to the source gets closer to and farther from the lens. If the source, lens, and observer are all in rectilinear motion, then by the Pythagorian theorem, $u(t) = [u_0^2 + (t-t_0)^2/t_E^2]^{1/2}$, where t_0 is the time of closest approach, $u_0 = u(t_0)$,

$$t_{\rm E} \equiv \frac{\theta_{\rm E}}{\mu_{\rm rel}}, \qquad \vec{\mu}_{\rm rel} \equiv \vec{\mu}_l - \vec{\mu}_s, \tag{7}$$

and $\vec{\mu}_{\rm rel}$ is the lens-source relative proper motion.

2.3. Observables

From a photometric microlensing event, one can then usually measure three parameters, t_0, u_0, and $t_{\rm E}$. Of these, the first two tell us nothing about the lens, and the third is related to M, $\pi_{\rm rel}$, and $\mu_{\rm rel}$ in a complicated way through equations (4) and (7). However, as mentioned above, there are actually seven (scalar) quantities that can in principle be observed in a microlensing event. These are: $t_{\rm E}$, $\theta_{\rm E}$, $\tilde{r}_{\rm E}$, ϕ (the angle of lens-source relative proper motion), and the source parallax and proper motion, π_s, and $\vec{\mu}_s$, which can be measured astrometrically after the event.

To date, there have been only a handful of measurements of $\theta_{\rm E}$, $\tilde{r}_{\rm E}$, and ϕ, no measurements of $\vec{\mu}_s$, and only estimates of π_s (although these are probably very good). However, all that could radically change in the next decade. I first discuss what it means that these quantities are "observable" and then briefly outline future prospects.

The angular Einstein ring $\theta_{\rm E}$ can be measured by scaling the event against some known "angular ruler" on the sky. The only such "ruler" to be used to date is the angular size of the source (Gould 1994a; Nemiroff & Wickramasinghe 1994; Witt & Mao 1994), which can be determined from the source flux and color, together with the empirical color/surface-brightness relation (van Belle 1999; Albrow et al. 2000a). So far, $\theta_{\rm E}$ has been measured for only 7 of the ~ 500 events detected to date (Alcock et al. 1997,2000; Albrow et al. 1999a,2000a; Afonso et al. 2000). Many other methods have been proposed (see Gould 1996 for a review; Han & Gould 1997), but none has been carried out.

The projected Einstein ring $\tilde{r}_{\rm E}$ can be measured by scaling the event against some known "physical ruler" in the plane of the observer. Three such "rulers" have been proposed, but the only one to be used to date is the Earth's orbit which induces a wobble in the light curve (Gould 1992b). Because this wobble depends on the Earth's motion being non-rectilinear, it is only significant if the event lasts more than a radian (i.e., $t_{\rm E} \gtrsim 58\,{\rm days}$), and such events are very rare. To date, only about a half dozen events have measured $\tilde{r}_{\rm E}$ (Alcock et al. 1995; Bennett et al. 1997; Mao 1999a). Another approach is to observe the event simultaneously from two locations ("parallax"). Since $\tilde{r}_{\rm E} \sim 5\,{\rm AU}$, it would be best if the second location were in solar orbit (Refsdal 1966; Gould 1994b).

The essential idea of a parallax satellite is illustrated in fig. 5 of Gould (1996). Basically, the Earth and the satellite each see a different microlensing event characterized respectively by $(t_{0,\oplus}, u_{0,\oplus}, t_{\rm E,\oplus})$, and $(t_{0,\rm sat}, u_{0,\rm sat}, t_{\rm E,sat})$. To zeroth order $t_{\rm E,sat} \simeq t_{\rm E,\oplus}$, but the differences in the other two components,

$$\Delta{\bf u} \equiv (\Delta t_0/t_{\rm E}, \Delta u_0) = (t_{0,\rm sat}/t_{\rm E}, u_{0,\rm sat}) - (t_{0,\oplus}/t_{\rm E}, u_{0,\oplus}), \tag{8}$$

give the displacement in the Einstein ring of the satellite relative to the Earth. That is,

$$\tilde{r}_{\rm E} = \frac{d_{\rm sat}}{\Delta u}, \tag{9}$$

where $d_{\rm sat}$ is the Earth-satellite separation (projected onto the plane perpendicular to the line of sight). There is actually a four-fold ambiguity in $\Delta u_0 = \pm(u_{0,\rm sat} \pm u_{0,\oplus})$ but this can be resolved, at least in principle, from the small difference in Einstein timescales which constrains Δu_0 by (Gould 1995),

$$w_\perp \Delta u_0 + w_\parallel \frac{\Delta t_{\rm E}}{t_{\rm E}} + d_{\rm sat}\frac{\Delta t_{\rm E}}{t_{\rm E}^2} = 0, \qquad (10)$$

where $w_\parallel$ and $w_\perp$ are the components of the Earth-satellite relative velocity respectively parallel and perpendicular to the Earth-satellite separation vector. Boutreux & Gould (1996) and Gaudi & Gould (1997) showed that the degeneracy could be broken with relatively modest satellite parameters. Unfortunately, no such satellite has been launched.

In a few specialized situations, it should be possible to obtain parallaxes using Earth-sized baselines (Hardy & Walker 1995; Holz & Wald 1996; Gould 1997; Gould & Andronov 1999; Honma 1999), but to date no such measurements have been made. All measurements of $\tilde{r}_{\rm E}$ simultaneously measure ϕ, and to date no other measurements of ϕ have been made.

2.4. Physical Parameters in Terms of Observables

It is convenient to replace $t_{\rm E}$ and ϕ together by $\vec{\mu}_{\rm E}$ whose direction is ϕ and whose magnitude is $\mu_{\rm E} \equiv t_{\rm E}^{-1}$, and to replace $\tilde{r}_{\rm E}$ by $\pi_{\rm E} \equiv {\rm AU}/\tilde{r}_{\rm E}$. Then equation (4) can be rewritten as,

$$\theta_{\rm E} = \sqrt{\kappa M \pi_{\rm rel}}, \qquad \pi_{\rm E} = \sqrt{\frac{\pi_{\rm rel}}{\kappa M}}, \qquad \kappa \equiv \frac{4G}{c^2\,{\rm AU}} \simeq 8.144\,\frac{\rm mas}{M_\odot}, \qquad (11)$$

and the physical parameters can be written in terms of the observables as,

$$M = \frac{\theta_{\rm E}}{\kappa \pi_{\rm E}}, \qquad \pi_l = \pi_{\rm E}\theta_{\rm E} + \pi_s, \qquad \vec{\mu}_l = \vec{\mu}_{\rm E}\theta_{\rm E} + \vec{\mu}_s. \qquad (12)$$

3. Astrometric Microlensing

While microlensing images cannot be resolved, the centroid of the images deviates from the source position by

$$\delta\vec{\theta} = \frac{A_+\vec{\theta}_{I,+} + A_-\vec{\theta}_{I,-}}{A_+ + A_-} - \vec{\theta}_s = \frac{\mathbf{u}}{u^2+2}\theta_{\rm E}, \qquad (13)$$

which reaches a maximum of $\theta_{\rm E}/\sqrt{8}$ when $u = \sqrt{2}$. See Figure 1. Since typically $\theta_{\rm E} \sim 300\mu{\rm as}$, $\delta\vec{\theta}$ is well within the range of detection of the Space Interferometry Mission (SIM) and perhaps ground-based interferometers as well. While it is not obvious from equation (13), if $\mathbf{u}(t)$ is rectilinear, then $\delta\vec{\theta}$ traces out an ellipse. The size of the ellipse gives $\theta_{\rm E}$ and its orientation gives ϕ. Hence, if microlensing were monitored astrometrically, it would be possible to *routinely* recover these two parameters (Boden, Shao & Van Buren 1998; Paczyński 1998). In fact, astrometric microlensing has a host of other potential applications including measurement of the lens brightness (Jeong, Han & Park 1999; Han &

Jeong 1999), removal of degeneracies due to blended light from unmicrolensed sources (Han & Kim 1999), and the detection and characterization of planetary (Safizadeh, Dalal & Griest 1999) and binary microlenses (Chang & Han 1999; Han, Chun & Chang 1999).

Because at late times the astrometric signature falls off as u^{-1} (eq. 13), compared to u^{-4} for the photometric signature (eq. 6), it could in principle be possible to measure $\tilde{r}_E$ astrometrically from the Earth's orbital motion, even for events with $t_E \ll 58$ days (Boden et al. 1998; Paczyński 1998). If practical, this would mean that all the observables listed in § 2.3 could be extracted from astrometric observations alone. Unfortunately, such measurements are not practical (Gould & Salim 1999).

Nevertheless, using SIM one can in fact extract all seven parameters, and therefore can accurately determine both the masses and distances of the lenses. SIM makes its astrometric measurements by centroiding the fringe, i.e., by *counting photons* as a function of fringe position. This means that SIM's *astrometric* measurements are simultaneously *photometric* measurements. Since SIM will be launched into solar orbit, it therefore can effectively act as a parallax satellite (Gould & Salim 1999). Moreover, SIM can break the degeneracy in Δu_0 in two ways: photometrically (according to eq. 10) and astrometrically (by measuring the angle ϕ associated with the astrometric ellipse, eq. 13).

An alternate approach to measuring $\tilde{r}_E$ would be to compare SIM and ground-based *astrometry* rather than *photometry* (Han & Kim 2000).

Another important application of astrometric micolensing is to measure the masses of nearby stars (Refsdal 1964; Paczyński 1995, 1998; Miralda-Escudé 1996). Equation (13) still effectively describes the astrometric deflection, but since typically $u \gg 1$, this equation can be more simply written as

$$\Delta\theta = \frac{\kappa M \pi_{\rm rel}}{\theta_{\rm rel}}, \qquad (14)$$

where $\theta_{\rm rel} \equiv |\vec{\theta}_l - \vec{\theta}_s|$. In this form, it is clear that the probability that a given lens will come close enough to a background source to allow a mass measurement of fixed precision in a fixed amount of time is

$$P \propto M \pi_l \mu_l N, \qquad (15)$$

where N is the density of background sources (Gould 2000a). Hence, the best place to look for such candidates is a proper-motion selected catalog near the Galactic plane. In fact, the selection of such candidates is a complex undertaking, but good progress is being made (Salim & Gould 2000).

4. Binary Lenses

Binary microlensing is one of the most active fields of theoretical investigation in microlensing today. In part this is due to the mathematical complexity of the subject and in part to the demands that are being placed on theory by new, very precise observations of binary events. Schneider & Weiss (1986) made a careful early study of binary lenses despite the fact that they never expected any to be detected (P. Schneider 1994, private communication), in order to learn about caustics in quasar macrolensing. Indeed caustics are the main new features of binaries relative to point lenses. These are closed curves in the source plane where a point source is infinitely magnified. The curves are composed of 3 or more concave segments that meet at cusps. Binary lenses can have 1, 2, or 3 closed caustic curves. If the two masses are separated by approximately an Einstein radius, then there is a single 6-cusp caustic. If they are separated by much more than an Einstein ring, then there are two 4-cusp caustics, one associated with each member of the binary. If the masses are much closer than an Einstein ring, there is a central 4-cusp caustic and two outlying 3-cusp caustics. Figure 3 shows two cases of the 6-cusp caustic, one close to breaking up into the two caustics characteristic of a wide binary and the other close to breaking up into the three caustics characteristic of a close binary. Witt (1990) developed a simple algorithm for finding these caustics. Multiple-lens systems can have even more complicated caustic structures (Rhie 1997; Gaudi, Naber, & Sackett 1998; Bozza 2000a,b).

4.1. Binary Lens Parameters

Recall that a point-lens light curve is defined by just three parameters, t_0, u_0, and t_E. These three generalize to the case of binaries as follows: u_0 is now the smallest separation of the source relative to the center of mass (alternatively geometric center) of the binary, t_0 is the time when $u = u_0$, and t_E is the timescale associated with the combined mass of the binary. At least three additional parameters are required to describe a binary lens: the angle α at which the source crosses the binary axis, the binary mass ratio q, and the projected separation of the binary in units of the Einstein ring. Several additional parameters may be required in particular cases. If caustic crossings are observed, then the infinite magnification of the caustic is smeared out by the finite size of the source, so one must specify $\rho_* = \theta_*/\theta_E$, where θ_* is the angular size of the source. If the observations of the crossing are sufficiently precise, one must specify one or more limb-darkening coefficients for each band of observation (Albrow et al. 1999a, 2000a; Afonso et al. 2000). Finally, it is possible that the binary's rotation is detectable in which case one or more parameters are required to describe that (Dominik 1998a; Albrow et al. 2000a). In addition, binary light curves often have data from several observatories in which case one needs two parameters (source flux and background flux) for each observatory.

4.2. Binary Lens Lightcurve Fitting

The problem of fitting binary-lens lightcurves is extremely complicated and is still very much under active investigation. There are actually three inter-related difficulties. First, as discussed in § 4.1, the parameter space is large. Secondly, it turns out the minima of χ^2 over this space are not well behaved. Thirdly, the evaluation of the magnification for a finite source straddling or near a caustic can be computationally time consuming. The combination of these three factors means that a brute force search for solutions can well fail or, worse yet, settle on a false minimum.

The ideas for tackling these problems go back to the detection of the first binary microlens OGLE-7 (Udalski et al. 1994). There are two major categories: ideas for improving efficiency in the evaluation of the magnification, and ideas

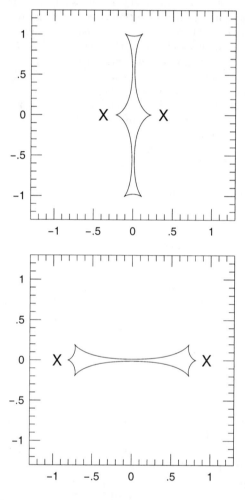

Figure 3. Two extreme examples of 6-cusp caustics generated by equal mass binaries. The tick marks are in units of Einstein radii. In each case, the crosses show the positions of the two components. The upper panel shows a relatively close binary with the components separated by $d = 0.76$ Einstein radii. For $d < 2^{-1/2}$ the caustic would break up into three caustics, a central 4-cusp caustic plus two outlying 3-cusp caustics. The lower panel shows a relatively wide binary with $d = 1.9$. For $d > 2$ the caustic would break up into two 4-cusp caustics.

for cutting down the region of parameter space that must be searched. For the first, various methods have been developed by Kayser & Schramm (1988), Bennett & Rhie (1996), Gould & Gaucherel (1997), Wambsganss (1997) and Dominik (1998b), although in fact all are still fairly time consuming. For the second, Mao & Di Stefano (1995) developed a densely sampled library of point-source binary microlensing events, each of which is characterized by cataloged "features" such as the number of maxima, heights of peaks, etc. They can then examine individual events, characterize their "features," and search their library for events that are consistent with these features. Di Stefano & Perna (1997) suggested that binary lenses could be fitted by decomposing the observed light curve into a linear combination of basis functions. The coefficients of these functions could then be compared to those fitted to a library of events in order to isolate viable regions of parameter space. This is essentially the same method as Mao & Di Stefano (1995), except that, rather than use gross features to identify similar light curves, one uses the coefficients of the polynomial expansion, which is more quantitative and presumably more robust.

Albrow et al. (1999b) developed a hybrid approach that both simplifies the search of parameter space and vastly reduces the computation time for individual light curves. It makes use of the fact that one of the very few things that is simple about a binary microlens is the behavior of its magnification very near to a caustic. A source inside a caustic will be imaged into five images, while outside the caustics it will be imaged into three images. Hence, at the caustic two images appear or disappear. These images are infinitely magnified. In the immediate neighborhood of a caustic (assuming one is not near a cusp), the magnification of the two new images diverges as $A_2 \propto (-\Delta u_\perp)^{-1/2}$, where $\Delta u_\perp$ is the perpendicular separation of the source from the caustic (in units of θ_E). On the other hand, the three other images are unaffected by the approach of the caustic, so $A_3 \sim$const. Hence, the total magnification is given by (Schneider & Weiss 1987)

$$A = A_2 + A_3 \simeq \left(-\frac{\Delta u_\perp}{u_r}\right)^{-1/2} \Theta(-\Delta u_\perp) + A_{cc}, \qquad (16)$$

where u_r is a constant that characterizes the approach to the caustic, A_{cc} is the magnification just outside the caustic crossing, and Θ is a step function. For a source of uniform brightness, or limb darkened in some specified way, one can therefore write a relatively simple expression for the total magnification as a function $\Delta u_\perp$ (Albrow et al. 1999b; Afonso et al. 2000).

4.3. Degeneracies

By fitting just the caustic-crossing data to a simple form based on equation (16), Albrow et al. (1999b) were able to reduce the search space from 7 to 5 dimensions and so effect a brute force search. This turned up a degeneracy between a wide-binary geometry and a close-binary geometry that both equally well accounted for the observed light curve. The Albrow et al. (1999b) data did not cover large parts of the light curve, but even when Afonso et al. (2000) combined these data with very extensive data sets from four other microlensing collaborations, the close/wide degeneracy survived. Simultaneously, Dominik (1999b) discovered a whole class of close/wide binary degeneracies whose roots lie deep in the lens equation itself.

This was unexpected. It was previously known that various observed light curves could be fit by several radically different binary geometries (Dominik 1999a). However, Han et al. (1999) showed that these geometries produced radically different astrometric deviations, so that the photometric degeneracy could be taken to be in some sense "accidental". Moreover, for all of the examples examined by Dominik (1999a) and Han et al. (1999), the data, while reasonably good, were substantially poorer in quality than those used by Afonso et al. (2000). Hence, it was plausible to hope that with better photometric data, the degeneracies could be resolved. However, since the Dominik (1999b) degeneracies are rooted in the lens equation, they may prove more intractible. For example, Gould & Han (2000) showed that, in contrast to the cases examined by Han et al. (1999), the wide and close models presented by Afonso et al. (2000) generated similar astrometric deviations, although they could be distinguished with sufficiently late-time data. On the other hand, Albrow et al. (2000b) showed that in at least one case this degeneracy is easily broken with photometric data alone.

5. Femtolensing

Femtolensing refers to interference effects in microlensing and derives its name from the very small angular scales that are usually required to produce such effects (Gould 1992a). To date, work on femtolensing has been almost entirely theoretical, although there has been at least one significant observational result.

As I mentioned in § 2.2, the magnification is given by the ratio of the area of the image to the area of the source, but this applies only to single images. Multiple images will in general interfere with one another. The reason that this can generally be ignored (and so one can write $A = A_+ + A_-$ as I did in eq. 6) is that usually the source is large enough that for some parts the two images interfere constructively, and for other parts they do so destructively, so that one can simply add intensities rather than amplitudes.

The validity of this approximation then depends on how rapidly the relative phase (proportional to the time delay) varies across the source. For a point lens and for $u \ll 1$, the time delay between the images is given by $\Delta t = 8GM/c^3(1+z_L)u$ where z_L is the redshift of the lens. The phase delay is therefore,

$$\Delta\phi = \frac{E\Delta t}{h} = 9.5 \times 10^9 \frac{M}{M_\odot} \frac{E}{\text{eV}} (1+z_L)u, \qquad (17)$$

where E is the energy of the photon. Hence, for ordinary Galactic microlensing observed in optical light, the interference effects are completely wiped out unless the source covers only a tiny part of the Einstein ring, $\rho_* \lesssim 10^{-10}$. Thus, applications of femtolensing require a search for unusual regions of parameter space. However, if such regions of parameter space can be identified, the effects can be dramatic: if the magnifications of the images are written in terms of $u_\pm$ (eq. 6), it is not difficult to show that,

$$\mathcal{A}_\pm = (A_+^{1/2} \pm A_-^{1/2})^2 = (1+4/u^2)^{\pm 1/2}, \qquad (18)$$

where $\mathcal{A}_\pm$ are the magnifications at constructive and destructive interference. Hence the ratio of the peaks to the troughs is $\mathcal{A}_+/\mathcal{A}_- = (1 + 4/u^2)$, which for typical $u \sim 0.5$ can be quite large. Hence, the interference pattern should manifest itself regardless of what intrinsic spectral features the source has.

Femtolensing was first discussed by Mandzhos (1981) and further work was done by Schneider & Schmidt-Burgk (1985) and Deguchi & Watson (1986). Peterson & Falk (1991) were the first to confront the problem of scales posed by equation (17). They considered radio sources (and so gained about 5 orders of magnitude relative to the optical) and advocated high signal-to-noise ratio observations that could detect the O(1/N) effects if the source subtends N fringes.

Gould (1992a) sought to overcome the huge factor in equation (17) by going to smaller M. He showed that asteroid-sized objects ($M \sim 10^{-16} M_\odot$) could femtolens gamma-ray bursts (i.e., $E \sim$ MeV). Plugging these numbers into equation (17) yields phase changes $\Delta\phi \sim 1$ over the entire Einstein ring. This is the only femtolensing suggestion that has ever been carried out in practice: Marani et al. (1999) searched BATSE and Ulysses data for femtolensing (as well as several other types of lensing) and used their results to place weak limits on cosmological lenses in this mass regime. Kolb & Tkachev (1996) showed that this type of observation can also be used to probe for axion clusters to determine if such axions make up the dark matter.

Ulmer & Goodman (1995) developed a formalism capable of going beyond the semi-classical approximation of previous investigations. They thereby found effects that are in principle observable even at optical wavelengths and solar masses, despite the seemingly pessimistic implications of equation (17). Jaroszynski, & Paczyński (1995) then showed that these results could have implications for microlensing observations of Huchra's Lens.

I close this review with a description of another potential application of femtolensing due to Gould & Gaudi (1997). The Einstein ring associated with a typical M star at ~ 20 pc is $\theta_E \sim 10$ mas, corresponding to $r_E \sim 0.2$ AU. Hence, the binary companions of such stars are likely to be several orders of magnitude farther away. The binary lens can then be approximated as a point lens perturbed by a weak shear due to the companion. This produces a Chang-Refsdal lens (Chang & Refsdal 1979, 1984), which is more thoroughly described by Schneider, Ehlers & Falco (1992). A source lying inside the caustic and near one of the cusps will produce four images near the Einstein ring, three highly magnified images on the same side of the Einstein ring and one moderately magnified image (which we will henceforth ignore) on the other. Depending on the details of the geometry, the three images can easily be magnified 10^6 times in one direction but are not much affected (indeed shrunk by a factor 2) in the other. Thus a $10^8 M_\odot$ black hole (Schwarzschild radius 2 AU) at the center of a quasar at 1 Gpc, could be magnified in one direction from 2 nas, to 2 mas, and so could easily be resolved with a space-based optical interferometer having a baseline of a few hundred meters.

The only problem, then, is how to get similar resolution in the other direction. Femtolensing provides the answer. Because the three images are not magnified in the direction perpendicular to the Einstein ring, they will each necessarily contain images not only of the black hole, but of considerable other junk along one-dimensional bands cutting through its neighborhood. The three bands from the three images will cut accross one another and will only intersect along a length approximately equal to the width of the bands, i.e., ~ 1 AU. If light from two of these images is brought together and dispersed in a spectrograph, then only the light from the intersecting region will give rise to interference fringes. The amplitude of these fringes will be set by the ratio of the length in the source plane over which the time delay between the images differs by 1 wavelength to the width of the bands. For typical parameters this could be a few percent. Since the $V \sim 22$ quasar will be magnified $\sim 10^6$ times to $V \sim 7$, this should not be difficult to detect.

There are a few difficulties that must be overcome to make this work in practice. First, the nearest position from which a dwarf star appears aligned with a quasar lies about 40 AU from the Sun. So while the huge "primary" of this "femtolens telescope" (the dwarf star) comes for free, getting the "secondary optics" aligned with the primary is a big job. Second, unlike other space missions that deliver payloads to 40 AU (e.g., the Voyagers), this package must be stopped at 40 AU so that it remains aligned with the dwarf star and quasar. Third, there are station keeping problems because the telescope will gradually fall out of alignment due to the Sun's gravity and will have to be realigned about every 10 hours. However, what sort of theorist would recoil from a few engineering challenges?

6. Conclusion

When microlensing experiments began in the early 1990s, few participants expected that there was much room for theoretical development at all. The physical effect (eq. 1) was completely understood, and the basic equations of microlensing had all been worked out. A decade later, microlensing theory has shown itself to be a very dynamic field. The original problems turned out to be much richer than expected, while new observations and new developments in instrumention have raised new problems. Thus, microlensing theory promises to remain vibrant.

Acknowledgements: This work was supported by grant AST 97-27520 from the NSF.

References

Afonso, C. et al. 2000, ApJ, 532, 000 (astro-ph/9907247)
Albrow, M. et al. 1999a, ApJ, 522, 1011
Albrow, M. et al. 1999b, ApJ, 522, 1022
Albrow, M. et al. 2000a, ApJ, 534, 000 (astro-ph/9910307)
Albrow, M. et al. 2000b, in preparation
Alcock, C., et al. 1995 ApJ, 454, L125
Alcock, C., et al. 1997 ApJ, 491, 436
Alcock, C., et al. 2000 ApJ, submitted (astro-ph/9907369)
Bennett, D.P. et al. 1997, BAAS, 191, 8303
Bennett, D.P. & Rhie S.H. 1996, ApJ, 472, 660

Boden, A.F., Shao, M., & Van Buren, D. 1998 ApJ, 502, 538
Bozza, V. 2000a, A&A, in press (astro-ph/9910535)
Bozza, V. 2000b, A&A, submitted (astro-ph/000287)
Boutreux, T., & Gould, A. 1996, ApJ, 462, 705
Chang, K. & Han, C. 1999, ApJ, 525, 434
Chang, K. & Refsdal, S. 1979, Nature, 282, 561
Chang, K. & Refsdal, S. 1984, A&A, 130, 157
Deguchi, S., & Watson, W.D. 1986, ApJ, 307, 30
Di Stefano, R., & Perna, R. 1997, ApJ, 488, 55
Dominik, M. 1998a, A&A, 329, 361
Dominik, M. 1998b, A&A, 333, L79
Dominik, M. 1999a, A&A, 341, 943
Dominik, M. 1999b, A&A, 349, 108
Einstein, A. 1936, Science, 84, 506
Froeschle, M., Mignard, F., & Arenou, F. 1997, Proceedings of the ESA Symposium "Hipparcos – Venice '97", p. 49, ESA SP-402
Gaudi, B.S., & Gould, A. 1997, 477, 152
Gaudi, B.S., Naber, R.M., & Sackett, P.D. 1998, 502, L33
Gould, A. 1992a, ApJ, 386, L5
Gould, A. 1992b, ApJ, 392, 442
Gould, A. 1994a, ApJ, 421, L71
Gould, A. 1994b, ApJ, 421, L75
Gould, A. 1995, ApJ, 441, L21
Gould, A. 1996, PASP, 108, 465
Gould, A. 1997, ApJ, 480, 188
Gould, A. 2000a, ApJ, submitted (astro-ph/9909455)
Gould, A. 2000b, ApJ, submitted (astro-ph/0001421)
Gould, A. & Andronov, N. 1999, ApJ, 516, 236
Gould, A. & Gaucherel, C. 1997, ApJ, 477, 580
Gould, A. & Gaudi, B.S. 1997, ApJ, 486, 687
Gould, A. & Han, C. 2000, ApJ, 538, 000 (astro-ph/00011930)
Gould, A., & Salim, S. 1999, ApJ, 524, 794
Griest, K. 1991, ApJ, 366, 412
Han, C., Chun, M.S., & Chang, K. 1999, ApJ, 526, 405
Han, C., & Gould, A. 1997, ApJ, 480, 196
Han, C., & Jeong, Y. 1999, MNRAS, 309, 404
Han, C., & Kim, H.-I. 2000, ApJ, 528, 687
Han, C., & Kim, T.-W. 1999, MNRAS, 305, 795
Hardy, S.J., & Walker, M.A. 1995, MNRAS, 276, L79
Holz, D.E., & Wald, R.M. 1996, ApJ, 471, 64
Honma, M. 1999, ApJ, 517, L35

Jaroszynski, M., & Paczyński, B. 1995, ApJ, 455, 433
Jeong, Y., Han, C., & Park, S.H. 1999, MNRAS, 304, 845
Kayser, R., & Schramm, T. 1998, A&A, 191, 39
Kolb, E.W., & Tkachev, I.I. 1996, ApJ, 460, L25
Mandzhos, A.V. 1981, Soviet Ast., 7, L213
Mao, S. 1999a, A&A, 350, L19
Mao, S. 1999b, in Gravitational Lensing, Recent Progress and Future Goals, Boston University, July 1999, ed. T.G. Brainerd and C.S. Kochanek, in press (astro-ph/9909302)
Mao, S., & Di Stefano, R. 1995, ApJ, 440 22
Marani, G.F., Nemiroff, R.J., Norris, J.P., Hurley, K., & Bonnell, J.T., 1999, ApJ, 512, L13
Miralda-Escudé, J. 1996, ApJ, 470, L113
Nemiroff, R.J. 1991, A&A, 247, 73
Nemiroff, R.J. & Wickramasinghe, W.A.D.T. 1994, ApJ, 424, L21
Paczyński, B. 1986, ApJ, 304, 1
Paczyński, B. 1995, Acta Astron., 45, 345
Paczyński, B. 1996, ARA&A, 34, 419
Paczyński, B. 1998, ApJ, 494, L23
Peterson, J.B., & Falk, T. 1991, ApJ, 374, L5
Refsdal, S. 1964, MNRAS, 128, 295
Refsdal, S. 1966, MNRAS, 134, 315
Rhie, S.H. 1997, ApJ, 484, 63
Roulet, E., & Mollerach, S. 1997, Physics Reports, 279, 68
Safizadeh, N., Dalal, N., & Griest, K. 1999, ApJ, 522, 512
Salim, S., & Gould, A., 2000 ApJ, 539, 000
Schneider, P., Ehlers, J., & Falco, E.E. 1992, Gravitational Lenses (Berlin: Springer)
Schneider, P., & Schmidt-Burgk, J. 1985, A&A, 148, 369
Schneider, P., & Weiss, A. 1986, A&A, 164, 237
Schneider, P., & Weiss, A. 1987, A&A, 171, 49
Udalski, A., Szymański, M., Pietrzyński, G., Kubiak, M., Woźniak, P., & Żebruń, K. 1998, Acta Astron., 48, 431
Ulmer, A., & Goodman, J. 1995, ApJ442, 67
van Belle, G.T. 1999, PASP, 111, 1515
Wambsganss, J. 1997, MNRAS, 284, 172
Witt, H.J. 1990, A&A, 236, 311
Witt, H.J. & Mao, S. 1994, ApJ, 429, 66

Discussion

Bennett: I'd like to point out that the degeneracies between the different binary fits only exist because the source flux is a free parameter. This can generally be determined with high resolution imaging, although follow-up spectroscopy may be required to resolve bright lens cases.

Gould: First, I agree that this degeneracy can be at least partially resolved with high-resolution photometry in this case. However, if the images reveal only a small contribution from a distinguishable source, it could still be that there is other "blended" light from the lens. Secondly, I meant to raise these binary degeneracies mainly to illustrate an area of ongoing research. At this point only MACHO 98-SMC-1 has really been studied exhaustively, so we don't know the full extent of binary degeneracies.

Crotts: What can you do when observing with the microlensing satellite to distinguish between the case of the lens passing between Earth and the satellite in the observer plane versus the one where the satellite and Earth are on the same side? How do you break this difference?

Gould: 1. There are (subtle) differences in event timescale and peak times in the two cases.
2. The astrometric lensing displacement vectors differ somewhat.
3. The lens geometries would be very different (one in the foreground, one in the background) and one might be dismissed based on a priori considerations.
How well any of these works is a technical question for each event, but there are ways to address this degeneracy.

Evans: When is the Chang-Refsdahl lens a good approximation to the binary lens equation?

Gould: Two conditions are necessary:
1. The mass ratio must be large (eg planetary lensing)
2. The separation between the star and the planet cannot be too close to one Einstein Radius.

Binney: There are plenty of low-mass stars out there and quite a few quasars. Is the probability of a transit entirely negligible?

Gould: Unfortunately, one must not only put the satellite on line with the dwarf-QSO line of sight, but also stop it there, and stopping the Earth would be a bit tougher.

Microlensing 2000: A New Era of Microlensing Astrophysics
ASP Conference Series, Vol. 239, 2000
John Menzies and Penny D. Sackett, eds.

Additional Information from Astrometric Gravitational Microlensing Observations

Cheongho Han

Department of Astronomy & Space Science, Chungbuk National University, Chongju, Korea 361-763, cheongho@ast.chungbuk.ac.kr

Abstract. Astrometric observations of microlensing events were originally proposed to determine the lens proper motion with which the physical parameters of lenses can be better constrained. In this proceeding, we demonstrate that besides this original usage astrometric microlensing observations can be additionally used in obtaining various important information about lenses. First, we demonstrate that the lens brightness can be determined with astrometric observations, enabling one to know whether the event is caused by a bright star or a dark lens. Second, we show that with additional information from astrometric observations one can resolve the ambiguity of the photometric binary lens fit and thus uniquely determine the binary lens parameters. Finally, we propose two astrometric methods that can resolve the degeneracy in the photometric lens parallax determination. Since one can measure both the proper motion and the parallax by these methods, the lens parameters of individual events can be uniquely determined.

1. Introduction

When a source is microlensed, it is split into two images. The flux sum of the individual images is greater than that of the unlensed source, and thus the source becomes brighter during the event. The sizes and brightnesses of the individual images change as the lens-source separation changes due to their transverse motion. Therefore, microlensing events can be detected either by photometrically monitoring the source brightness changes or by directly imaging the two separated images. However, with current instruments direct imaging of the separate images is impossible due to the low precision of the instrument. As a result, current microlensing observations have been and are being carried out only by using the photometric method (Aubourg et al. 1993; Alcock et al. 1993; Udalski et al. 1993; Alard & Guibert 1997).

However, if an event is astrometrically observed by using the planned high precision interferometers from space-based platform, e.g. the *Space Interferometry Mission* (SIM), and ground-based interferometers soon available on 8-10 m class telescope, e.g. the Keck and the Very Large Telescope, one can measure the shift of the source star image centroid caused by microlensing. The astrometric centroid shift vector as measured with respect to the position of the unlensed

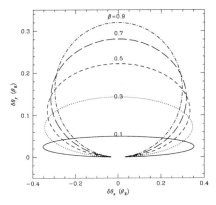

Figure 1. Trajectory of the source star image centroid shifts for several example microlensing events with various lens-source impact parameters. The directions of x- and y-axis are parallel and normal to the lens-source proper motion, respectively.

source is related to the lens parameters by

$$\delta\boldsymbol{\theta}_{c,0} = \frac{\theta_E}{u^2 + 2}\left[\left(\frac{t - t_0}{t_E}\right)\hat{\mathbf{x}} + \beta\hat{\mathbf{y}}\right], \tag{1}$$

where θ_E is the angular Einstein ring radius, t_E is the time required for the source to cross θ_E (Einstein time scale), t_0 is the time of the closest lens-source approach (and thus the time of maximum amplification), and β is the separation at this moment (i.e. impact parameter). The notation $\mathbf{x}$ and $\mathbf{y}$ represent the unit vectors with their directions that are parallel and normal to the lens-source proper motion. If one defines $x = \delta\theta_{c,x}$ and $y = \delta\theta_{c,y} - \beta\theta_E/2(\beta^2 + 2)$, equation (1) becomes

$$x^2 + \frac{y^2}{q^2} = a^2, \tag{2}$$

where

$$a = \frac{\theta_E}{2(\beta^2 + 2)^{1/2}}, \tag{3}$$

and

$$q = \frac{\beta}{(\beta^2 + 2)^{1/2}}. \tag{4}$$

Therefore, during the event the image centroid traces out an elliptical trajectory (hereafter astrometric ellipse) with a semi-major axis a and an axis ratio q. In Figure 1, we present astrometric ellipses for several example microlensing events with various lens-source impact parameters.

The greatest importance of astrometric microlensing observation is that one can determine θ_E from the observed astrometric ellipse (Høg, Novikov & Polarev 1995; Walker 1995; Paczyński 1998; Boden, Shao, & Van Buren 1998). This is because the size (i.e. semi-major axis) of the astrometric ellipse is directly proportional to θ_E [see equation (3)]. Once θ_E is determined, the lens proper motion

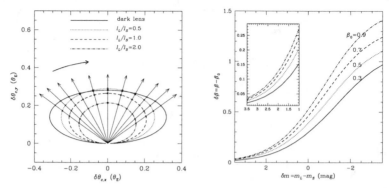

Figure 2. Left part: Astrometric behavior of bright lens events. The ellipses represents the trajectories of image centroid shift for events caused by bright lenses with various lens/source flux ratio ℓ_L/ℓ_S. The straight arrows represent the position vectors of the image centroid at different times during events. The curved arrow represents the direction of centroid motion with the progress of time. All example events have the same impact parameter of $\beta = 0.5$. Right part: Difference between the impact parameters determined from the centroid shift trajectory, β, and the angular speed curve, β_0, as a function of lens-source brightness difference.

is determined by $\mu = \theta_E/t_E$ with the independently determined t_E from the light curve. While the photometrically determine t_E depends on the three physical lens parameters of the lens mass (M), location (D_{ol}), and the transverse motion (v), the astrometrically determined μ depends only on the two parameters of M and D_{ol}. Therefore, by measuring μ one can significantly better constrain the nature of lens matter. However, we note that to completely resolve the lens parameter degeneracy, it is still required to additionally determine the lens parallax (see more details in § 4).

In this proceeding, we demonstrate that besides this original usage astrometric microlensing observations can be additionally used in obtaining various important information about lenses. First, we show that the lens brightness can be determined with astrometric observations, enabling one to know whether the event is caused by a bright star or a dark lens (§ 2). Second, we demonstrate that additional astrometric microlensing observations allow one to uniquely determine the binary lens parameters (§ 3). Finally, we propose two astrometric methods that can uniquely determine the lens parallax, with which one can completely break the lens parameter degeneracy along with the measured proper motion (§ 4).

2. Lens Brightness Determination

If an event is caused by a bright lens (i.e. star), the centroid shift trajectory is distorted by the brightness of the lens. The lens brightness affects the centroid shift trajectory in two ways. First, the lens makes the image centroid further shifted toward the lens. Second, the bright lens makes the reference of centroid shift measurements changed from the position of the unlensed source to the one between the source and the lens. By considering these two effects of the bright lens, the resulting centroid shift vector is computed by

$$\delta\boldsymbol{\theta}_c = \frac{1 + f_L + f_L[(u^2 + 2) - u(u^2 + 4)^{1/2}]}{(1 + f_L)[1 + f_L u(u^2 + 4)^{1/2}/(u^2 + 2)]} \delta\boldsymbol{\theta}_{c,0}, \quad (5)$$

where $f_L = \ell_L/\ell_S$ is the flux ratio between the lens and the source star.

In the left part of Figure 2, we present the trajectories of astrometric centroid shifts for events caused by bright lenses with various brightnesses. From the figure, one finds that the trajectories are also ellipses like those of dark lens events. As seen from the view of identifying bright lenses from the distorted trajectories, this is a bad news because one cannot identify whether the event is caused by a bright lens or not just from the shape of the trajectory (Jeong, Han, & Park 1999). One also finds that as the lens becomes brighter, the observed astrometric ellipse becomes rounder and smaller (measured by a).

Fortunately, identification of bright lenses is possible by measuring the angular speed (ω) of the image centroid motion around the unlensed source position (Han & Jeong 1999). In the left part of Figure 2, we present the position vectors (arrows with straight lines) of the image centroid at different times during events for both the dark and bright lens events. From the figure, one finds that the position vector at a given moment directs towards the same direction for both the dark and bright lens events, implying that ω is the same regardless of the lens brightness. The angular speed does not depend on the lens brightness because lens always lies on the line connecting the two images, and thus additional shift caused by the bright lens occurs along this line. As a result, although the amount of shift changes due to lens brightness, the direction of the shift does not change. Since the angular speed is related to the lensing parameters of $(\beta, t_{\rm E}, t_0)$ by

$$\omega(t) = \frac{\beta t_{\rm E}}{(t - t_0)^2 + \beta^2 t_{\rm E}^2}, \quad (6)$$

these parameters can be determined from the observed angular speed curve. Note that these parameters are the same regardless of the lens brightness because the angular speed curve is not affected by the lens brightness. By contrast, the the impact parameter determined from the shape of the observed centroid shift trajectory [see equation (4)] differs from the true value because the shape of the astrometric ellipse for a bright lens event differs from that of a dark lens event. Then, if an event is caused by a bright lens, the impact parameter determined from the observed centroid shift trajectory, β, will differ from that determined from the angular speed curve, β_0. Therefore, by comparing β and β_0, one can identify the bright lens and measure its flux. In the right part of Figure 2, we present $\delta\beta = \beta - \beta_0$ as a function of lens-source brightness difference in magnitudes.

3. Resolving Binary-Lens Parameter Degeneracy

If an event is caused by a binary lens, the resulting light curve deviates from that of a single lens event. Detecting binary lens events is important because one

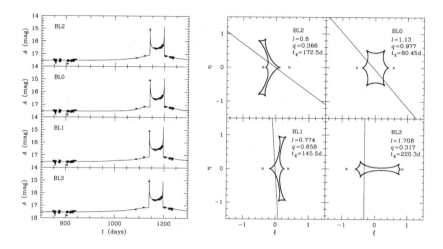

Figure 3. The ambiguity of the photometric binary lens fit. Left part: The observed light curve (dots with error bars) of the binary lens event OGLE-7 and several example model fits. Right part: The binary lens geometries for the individual model fits and their parameters. The 'x' marks represent the lens locations and the caustics and the source trajectories are marked by solid curves and straight lines.

can determine important binary lens parameters such as the mass ratio (q) and separation (ℓ). These parameters are determined by fitting model light curves to the observed one.

For many cases of binary lens events, however, it is difficult to uniquely determine the solutions of the binary lens parameters with the photometrically constructed light curves alone. In Figure 3, we illustrate this ambiguity of the photometric binary lens fit. In the left part of the figure, we present the observed light curve of the binary lens event OGLE-7 (dots with error bars, Udalski et al. 1994) and several example model light curves (solid curves) obtained from the fit to the observed light curve by Dominik (1999). In the right part of the figure, we also present the binary lens system geometries for the individual solutions responsible for the model light curves. The binary lens parameters (ℓ, q, and t_E) for each model are marked in the corresponding panel. From the figure, one finds that despite the dramatic differences in the binary lens parameters between different solutions, the resulting light curves fit the observed light curve very well, implying that unique determination of lens parameters is difficult by using the photometrically measured light curve alone.

However, the binary lens parameter degeneracy can be lifted if events are additionally observed astrometrically (Han, Chun, & Chang 1999). To demonstrate this, we compute the expected astrometric centroid shifts of the binary lens events resulting from the lens parameter solutions responsible for the model light curves in Figure 3, and the resulting trajectories are presented in Figure 4. From the figure, one finds that the trajectories are dramatically different from each other. Therefore, with the additional information provided by the as-

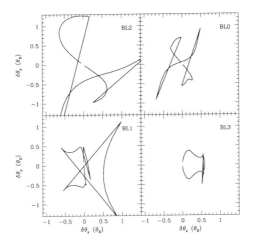

Figure 4. The astrometric centroid shifts of the binary lens events resulting from the lens parameter solutions responsible for the model light curves in Figure 3.

trometric microlensing observations, one can completely resolve the ambiguity of the photometric binary lens fit and thus uniquely determine the binary lens parameters.

4. Resolving Parallax Degeneracy

Although the astrometrically determined μ better constrains the lens parameters than the photometrically determined t_E does, μ still results from the combination of the lens mass and location, and thus the lens parameter degeneracy is not completely resolved. To completely resolve the lens parameter degeneracy, it is required to determine the transverse velocity projected on the source plane ($\tilde{v}$, hereafter simply projected speed). Determination of $\tilde{v}$ is possible by measuring the lens parallax Δu from photometric observations of the source light variations from two different locations, one from the ground and the other from a heliocentric satellite (Gould 1994, 1995). Once both μ and $\tilde{v}$ are determined, the individual lens parameters are determined by

$$M = \left(\frac{c^2}{4G}\right) t_E^2 \tilde{v}\mu, \quad (7)$$

$$D_{ol} = \frac{D_{os}}{\mu D_{os}/\tilde{v} + 1}, \quad (8)$$

$$v = \frac{1}{[\tilde{v}^{-1} + (\mu D_{os})^{-1}]^{-1}}. \quad (9)$$

However, the elegant idea of lens parallax measurements proposed to resolve the lens parameter degeneracy suffers from its own degeneracy. The parallax degeneracy is illustrated in Figure 5. In the upper panel, we present two light

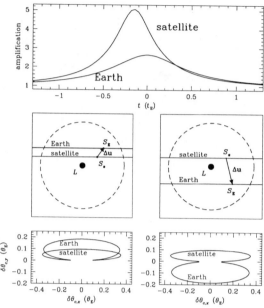

Figure 5. Degeneracy in the determination of parallax and astrometric resolution of the degeneracy. The two curves in the upper panel represent the light curves of an event observed from both the ground and the satellite. The two possible lens system geometries that can produce the light curves in the upper panel are presented in the middle panels. The dotted circle represents the Einstein ring. The vector $\Delta \boldsymbol{u}$ connecting the two points (S_S for the source seen from the satellite and S_E seen from the Earth) on the individual trajectories (straight lines) represent the displacements of the source positions (i.e. parallaxes) observed at a given time. In the lower panels, we present two sets of astrometric ellipses as seen from both the Earth and the satellite corresponding to the source trajectories in the middle panels.

curves of an event observed from the Earth and the satellite. Presented in the middle panels are the two possible lens system geometries that can produce the light curves in the upper panel. From the figure, one finds that depending on whether the source trajectories as seen from the Earth and the satellite are on the same or opposite sides with respect to the lens, there can be two possible values of Δu.

Astrometric microlensing observations are useful in resolving the degeneracy in parallax determination (Han & Kim 2000). The first method is provided by simultaneous *astrometric* observations from the ground and the satellite instead of *photometric* observations. Note that the SIM will have a heliocentric orbit, and thus can be used for this purpose. In the lower panels of Figure 5, we present two sets of the astrometric ellipses as seen from the Earth and the satellite that are expected from the corresponding two sets of source trajectories in the middle panels. One finds that these two sets of astrometric ellipses have opposite orientations, and thus can be easily distinguished from one another.

The parallax degeneracy can also be resolved if the event is astrometrically observed on one site and photometrically observed on a different site, instead of simultaneous astrometric observations from the ground and the satellite. This is possible because astrometric observations allow one to determine the lens-source proper motion $\boldsymbol{\mu}$. Then with the known Earth-satellite separation vector, which is parallel to $\Delta \boldsymbol{u}$, one can uniquely determine the angle between $\boldsymbol{\mu}$ and $\Delta \boldsymbol{u}$, allowing one to select the right solution of u.

5. Summary

In this proceeding, we demonstrate various additional usages of astrometric microlensing observations besides the original usage of the lens proper motion determination. These are summarized as follows.

1. By astrometrically observing a microlensing event caused by a bright lens, one can identify the bright lens and measure its flux.

2. With additional information from astrometric observations one can resolve the ambiguity of the photometric binary lens fit and thus uniquely determine the binary lens parameters.

3. With application of the two proposed astrometric methods, the degeneracy in the photometric lens parallax determination can be resolved, allowing one to completely break the lens parameter degeneracy along with simultaneously determined lens proper motion.

References

Alard, C., & Guibert, J. 1997, A&A, 326, 1
Alcock, C., et al. 1993, Nature, 365, 621
Aubourg, E., et al. 1993, Nature, 365, 623
Boden, A. F., Shao, M., & Van Buren, D. 1998, ApJ, 502, 538
Dominik, M. 1999, A&A, 341, 943
Gould, A. 1994, ApJ, 421, L75
Gould, A. 1995, ApJ, 441, L21
Han, C., & Kim, H.-I. 2000, ApJ, 528, 687
Han, C., & Jeong, Y. 1999, MNRAS, 309, 404
Han, C., Chun, M.-S., & Chang, K. 1999, ApJ, 526, 405
Høg, E., Novikov, I. D., & Polarev, A. G. 1995, A&A, 294, 287
Jeong, Y., Han, C., & Park, S.-H. 1999, ApJ, 511, 569
Paczyński, B. 1998, ApJ, 494, L23
Udalski, A., et al. 1993, Acta Astron., 43, 289
Udalski, A., et al. 1994, ApJ, 436, L103
Walker, M. A. 1995, ApJ, 453, 37

Discussion

Bennett: Your "angular speed curve" refers to an observation at $t = \infty$ to define the center of your coordinate system. Do you know how long you must wait to make these measurements in realistic cases?

Han: I think there will be two ways to find the center of the coordinates. First, by fitting astrometric ellipses, one can determine the origin. Secondly, one should wait for a long time (several years) so that the centroid returns to the origin. This will, however, take a long time.

Zinnecker: In practice, VLT or Keck astrometric observations will be more difficult than theorists take for granted! It involves tricky things such as dual-beam laser guide star systems and two large telescopes with fully functional adaptive optics before interferometric beam combination.

Bennett: The Keck Interferometer is supposed to require a $K \leq 13$ guide star within 30 arcsec. A typical microlensed source in the Galactic Bulge will have such a star within 12 arcsec.

Han: I agree. The microlensing fields are crowded regions, so one can easily find nearby bright reference stars.

Vermaak: Is it fair to compare VLT astrometry to current photometry with 1- to 2-m class telescopes?

Han: In the Bulge, photometric S/N is determined mostly by crowding, so a large telescope should not improve the precision.

Gaudi: I doubt that would be the case for difference imaging.

Binney: What limits do atmospheric effects place on the precision of your measurements?

Sackett: For photometric lensing, one monitors all stars in the same field simultaneously. By referencing the source photometry to these other stars, most variations in transparency can be removed. The image subtraction technique can reduce the effects of seeing.

Han: I have little expertise in observation. I appreciate Dr Sackett's answer.

Glicenstein: Is orbital motion of a binary important?

Han: It is unimportant.

Gaudi: Orbital motion does not affect the detectability of close ($b \lesssim 0.2$) binaries, because the regions of significant deviation are roughly circular in shape and rotation causes these circular regions to rotate.

Evans: Why is there no intermediate regime where the effect of orbital motion is detectable in the plane of astrometric observables?

Han: I know the answer to that question. Except for some extreme cases, orbital motion in both photometry and astrometry, is not important. I investigated the influence of orbital motion several months ago, so the equation is not at the top of my head, but I remember the result. Orbital motion is not important.

Microlensing Observations

K. Cook

Lawrence Livermore National Laboratory, MS L-413, P.O.Box 808, Livermore, CA 94550, USA

Abstract. Microlensing surveys are generally mounted to probe the lens population. The observation of microlensing is challenging because the probability of a background source being microlensed at a given time is generally less than 1-in-one-million for most lines of sight. In spite of this, a number of groups have mounted successful microlensing surveys and the study of microlensing has moved beyond its mere detection into more detailed characterization. I will present an overview of some of the techniques and technologies which have been developed to better detect and characterize microlensing events. Because microlensing is typically detected in crowded fields, significant developments in crowded field photometry and techniques to account for confusion and blending have been and continue to be made. Accurate photometry and/or astrometry has the potential to determine the most interesting parameters of microlensing such as the lens mass and distance, particularly for certain types of exotic microlensing. Microlensing studies can also probe for planets around the lensing objects, allow detailed study of the source star and even test gravitational theory. I will detail efforts toward these ends and plans for the future study of microlensing.

Microlensing 2000: A New Era of Microlensing Astrophysics
ASP Conference Series, Vol. 239, 2000
John Menzies and Penny D. Sackett, eds.

The EROS Microlensing Alert System

J. -F. Glicenstein (EROS collaboration)

DSM/DAPNIA/SPP, CEA Saclay, F-91191 Gif-sur-Yvette

Abstract. The EROS2 microlensing alert ("trigger") system is described. The system provides alerts on microlensing events towards both the Magellanic Clouds and the Galactic Center. More than fifteen million stars are being monitored on these fields. The system has been fully operational since August 1999. A few microlensing candidates have already been detected towards the Galactic Center.

1. Introduction

EROS is searching for compact halo objects (the so-called MACHOs) with microlensing. Microlensing candidates have been detected by the EROS (Lasserre et al. 2000) and MACHO (Alcock et al. 2000) collaborations towards the LMC and the SMC. These microlensing events could be caused by MACHOs, but also by ordinary stars from the Clouds ("self-lensing" events). Two microlensing candidates have been detected towards the SMC, in 1997 (SMC-97-1) and 1998 (SMC-98-1). Event SMC-98-1, a lensing by a binary, is a "self-lensing" event (Afonso et al 2000). The absence of "parallax" (the effect of the Earth motion is not detected) is one of the hints in favor of SMC-97-1 not being a lensing by a MACHO. It has been suggested by Gould (Gould 1998) that the "parallaxes" (or limits thereon) of LMC events could be measured if the photometry was accurate at the 1% level. This is possible only if the microlensing candidate is detected in real-time. The main purpose of the EROS trigger system is to provide alerts on the Magellanic Clouds which could lead to parallax measurements.

EROS is also taking data towards the Galactic Disk and the Galactic Bulge. The goal is to separate the contributions of disk and bulge lenses to the optical depth towards the Galactic Center by studying the microlensing optical depth as a function of galactic longitude. Only bright stars are being monitored in the Galactic Bulge (a large fraction of which are clump giants), as advocated in (Gould 1995). A fraction of the Galactic Center fields are also monitored by the EROS trigger system and can be used to search for exotic lenses or planets.

2. Online photometry

The EROS data acquisition system has been described in (Glicenstein & Gros 1997). Images are taken in two wide optical passbands V_{EROS} ($\lambda_{center} \simeq$ 560 nm), and R_{EROS} ($\lambda_{center} \simeq$ 760 nm). After being read out on VME crates, the images are sent through a local network connection to two DEC/OSF1 computers. After correcting for electronic offsets and flat fields, the images are stored on local disks and copied to DLT tapes.

The online photometry runs automatically every day on two dedicated DEC/OSF1 computers. The photometry code is PEIDA++ (Ansari 1996). It makes use of a catalog of stars which was created from a template image. The typical photometric accuracy for a clump giant is less than 5%. Fifty photometry points are stored for every star to help taking the trigger decision.

3. Alert system

A measurement of the flux $\Phi(t)$ of a star at a given time t deviates from the baseline flux Φ_b because of photometric errors. The typical deviation from Φ_b is estimated by

$$\sigma_\Phi = \sqrt{<\Phi(t)^2 - \Phi_b^2>} \qquad (1)$$

where the average is taken over ~ 50 measurements. Both Φ_b and σ_Φ have been determined offline with data taken during the first year of EROS2.

The trigger program runs automatically every day. It analyzes the output of the photometry program and looks for a set of trigger conditions. A typical trigger condition is that

$$\Phi(t) - \Phi_b \geq N\sigma_\Phi \qquad (2)$$

for more than M consecutive measurements of $\Phi(t)$ in both colors. More sophisticated trigger conditions, involving moving averages or filtering have also been tried (Glicenstein & Gaucherel 1996, Gaucherel 1997, Mansoux 1997). The actual values of N,M are field-dependent and obtained by simultaneously maximizing the efficiency for finding a microlensing event and minimizing the number of false alarms. Typical values are $N = M = 4$. The typical trigger efficiency is $\sim 50\%$ for a simulated microlens with an Einstein radius crossing time $t_E = 30$ days and a magnification $A = 2$, with less than two spurious alerts per million stars per day.

With the present setting, the EROS trigger system can measure the photometry (in the two passbands) and check the trigger conditions for twenty million stars per day.

The light curves of the alerts are scanned by eye. If the alert is not an obvious background (long period variable or systematic effect of the photometry), a microlensing fit is performed. The position of the source star in the colour-magnitude diagram is also checked. Figure 1 shows a trigger which was rejected after close examination. It fits quite well a microlensing light curve in the V_{EROS} passband, but not so well in the R_{EROS} passband. The source star has a red colour and is located in an underpopulated part of the colour-magnitude diagram. This type of background is hard to reject when data in only one color is available.

If the trigger passes all the checks, it is accepted as a microlensing alert. The coordinates of the source star, a finding chart, and data points are written to the EROS microlensing alerts page:

http://www-dapnia.cea.fr/Spp/Experiences/EROS/alertes.html

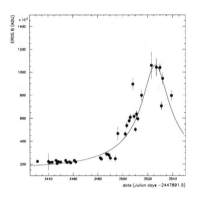 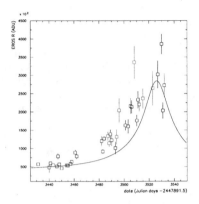

Figure 1. A source star monitored by EROS towards the Galactic Bulge which triggered the EROS alert system, but didn't end up as a microlensing candidate. The V_{EROS} measurements (left) are plotted with dots and the R_{EROS} measurements (right) are plotted with squares. A microlensing fit with $t_E = 64.5$ days and $u_o = 0.16$ (solid line) is superimposed to the data. The source star has $V_J \sim 17.7$ and $R_c \sim 16.4$ and is located in an underpopulated region of the colour-magnitude diagram.

An alert is also sent by e-mail to a distribution list[1].

4. Preliminary results

The system has been running since August 1999. Four and half a million stars (almost all the EROS2 catalog) are being monitored in the SMC and twelve million in the LMC. Stars monitored in the LMC are distributed over a large area of 17.5 square degrees, to help discriminate between "self-lensing" and halo microlenses (Palanque-Delabrouille et al. 1998). No microlensing candidate has been found yet, but two novæ reported in IAUC telegrams 7239 and 7286, triggered the system in the SMC.

Four and half a million bright stars are being monitored towards the Galactic Bulge. Eight microlensing candidates have been found during the 1999 season and six, as of writing, during the 2000 season. One of the 1999 microlensing alerts is shown in figure 2.

5. Conclusion

EROS has implemented an alert system on Magellanic Clouds and Galactic Center fields. The system has been triggered by a few microlensing candidates

---

[1]Requests for suscription should be sent to *glicens@hep.saclay.cea.fr*

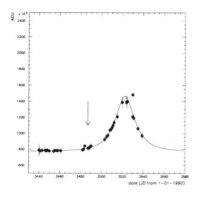 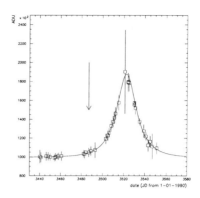

Figure 2. Microlensing candidate EROS 99-BLG-1. The two graphs show the light curves respectively in the V_{EROS} (left) and R_{EROS} (right) passbands. The arrows point to the date of trigger.

from the Galactic Bulge sources. No candidate has been observed towards the Magellanic Clouds, but two novæ were detected in 1999.

References

Afonso, C., et al. (EROS,MACHO/GMAN,OGLE,PLANET collaborations), 2000,ApJ, 532,340
Alcock, C., et al. (MACHO collaboration), 2000, ApJ, submitted (astro-ph/0001272)
Ansari, R., 1996, Vistas in Astronomy, 40, 519
Gaucherel, C., 1997, PhD thesis, Université Paris XI, SPP 97-1002
Glicenstein, J-F. & Gaucherel, C., 1996, EROS internal note 34
Glicenstein, J-F. & Gros, M., 1998, IEEE Transactions On Nuclear Science, 45, 1830
Gould, A., 1995, ApJ, 447, 491
Gould, A., 1998, ApJ, 506, 253
Lasserre, T., et al. (EROS collaboration), 2000, A&A, 355, L39
Mansoux, B., 1997, PhD thesis, Université Paris VII, LAL 97-19
Palanque-Delabrouille, N., et al. (EROS collaboration), 1998, A&A, 332, 1

Discussion

Cook: Is the CMD presented with the 99 Bulge Alert representative of those stars in the alert system or all stars photometered?
Glicenstein: The whole CMD shown is monitored.

Wambsganss: What is the sampling frequency for your different fields? Is it the same for all fields in the LMC, SMC and Bulge or do you select fields with higher frequency?

Glicenstein: The sampling frequencies are different for the different directions. This is determined by the offline analysis strategy and not by online wiggle consideration. In general, we try to have a uniform coverage of the fields in a given program (LMC, SMC...).

Graff: In practice, the sampling frequency can vary widely due to adverse observing conditions. Some "high-priority" fields are taken regularly, others fall to the bottom of the stack.

Glicenstein: I agree with this comment.

Gould: Is the 20-million limit set fundamentally by disk space?

Glicenstein: No, it is limited by CPU power. But disk space is also an important restriction, though less severe.

Menzies: You mentioned that your photometric pass bands are non-standard. Are you doing any work on the transformation of your magnitudes to some standard system?

Glicenstein: A lot of work has been done to provide transformation equations from our passbands to standard colors, but the accuracy of the transformations is low (~ 0.1 mag) and standard stars are rarely taken.

Difference Imaging Analysis of the MOA Image Data Base

Ian Bond (MOA Collaboration)

University of Auckland, Private Bag 92019, Auckland, New Zealand

Abstract. The MOA Collaboration has been conducting survey observations of selected fields in the Galactic Bulge and Magellanic Clouds using a 24 Megapixel CCD mosaic camera. During the 1998 and 1999 Austral Winters, a number of Bulge fields covering a 6 degree square region were monitored several times per night. The resulting data set is being analysed using difference imaging photometry. Subsequently, MOA plans to issue microlensing alerts in the winter 2000 season based on on-line analysis using difference imaging.

1. Introduction

The MOA (Microlensing Observations in Astrophysics) Collaboration comprises a number of institutes in Japan and New Zealand and undertakes observations at the Mt John Observatory in the South Island of New Zealand. From June 1996 to July 1998, a mosaic camera comprising 9 TI 1000×1018 pixel CCD chips (MOACamI) was employed (Abe et al. 1996). Afterwards a new mosaic camera comprising 3 SITe 2048×4096 pixel CCD chips (MOACamII) was commissioned and observations with this camera are continuing at present (Yanagisawa et al. 1999).

The aims of the project are to study microlensing events through follow-up observations of alerts issued by the MACHO, EROS, and OGLE groups and survey observations of selected fields towards the Magellanic Clouds and Galactic Bulge. Until now, all reductions of CCD images have been based on "profile-fitting" analysis on resolved stars using the DoPHOT and DAOPHOT packages. This approach has served us well but there are disadvantages. Extracting reliable photometric measurements of stars in crowded fields is very difficult. Furthermore, one usually requires stars to be resolved on some reference image making it difficult to detect microlensing events which may not be detectable at minimum light but cross the detection threshold during times of magnification. An analysis system based on image subtraction promises to overcome these problems.

2. Image Subtraction Method

Before subtracting one image from another it is necessary to adjust the seeing of one image to match that of the other. Two images taken under conditions of

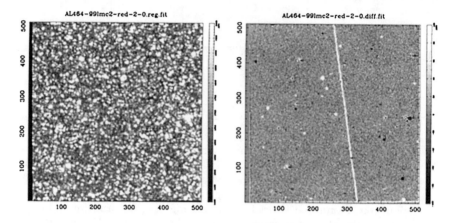

Figure 1. Illustration of difference imaging. After subtracting a reference image from the left image, one obtains the corresponding difference image on the right.

different seeing are related through a convolution kernel, k, as follows:

$$i(x,y) = r \star k(x,y) + b(x,y) \qquad (1)$$

where $r(x,y)$ denotes a "reference" image and $i(x,y)$ denotes a "current" image. The additional term $b(x,y)$ denotes the possibly spatially varying differential sky background between the two images.

There are two approaches in use to solve for the kernel. One is based on direct Point Spread Function (PSF) matching, where a PSF model is constructed for each image and the kernel is then found by deconvolution techniques (Phillips & Davis 1995; Tomaney & Crotts 1996). The other approach is to directly model the kernel itself using a number of basis analytical functions (Alard & Lupton 1998).

In the analysis system I am implementing for the MOA project, I essentially follow the approach of Alard & Lupton (1998). The kernel may be modelled using a linear combination of basis functions as follows:

$$k(x,y) = a_p k_p(x,y) + \sum a_i k_i(x,y) \qquad (2)$$

where $k_p(x,y)$ is the "primary" basis function and $k_i(x,y)$ are "secondary" basis functions. If the point spread functions on the two images can be approximated by a Gaussian profile, then the convolution kernel can also be approximated by a Gaussian. Subsequently, for the primary basis function, I adopt a generalised two-dimensional elliptical Gaussian profile whose shape parameters depend upon the shapes of the point spread function profiles on the two images. The "secondary" basis functions model the fine structure in the shape of the kernel which would arise if the point spread function profiles deviate from the idealised Gaussian profile. The secondary basis functions are constructed using Gaussian functions modified by polynomial terms (Alard & Lupton 1998).

A solution to the convolution kernel can be found using linear techniques to solve for the coefficients of the basis functions. A subtracted image may then be constructed using equation (1). An example is shown in Figure 1. This procedure can be extended to the situation where the convolution kernel varies across the image. In this case, the coefficients in equation (2) may themselves be modelled as a linear combination of polynomial basis functions. The size of the linear system becomes large, but one can utilise an accelerated summation technique to build the system (Alard 1999).

3. MOA Image Data Base

From 1998, the MOA Collaboration has been monitoring a number of fields towards the Galactic Bulge and Magellanic Clouds with the MOACamII. A complete field of view over the three-chip mosaic comprises $0.9° \times 1.3°$.

The MACHO collaboration has re-analysed a small subset of their data base of images obtained by observations towards the Galactic Bulge. Their results are encouraging in that they find twice as many microlensing events as they did with the initial analysis (Alcock et al. 1999). The MOA survey observations are now being re-analysed using image subtraction. The image data base corresponding to the Galactic Bulge observations have been put through the difference imaging pipeline. Each observation image now has a corresponding difference image. The next stage is to search this data base for microlensing events.

4. Plans for MOA Alerts

During the Austral Winter 2000 season, the MOA Collaboration will endeavour to issue alerts of microlensing events based on online analysis using difference imaging. This requires considerable computing resources. The analysis will be done at the observatory site using a 4 CPU Sun E450 server with 90 GB internal disc space. Additionally there is 100 GB of external disc space which is continually being expanded. DLT tapes are used for data archiving.

Of particular interest are high magnification microlensing events which may not be visible at their times of baseline intensities. Such events would be rising very rapidly during times of high magnification above the detection threshold. Therefore, rather than survey a large area of the Galactic Bulge once per night a smaller $10° \times 10°$ area is being monitored several times per night.

References

Abe, F. et al. (MOA Collaboration) 1996, in Proc. 12th IAP Conf., Variable Stars and Astrophysical Returns of Microlensing Surveys, eds. R. Ferlet, J-P. Maillard & B. Raban (Gif-sur-Yvette: Editions Frontieres), p. 75

Alard, C. 1999, A&AS, in press (astro-ph/9903111)

Alard, C. & Lupton, R.H. 1998, ApJ, 503, 325

Alcock, C. et al. (MACHO Collaboration) 1999, ApJ, 521, 602

Phillips, A.C. & Davis, L.E. 1995, in ASP Conf. Ser., Vol. 77, Astronomical Data Analysis Software and Systems IV, ed R.A. Shaw, H.E. Payne, & J.J.E. Hayes (San Francisco: ASP), 297

Tomaney, A. & Crotts, A. 1996, AJ, 112, 287

Yanagisawa, T. et al. (MOA Collaboration) 1999, Experimental Astronomy, submitted

Discussion

Crotts: How much CPU time does variable-PSF DIA code take, per 2K × 4K image?

Bond: About 15–20 CPU minutes, but most time is taken in the last convolution calculation, not the variable PSF determination.

Gould: What fitting functions are used?

Bond: The primary basis function takes into account the approximate seeing differences. The additional basis functions then give a better approximation to the real kernel. If the two images have similar seeing, a δ-function + secondary basis function set can be used.

Kerins: To implement an alert system you need to choose one of your initial images as a reference image. Are you guaranteed of having an image of sufficient quality to use as a reference?

Bond: Yes. Before starting the alert system, we will already have observed a number of selected fields which can be stacked to form a high signal-to-noise image. Also we have a large data base of images from 1998 and 1999 from which we can select images to form a reference.

Bennett: Why are your Bulge fields so far from the Galactic center?

Bond: We are currently revising our selection of Galactic Bulge fields.

Glicenstein: The output of difference imaging gives variable stars and microlenses. How do you distinguish one from the other?

Bond: The light curves will be investigated individually.

Menzies: You get relative magnitudes from your image subtraction procedure. How do you convert them to a true magnitude scale?

Bond: Actually we get flux differences from image subtraction. If the flux of the star of interest on the reference frame is known, then the total flux can be found which can then be converted to a magnitude scale. On the other hand, if the reference flux cannot be known (if the star is highly blended for example) we can only make a δ-flux light curve. This is better than no light curve at all, which would be the case if profile-fitting techniques were applied.

Baryonic Dark Matter in Galaxies

B. J. Carr
Astronomy Unit, Queen Mary & Westfield College, Mile End Road, London E1 4NS, UK

Abstract. Cosmological nucleosynthesis calculations imply that many of the baryons in the Universe must be dark. We discuss the likelihood that some of these dark baryons may reside in the discs or halos of galaxies. If they were in the form of compact objects, they would then be natural MACHO candidates, in which case they are likely to be the remnants of a first generation of pregalactic or protogalactic Population III stars. Various candidates have been proposed for such remnants – brown dwarfs, red dwarfs, white dwarfs, neutron stars or black holes – and we review the many types of observations (including microlensing searches) which can be used to constrain or exclude them.

1. Introductory Overview

Evidence for dark matter has been claimed in many different contexts. There may be *local* dark matter in the Galactic disc, dark matter in the *halos* of our own and other galaxies, dark matter associated with *clusters* of galaxies and finally – if one believes that the total cosmological density has the critical value – smoothly distributed *background* dark matter. Since dark matter probably takes as many different forms as visible matter, it would be simplistic to assume that all these dark matter problems have a single solution. The local dark matter is almost certainly unrelated to the other ones and, while the halo and cluster dark matter would be naturally connected in many evolutionary scenarios, there is a growing tendency to regard the unclustered background dark matter as different from the clustered component.

As emphasized in a recent review by Turner (1999), there are also many different types of dark matter candidates. The latest supernovae measurements indicate that the cosmological expansion is accelerating, which means that the total density must be dominated by some form of energy with negative pressure (possibly a cosmological constant). Theorists therefore now distinguish between this exotic (unclustered) dark energy and the ordinary *matter* component (with positive pressure). The combination of the supernovae and microwave background observations suggest that the density parameters of these two components are $\Omega_X = 0.8 \pm 0.2$ and $\Omega_M = 0.4 \pm 0.1$ respectively. This is compatible with the total density parameter being 1, as expected in the inflationary scenario, but does not definitely require this.

The matter density may itself be broken down into different components (Turner 1999). Large-scale structure observations suggest that there must be *cold* dark matter (i.e. "Weakly Interacting Massive Particles" or WIMPs) with a density parameter $\Omega_C = 0.3 \pm 0.1$, the latest determination of the neutrino mass by the Super-Kamiokande experiment requires that there is *hot* dark matter with a density parameter $\Omega_H \approx 0.01$, and we will see that Big Bang nucleosynthesis calculations require that the baryonic matter (be it visible or dark) has a density parameter $\Omega_B = 0.05 \pm 0.005$. It is remarkable, not only that each of these components seems to be needed, but also that their densities are all within one or two orders of magnitude of each other:

$$1 > \Omega_X \sim \Omega_C \sim \Omega_H \sim \Omega_B > 0.01. \tag{1}$$

Why this should be remains a mystery. Note that all the Ω values given above assume a Hubble parameter $H_o = 65$ km s^{-1} Mpc^{-1}.

This paper will focus exclusively on baryonic dark matter, because that is the form of dark matter most relevant to microlensing observations. We will address four issues: (1) What is the evidence that some of the baryons in the Universe are dark? (2) What are the reasons for believing that some of these dark baryons are in galaxies (i.e. in the form of MACHOs)? (3) What are the microlensing signatures of baryonic dark matter? (4) What constraints do these and other observations place on MACHO candidates? However, it is important to place these considerations in the broader context discussed above.

2. Evidence for Baryonic Dark Matter

The main argument for both baryonic and non-baryonic dark matter comes from Big Bang nucleosynthesis calculations. This is because the success of the standard picture in explaining the primordial light element abundances only applies if the baryon density parameter lies in the range (Copi et al. 1995)

$$0.007h^{-2} < \Omega_B < 0.022h^{-2} \tag{2}$$

where h is the Hubble parameter in units of 100 km s^{-1} Mpc^{-1}. For comparison, the latest measurements of the primordial deuterium abundance imply a much tighter constraint (Burles et al. 1999):

$$0.018h^{-2} < \Omega_B < 0.020h^{-2}. \tag{3}$$

In any case, the upper limit implies that Ω_B is well below 1, which suggests that no baryonic candidate could provide the matter density required by large-scale structure observations. This conclusion also applies if one invokes inhomogeneous nucleosynthesis since one requires $\Omega_B < 0.09h^{-2}$ even in this case (Mathews et al. 1993). On the other hand, the value of Ω_B allowed by equations (2) and (3) almost certainly exceeds the density of visible baryons Ω_V. A careful inventory by Persic & Salucci (1992) shows that the density in galaxies and cluster gas is

$$\Omega_V \approx (2.2 + 0.6h^{-1.5}) \times 10^{-3} \approx 0.003 \tag{4}$$

where the last approximation applies for reasonable values of h. This is well below the lower limits allowed by equations (2) and (3), so it seems that one needs *both* non-baryonic and baryonic dark matter.

The claim that some of the nucleosynthetic baryons must be dark is such an important one that it is important to consider whether it can be circumvented by decreasing Ω_B or increasing Ω_V in some way. Both of these possibilities have been advocated.

A few years ago the anomalously high deuterium abundance measured in intergalactic Lyman-α clouds suggested that the nucleosynthesis value for Ω_B could be lower than the standard one (Carswell et al. 1994, Songaila et al. 1994, Rugers & Hogan 1996, Webb at al. 1997). However, the evidence for this has always been strongly disputed (Tytler et al. 1996). In particular, recent studies of the quasars Q 1937-1009 and Q 1009+2956 suggest that the deuterium abundance is 3.3×10^{-5}. This corresponds to $\Omega_B h^2 = 0.019$ (Burles & Tytler 1998a,b), which is in the middle of the range given by eqn (3).

The possibility that Ω_V could be larger than indicated by eqn (4) is much harder to refute. Certainly one can now make a more precise estimate for some of the components of Ω_V than Persic & Salucci. A recent review by Hogan (1999) replaces the factors in brackets in eqn (4) by 2.6 for spheroid stars, 0.86 for disc stars, 0.33 for neutral atomic gas, 0.30 for molecular gas and 2.6 for the ionized gas in clusters. This assumes $H_o = 70$ km s^{-1} Mpc^{-1}. However, this may still not account for all components. For example, there may be some baryons in low surface brightness galaxies (De Blok & McGaugh 1997) or dwarf galaxies (Loveday et al. 1997) or in a hot intergalactic medium. The last possibility is emphasized by Cen & Ostriker (1999), who find that – in the context of the CDM scenario – half the mass of baryons should be in warm ($10^5 - 10^7$ K) intergalactic gas at the present epoch.

Another possibility is that the missing baryons could be in gas in groups of clusters. This has been emphasized by Fukugita et al. (1998), who argue that plasma in groups (which would be too cool to have appreciable X-ray emission) could provide all of the cosmological nucleosynthesis shortfall. Indeed the review by Hogan (1999) suggests that the ionized gas in groups could have a density parameter as high as 0.014. However, it must be stressed that this estimate is based on an extrapolation of observations of rich clusters and there is no direct evidence for this.

3. Is There Baryonic Dark Matter in Galaxies?

Which of the dark matter problems identified in Section 1 could be baryonic? Certainly the dark matter in galactic discs could be – indeed this is the only dark matter problem which is definitely baryonic. Even if all discs have the 60% dark component envisaged for the Galaxy by Bahcall et al. (1992), this only corresponds to $\Omega_d \approx 0.001$, well below the nucleosynthesis bound. However, the Bahcall et al. claim is strongly disputed by Kuijken & Gilmore (1989a,b) and Flynn & Fuchs (1995). Indeed recent Hipparcos observations suggest that the dark disc fraction is below 10% (Crézé et al. 1998).

The more interesting question is whether the halo dark matter could be baryonic. If the Milky Way is typical, the density associated with halos would be $\Omega_h \approx 0.03 h^{-1}(R_h/100\text{kpc})$, where R_h is the (uncertain) halo radius, so the upper limit in eqn (2) implies that *all* the dark matter in halos could be baryonic providing $R_h < 70 h^{-1}$kpc. This is marginally possible for our galaxy (Fich

& Tremaine 1991), in which case the dark baryons could be contained in the remnants of a first generation of "Population III" stars (Carr 1994). This corresponds to the "Massive Compact Halo Object" or "MACHO" scenario and has attracted considerable interest as a result of the LMC microlensing observations. Even if halos are larger than $70h^{-1}$kpc, and studies of the kinematics of other spirals by Zaritsky et al. (1993) suggest that they could be as large as $200h^{-1}$kpc, a considerable fraction of their mass could still be in stellar remnants.

Probably the only direct evidence that there are dark baryons in galaxies comes from studying the density profiles and rotation curves in dwarf galaxies. It is well known that the presence of dark matter in dwarf galaxies requires that it cannot be entirely hot. However, if the dark matter consisted only of WIMPs, one would expect it to have the standard Navarro-Frenk-White density profile (Navarro et al. 1997) and Burkert & Silk (1999) claim that the profile for DDO 154 is very different from this. They argue that the measurements indicate the presence of a centrally condensed baryonic dark component, having about 25% of the total dark matter density. Another possible signature of a baryonic halo would be *flattening* of the halo since more dissipation would be expected in this case. For our galaxy observations are best fit by an axis ratio $q \approx 0.6$, which constrains the fraction of the halo in MACHOs but does not necesssarily require them (Samurovic et al. 1999).

On theoretial grounds one would *expect* the halo dark matter to be a mixture of WIMPs and MACHOs. For since the cluster dark matter must be predominantly cold, one would expect at least some of it to clump into galactic halos. The relative densities of the two components would depend sensitively on the formation epoch of the Population III stars. If they formed pregalactically, one would expect the halo ratio to reflect the cosmological background ratio (viz. $\Omega_B/\Omega_C \approx 0.1$). However, if they formed protogalactically, the ratio could be larger since the baryons could have dissipated and become more centrally concentrated than the WIMPs.

In order to distinguish between the pregalactic and protogalactic scenarios, it is important to gain independent evidence about the formation epoch of the putative MACHOs. At moderate redshifts one can obtain a lower limit to the baryon density by studying Lyman-α clouds. The simulations of Weinberg et al. (1997) suggest that the density parameter of the clouds must be at least $0.017h^{-2}$ at a redshift of 2 and this is already close to the upper limit given by eqn (3). By today some of these baryons might have been transformed into a hot intergalactic medium or stars but this suggests that there was little room for any dark baryons before $z = 2$. On the other hand, we will see in Section 5.4 that background light constraints require that any massive Population III stars must have formed much earlier than this.

4. Lensing Constraints

One of the most useful signatures of compact objects is their gravitational lensing effects. Indeed it is remarkable that lensing could permit their detection over the entire mass range $10^{-7} M_\odot$ to $10^{12} M_\odot$. There are two distinct lensing effects and these probe different but nearly overlapping mass ranges: macrolensing (the multiple-imaging of a source) can be used to search for objects larger than

$10^5 M_\odot$, while microlensing (modifications to the intensity of a source) can be used for objects smaller than $10^3 M_\odot$. The current constraints on the density parameter of compact objects in various mass ranges are brought together in Figure 1, shaded regions being excluded. Although the focus of this meeting is mainly on microlensing, it is useful to bring all the limits together. Most of these limits presume a cosmological distribution of lenses but they are still applicable for lenses confined to halos provided the halos cover the sky.

Macrolensing by Compact Objects. If one has a population of compact objects with mass M and density parameter Ω_c, then the probability of one of them multiply-imaging a source at a cosmological redshift and the separation between the images are roughly

$$P \approx 0.1\,\Omega_c, \qquad \theta \approx 6 \times 10^{-6} (M/M_\odot)^{1/2} h^{1/2} \text{ arcsec} \qquad (5)$$

(Press & Gunn 1973). One can therefore use upper limits on the frequency of macrolensing for different image separations to constrain Ω_c as a function of M. In particular, in the context of quasars, VLA observations imply $\Omega_c(10^{11} - 10^{13} M_\odot) < 0.4$ (Hewitt 1986) and HST data imply $\Omega_c(10^{10} - 10^{12} M_\odot) < 0.02$ (Surdej et al. 1993). To probe smaller scales, one can use high resolution radio sources: Kassiola et al. (1991) have invoked lack of lensing in 40 VLBI objects to infer $\Omega_c(10^7 - 10^9 M_\odot) < 0.4$, while a study of VLBA sources leads to a limit $\Omega_c(10^6 - 10^8 M_\odot) < 0.03$ (Henstock 1996). Other techniques (eg. speckle interferometry) could strengthen these constraints, as indicated by the broken lines in Figure 1. It should be stressed that the expression for P in eqn (5) has some dependence on the cosmological model. Indeed one of the important recent uses of macrolensing searches is to constrain the cosmological constant, the observations requiring $\Omega_\Lambda < 0.7$.

Microlensing in Macrolensed Sources. If a galaxy is suitably positioned to image-double a quasar, then there is also a high probability that an individual halo object will traverse the line of sight of one of the images and this will give intensity fluctuations in one but not both images (Gott 1981). The effect would be observable for objects bigger than $10^{-4} M_\odot$ but the timescale of the fluctuations $\sim 40 (M/M_\odot)^{1/2}$y would make them detectable only for $M < 0.1\ M_\odot$. There is already evidence of this effect for the quasar Q2237+0305 (Irwin et al. 1989), the observed timescale for the variation in the luminosity of one of the images possibly indicating a mass below $0.1\ M_\odot$ (Webster et al. 1991). However, because the optical depth is high, the mass estimate is very uncertain and a more recent analysis suggests that it could be in the range $0.1 - 10\ M_\odot$ (Lewis et al. 1998), in which case the lens could be an ordinary star. The absence of this effect in the quasar Q0957+561 has also been used to exclude MACHOs with mass in the range $10^{-7} - 10^{-3} M_\odot$ from making up all of the halo of the intervening galaxy, although the precise limit has some dependence on the quasar size (Schmidt & Wambsganss 1998). Another application of this method is to seek microlensing in a compact radio source which is macrolensed by a galaxy. Indeed Koopmans & Bruyn (2000) claim to have detected this effect for the CLASS gravitational lens B1600+434. The inferred lens mass is around $0.5\ M_\odot$, comparable to the mass implied by the LMC data. This result is discussed in more detail at this meeting.

Microlensing of Quasars. More dramatic but rather controversial evidence for the microlensing of quasars comes from Hawkins (1993, 1996, 1999), who has been monitoring 300 quasars in the redshift range $1-3$ for nearly 20 years using a wide-field Schmidt camera. He finds quasi-sinusoidal variations with an amplitude of 0.5 magnitudes on a timescale 5 yr and attributes this to lenses with mass $\sim 10^{-3} M_\odot$. The crucial point is that the timescale decreases with increasing redshift, which is the opposite to what one would expect for intrinsic variations, although this has been disputed (Alexander 1995, Baganoff & Malkan 1995). The timescale also increases with the luminosity of the quasar and he explains this by noting that the variability timescale should scale with the size of the accretion disc (which should itself correlate with luminosity). A rather worrying feature of Hawkins's claim is that he requires the density of the lenses to be close to critical (in order that the sources are transited continuously), so he has to invoke primordial black holes which form at the quark-hadron phase transition (Crawford & Schramm 1982). However, this requires fine-tuning since the fraction of the Universe going into black holes at this transition must only be about 10^{-9}. As discussed in Section 5.5, Walker (1999) has proposed that Hawkins's lenses might also be Jupiter-mass gas clouds.

Line-Continuum Effects for Quasars. In some circumstances, the continuum part of the quasar emission will be microlensed but not the line part. This is because only the continuum region may be small enough to act as a point-like source. (For a lens at a cosmological distance the Einstein radius is $0.05(M/M_\odot)^{1/2} h$ pc, whereas the sizes of the optical continuum and line regions are of order 10^{-4} pc and 1 pc respectively.) This would decrease the equivalent width of the emission lines, so in statistical studies of many quasars one would expect the characteristic equivalent width of quasar emission lines to decrease as one goes to higher redshift because there would be an increasing probability of having an intervening lens. Dalcanton et al. (1994) compared the equivalent widths for a high and low redshift sample of quasars and found no difference. They inferred

$$\Omega_c(0.001-60\ M_\odot) < 0.2, \quad \Omega_c(60-300\ M_\odot) < 1, \quad \Omega_c(0.01-20\ M_\odot) < 0.1. \quad (6)$$

The mass limits come from the fact that the amplification of even the continuum region would be unimportant for $M < 0.001\ M_\odot$, while the amplification of the line regions would be important (cancelling the effect) for $M > 20\ M_\odot$ if $\Omega_c = 0.1$ or $M > 60\ M_\odot$ if $\Omega_c = 0.2$ or $M > 300\ M_\odot$ if $\Omega_c = 1$. These limits are indicated in Fig.1 and are marginally incompatible with Hawkins's claim that $\Omega_c(10^{-3} M_\odot) \sim 1$.

Lensing of Quasars by Dark Objects in Clusters. If halos are made of MACHOs, one would also expect some of these to be spread throughout a cluster of galaxies. This is because individual galaxies should be stripped of some of their outer halo as a result of collisions and tidal interactions. This method is sensitive to MACHOs in the mass range $10^{-6} - 10^{-3}\ M_\odot$. Tadros et al. (1998) have therefore looked for the microlensing of quasars by MACHOs in the Virgo cluster: four months of observations of 600 quasars with the UK Schmidt telescope have yielded no candidates and this already implies that less than half the mass of Virgo is in $10^{-5}\ M_\odot$ objects. A more extensive follow-up campaign is

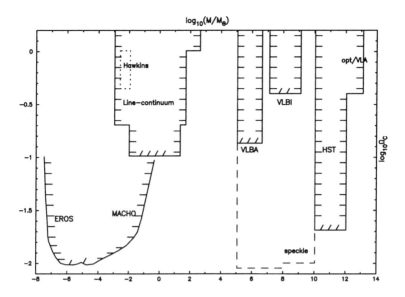

Figure 1. Lensing constraints on density parameter for compact objects

currently underway. This technique would also be sensitive to cold molecular clouds of the kind advocated by Walker (1999).

Microlensing of Supernovae by Halos. In principle galactic halos could produce luminosity variations in high redshift supernovae, many of which are now routinely detected as a result of supernova searches. As pointed out by Metcalf & Silk (1999), a particularly interesting aspect of this effect is that it could discriminate between what they term "macroscopic" and "microscopic" dark matter, corresponding to MACHO and WIMP halos respectively. This is because the distribution of amplifications would be different in these two cases, although the shape of the distribution is also sensitive to the cosmological and halo model.

Microlensing of Stars in LMC. Attempts to detect microlensing by objects in our own halo by looking for intensity variations in stars in the Magellanic Clouds and the Galactic Bulge have now been underway for a decade (Paczyński 1996). This method is sensitive to lens masses in the range $10^{-7} - 10^2 M_\odot$ but the probability of an individual star being lensed is only $\tau \sim 10^{-6}$, so one has to look at many stars for a long time (Paczyński 1986a,b). The duration and likely event rate are

$$P \sim 0.2(M/M_\odot)^{1/2} y, \quad \Gamma \sim N\tau P^{-1} \sim (M/M_\odot)^{-1/2} y^{-1} \qquad (7)$$

where $N \sim 10^6$ is the number of stars. As discussed elsewhere in this meeting, the MACHO group currently has 13-17 LMC events and the durations span the range 34 – 230 days (Alcock et al. 2000). For a standard halo model, the data suggest an average lens mass of around 0.5 $M_\odot$ and a halo fraction of 0.2, with the 95% confidence ranges being $0.15 - 0.9$ $M_\odot$ and $0.08 - 0.5$. The mass is comparable with the earlier estimates but the fraction is somewhat smaller (Alcock et al. 1997). This might appear to indicate that the MACHOs are white dwarfs but, as discussed in Section 5.2, this would seem to be excluded on astrophysical grounds, so this presents a dilemma for MACHO enthusiasts. One possible resolution is to invoke a less conventional candidate; for example, primordial black holes forming at the quark-hadron phase transition might have the required mass (Jedamzik 1997) and the microlensing implications of this scenario have been studied by Green (1999). Perhaps the most important result of the LMC searches is that they *eliminate* many candidates. Indeed the combination of the MACHO and EROS results already excludes objects in the mass range $5 \times 10^{-7} - 0.002$ $M_\odot$ from having more than 0.2 of the halo density (Alfonso et al. 1997).

Microlensing of Stars in M31. The LMC studies are complemented by searches for microlensing of stars in M31. In this case, the sources are too distant to resolve individually (i.e. there are many stars per pixel), so a lensing event is observed only if the amplification is large enough for the source to stand out above the background, but observations of the LMC already demonstrate the efficacy of the method (Melchior et al. 1999). For sources in M31 the halo objects may reside in our own galaxy or M31 but the crucial point is that one expects an asymmetry between the far and near side of the disc. Two groups have been involved in this work: the AGAPE collaboration (Ansari et al. 1997), who use the "pixel" method, and the VATT-Columbia collaboration, who use "differential image photometry" (Crotts & Tomaney 1997). The AGAPE team have been monitoring seven fields in M31 in red and blue and have already detected one good lensing candidate (Ansari et al. 1999). The important theoretical implications of this approach are considered by Kerins et al. (2000) and this is discussed further at this meeting.

5. Constraints on MACHO candidates

Although one cannot state definitely that MACHOs exist, one can already place important constraints on the possible candidates. In this section we will discuss each candidate in turn, focussing particularly on brown dwarfs, red dwarfs, white dwarfs and black holes. The combined constraints, including the microlensing ones discussed above, are indicated in Figure 2. This shows which candidates are excluded by various types of observational signature. A cross indicates that exclusion is definite, while a question mark indicates that it is tentative. Candidates associated with one or more crosses should clearly be rejected but those with question marks alone may still be viable. Although no candidate is entirely free of crosses or questions marks, the title of a recent paper by Freese et al. (1999), "Death of Stellar Baryonic Dark Matter", suggesting that there are no viable MACHO candidates, may be overly pessimistic.

5.1. Brown Dwarfs

Objects in the range $0.001 - 0.08$ $M_\odot$ would never burn hydrogen and are termed "brown dwarfs" (BDs). They represent a balance between gravity and degener-

	SB	BD	RD	WD	NS	BH	VMO	SMO
Source Counts		?	×	?				
Halo Light			?					
Background Light				×	×	×	×	
Enrichment				×	×	×	?	
Lensing	×	?	?					
Encounters	×							
Dynamical Effects							?	×

Figure 2. Constraints on and exclusions of MACHO candidates.

acy pressure. Objects below 0.001 $M_\odot$, being held together by intermolecular rather than gravitational forces, have atomic density and are sometimes termed "snowballs" (SBs). However, such objects would have evaporated within the age of the Universe if they were smaller than $10^{-8} M_\odot$ (De Rujula et al. 1992) and there are various encounter constraints for snowballs larger than this (Hills 1986).

It has been argued that objects below the hydrogen-burning limit may form efficiently in pregalactic or protogalactic cooling flows (Ashman & Carr 1990, Thomas & Fabian 1990) but the direct evidence for such objects remains weak. While some BDs have been found as companions to ordinary stars, these can only have a tiny cosmological density and it is much harder to find isolated field BDs. The best argument therefore comes from extrapolating the initial mass function (IMF) of hydrogen-burning stars to lower masses than can be observed directly. The IMF for Population I stars ($dN/dm \sim m^{-\alpha}$ with $\alpha < 1.8$) suggests that only 1% of the disc could be in BDs (Kroupa et al. 1995). However, one might wonder whether α could be larger, increasing the BD fraction, for zero-metallicity stars. Although there are theoretical reasons for entertaining this possibility, earlier observational claims that low metallicity objects have a steeper IMF than usual are now discredited. Indeed observations of Galactic and LMC globular clusters (Elson et al. 1999) and dwarf spheroidal field stars (Feltzing et al. 1999) suggest that the IMF is *universal* with $\alpha < 1.5$ at low masses (Gilmore 1999). This implies that the BD fraction is much less than 1% by mass. However, it should be stressed that nobody has yet measured the IMF in the sites which are most likely to be associated with Population III stars.

We have seen that the LMC microlensing results would now seem to exclude a large fraction of BDs in our own halo. Although Honma & Kan-ya (1998) have presented 100% BD models, these would require falling rotation curves and most theorists would regard these as rather implausible. Another exotic possibility, suggested by Hansen (1999), is "Beige Dwarfs" in the mass range $0.1 - 0.3$ $M_\odot$. Such objects are larger than the traditional BD upper limit but they are supposed to form by sufficiently slow accretion that they never ignite their nuclear fuel.

5.2. Red Dwarfs

Discrete source counts for our own Galaxy suggest that the fraction of the halo mass in low mass hydrogen burning stars – red dwarf (RDs) – must be less than 1% (Bahcall et al. 1995, Gould et al. 1998, Freese et al. 2000). These limits might be weakened if the stars were clustered (Kerins 1997) but not by much. For other galaxies, the best constraint on the red dwarf fraction comes from upper limits on the halo red light and such studies go back several decades (Boughn & Saulson 1983).

The discovery of red light around NGC 5907 by Sackett et al. (1994), apparently emanating from low mass stars with a density profile like that of the halo, was therefore a particularly interesting development. This detection was confirmed in V and I by Lequeux et al. (1996) and in J and K by James & Casali (1996). However, the suggestion that the stars might be of primordial origin (with low metallicity) was contradicted by the results of Rudy et al. (1997), who found that the color was indicative of low mass stars with solar metallicity. Furthermore, it must be stressed that the red light has only been observed within a few kpc and no NIR emission is detected at 10-30 kpc (Yost et al. 1999). Both these points go against the suggestion that the red light is associated with MACHOs.

Recently it has been suggested that the red light seen in NGC 5907 is more likely to derive from a disrupted dwarf galaxy, the stars of which would naturally follow the dark matter profile (Lequeux et al. 1998), or to be a ring left over from a disrupted dwarf spheroidal galaxy (Zheng et al. 1999). However, in this case one would expect of order a hundred bright giants for a standard IMF, whereas NICMOS observations find only one (Zepf et al. 1999). This requires that either the galaxy is much further away than expected (24 Mpc) or it has a very low metallicity or the dwarf-to-giant ratio is very large (requiring a very steep IMF with $\alpha > 3$). There is clearly still a mystery here. In any case, NGC 5907 does not seem to be typical since ISO observations of four other edge-on bulgeless spiral galaxies give no evidence for red halos (Gilmore & Unavane 1998).

5.3. White Dwarfs

A few years ago white dwarfs (WDs) were regarded as rather implausible dark matter candidates. One required a very contrived IMF, lying between 2 $M_\odot$ and 8 $M_\odot$, in order to avoid excessive production of light or metals (Ryu et al. 1990); the fraction of WD precursors in binaries would be expected to produce too many type-Ia supernovae (Smecker & Wyse 1991); and the halo fraction was constrained to be less than 10% in order to avoid the luminous precursors contradicting the upper limits from galaxy counts (Charlot & Silk 1995). The observed WD luminosity function also placed a severe lower limit on the age of any WDs in our own halo (Tamanaha et al. 1990).

More recent constraints have strengthened these limits. A study of CNO production suggests that a halo comprised entirely of WDs would overproduce C and N compared to O by factors as large as 60 (Gibson & Mould 1997) and, although one might be able to circumvent this constraint in some circumstances, a similar limit comes from considering helium and deuterium production (Field et al. 2000). Extragalactic background light limits now require that the halo WD fraction be less than 6% (Madau & Pozzetti 1999) and the detection of TEV γ-rays from the the galaxy Makarian 501 (which indirectly constrains the infrared background) requires that the WD density parameter be less than $0.002h^{-1}$ (Graff et al. 1999).

The "many nails in the coffin" of the WD scenario are confounded by the results of the LMC microlensing observations, the lens mass estimate for which suggests that WDs are the most plausible explanation. Not surprisingly, therefore, theorists have been trying to resuscitate the scenario. At least some of the afore-mentioned limits must be reconsidered in view of recent claims by Hansen (1998) that metal-poor old WDs with hydrogen envelopes could be much bluer and brighter than previously supposed, essentially because the light emerges from deeper in the atmosphere.

This suggestion has been supported by HST observations of Ibata et al. (1999), who claim to have detected five candidates of this kind. The objects are blue and isolated and show high proper motion. They infer that they are 0.5 $M_\odot$ hydrogen-atmosphere WDs with an age of around 12 Gyr. Three such objects have now been identified spectroscopically (Hodgkin et al. 2000, Ibata et al. 2000), so this possibility must be taken very seriously. However, this does not circumvent the nucleosynthetic arguments against WDs.

5.4. Black Hole Remnants

Stars bigger than 8 $M_\odot$ would leave neutron star (NS) remnants, while those in some range above about 20 $M_\odot$ would leave black hole (BH) remnants. However, neither of these would be plausible candidates for either the disc or halo dark matter because their precursors would have unacceptable nucleosynthetic yields. Stars larger than 200 $M_\odot$ are termed "Very Massive Objects" or VMOs and might collapse to black holes without these nucleosynthetic consequences (Carr et al. 1984). However, during their main-sequence phase, such VMOs would be expected to generate a lot of background light. By today this should have been shifted into the infrared or submillimetre band, as a result of either redshift effects or dust reprocessing, so one would expect a sizeable infrared/submillimetre cosmic background (Bond et al. 1991). Over the last few decades there have been several reported detections of such a background but these have usually turned out to be false alarms. COBE does now seem to have detected a genuine infrared background (Fixsen et al. 1998) but this can probably be attributed to ordinary Population I and II stars. In any case, the current constraints on such a background strongly limit the density of any VMOs unless they form at a very high redshift. For this reason massive Population III stars would need to be pregalactic rather than protogalactic.

Stars larger than 10^5 $M_\odot$ – termed "Supermassive Objects" or SMOs – would collapse directly to black holes without any nucleosynthetic or background light production. However, supermassive black holes would still have noticeable

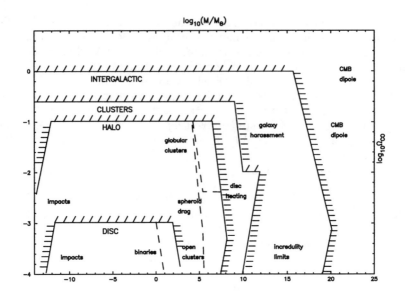

Figure 3. Dynamical constraints on the density parameter for black holes of mass M located in the Galactic disc, the Galactic halo, clusters of galaxies and intergalactic space.

lensing effects, as discussed in Section 4, and dynamical effects. The latter have been investigated by many authors and are reviewed in detail by Carr & Sakellariadou (1998). The constraints on black holes in our own disc – due to the disruption of open clusters – and in our own halo – due to the heating of disc stars, the disruption of globular clusters and dynamical friction effects – are indicated by the shaded regions in Figure 3. Although it has been claimed that there is positive evidence for some of these effects, such as disc heating (Lacey & Ostriker 1985), the interpretation of this evidence is not clear-cut. It is therefore more natural to regard these dynamical effects as merely imposing an upper limit on the density of black holes or indeed any other type of compact object.

The limits in Figure 3 are expressed in terms of the density parameter, taken to be 0.001 for the disc and 0.1 for the halo. The figure also shows the dynamical constraints for black holes in clusters of galaxies or intergalactic space, although this goes beyond the context of the present discussion. It should be stressed that many of these limits would also apply if the black holes were replaced by "dark clusters" of smaller objects, a scenario which has been explored by many authors (Carr & Lacey 1987, Ashman 1990, Kerins & Carr 1994, De Paolis et al. 1995, Moore & Silk 1995).

5.5. Cold Clouds

The suggestion that the halo dark matter could be in cold clouds was first made by Pfenniger et al. (1994). They envisaged the clouds having a mass of around 10^{-3} $M_\odot$ and being distributed in a disc, which now seems dynamically rather implausible, but several people have proposed a similar scenario with a spheroidal halo of clouds (De Paolis et al. 1995). Walker (1999) argues that such clouds could explain both the "Extreme Scattering Events" detected by radio observations in our own galaxy and the quasar microlening events claimed by Hawkins. Indeed Walker & Wardle (1999) advocate a model in which the halo entirely comprises such clouds, with visible stars being formed as a result of collisions between these clouds. They claim that this scenario naturally produces various observed features of galaxies. Although clouds of 10^{-3} $M_\odot$ could not explain the LMC microlensing events, Draine (1998) argues that such clouds might still produce the apparent microlensing events through gas absorption effects. The interaction of cosmic rays with such clouds would also produce an interesting γ-ray signature (Sciama 2000, De Paolis et al. 1999).

6. Conclusions

Although it is premature to assess the importance of baryonic dark matter definitively, there have been many interesting developments in this field in the last few years and various conclusions can be drawn.

(1) Although cosmological nucleosynthesis calculations suggest that many baryons are dark, one cannot be sure that the dark baryons are inside galaxies. The more conservative conclusion would be that they are contained in an intergalactic medium or in gas within groups or clusters of galaxies.

(2) Over the years there have been several observational claims of effects which seem to indicate the existence of MACHOs but these have usually turned out to be false alarms. For example, the discovery of a red halo around NGC 5907 is suggestive but we have seen that its interpretation is far from clear.

(3) Currently the only *positive* evidence for MACHOs comes from microlensing observations. The LMC results suggest that white dwarfs may be the best MACHO candidate but it must be stressed that the mass estimate upon which this inference is based is sensitive to assumptions about the halo model. In any case, the large number of arguments which have been voiced against white dwarfs in the past cannot be brushed aside too cavalierly. The detection of microlensing in a compact radio source also gives a mass in the white dwarf range but the Hawkins result requires a much smaller mass. It would perhaps be strange to have two distinct populations.

(4) We have seen that there are many important *constraints* on MACHO candidates, not only from microlensing but also from a wide variety of other astrophysical effects. These constraints are summarized in Figure 2. Until there is a definite detection, therefore, the best strategy is to proceed by *eliminating* candidates, on the Sherlock Holmes principle that whatever candidate remains, however implausible, must be correct.

(5) What is clearly missing from current speculation is a good cosmological scenario for the formation of the MACHOs. There is considerable uncertainty as to whether they form pregalactically (as suggested by background light constraints) or more recently (as suggested by observations of Lyman-α clouds). There is also ambiguity as to whether they comprise a thick disc, as proposed by Gates & Gyuk (2000), or a spheroidal halo.

References

Alcock, C. et al. 1997, ApJ, 486, 697
Alcock, C. et al. 2000 (astro-ph/0001272)
Alexander, T. 1995, MNRAS, 274, 909
Alfonso, A. et al. 1997, ApJ, 99, L12
Ansari, R. et al. 1997, A&A, 324, 843
Ansari, R. et al. 1999, A&A, 344, L49
Ashman, K.A. 1990, MNRAS, 247, 662
Ashman, K.A. & Carr, B.J. 1991, MNRAS, 249, 13
Baganoff, F.K. & Malkan, M.A. 1995, ApJ, 444, L13
Bahcall, J.N., Flynn, C. & Gould, A. 1992, ApJ, 389, 234
Bahcall J.N. et al. 1995, ApJ, 435, L31
Bond, J.R., Carr, B.J. & Hogan, C.J. 1991, ApJ, 367, 420
Boughn, S.P. & Saulson, P.R. 1983, ApJ, 265, L55
Burkert, A. & Silk, J. 1999 (astro-ph/9904159)
Burles, S. & Tytler, D. 1998a, ApJ, 499, 699
Burles, S. & Tytler, D. 1998b, ApJ, 507, 732
Burles, S. et al. 1999, Phys. Rev. Lett., 82, 4176
Carr, B.J. 1994, ARA&A, 32, 531
Carr, B.J. & Lacey, C.G. 1987, Ap.J. 316, 23
Carr, B.J. & Sakellariadou, M. 1998, ApJ, 516, 195
Carr, B.J., Bond, J.R. & Arnett, W.D. 1984, ApJ, 277, 445
Carswell R.F. et al. 1994, MNRAS, 268, L1
Cen, R. & Ostriker, J.P. 1999, ApJ, 514, 1
Charlot, S. & Silk, J. 1995, ApJ, 445, 124
Copi, C.J., Schramm, D.N. & Turner, M.S. 1995, Phys.Rev.Lett., 75, 3981
Crézé, M. et al. 1998, A&A, 329, 920
Crawford, M. & Schramm, D.N. 1982, Nature, 298, 538
Crotts, A.P.S. & Tomaney, A.B. 1997, ApJ, 473, L87
Dalcanton, J. et al. 1994, ApJ, 424, 550
De Blok, W.J.G. & McGaugh, S. 1997, MNRAS, 290, 533
De Paolis, F. et al. 1995, A&A, 295, 567; 299, 647
De Paolis, F., Ingrosso, G., Jetzer, Ph. & Roncadelli, M. 1999, ApJ, 510, L103
De Rujula, A., Jetzer, Ph. & Masso, E. 1992, A&A, 254, 99
Draine, B.T. 1998, ApJ, 509, L41
Elson, R.A. et al. 1999, in Stellar Populations in the Magellanic Clouds, in press
Feltzing, S., Gilmore, G. & Wyse, R.F.G. 1999, ApJ, 516, L17
Fich, M. & Tremaine, S. 1991, ARA&A, 29, 409
Field, B.D., Freese, K. & Graff, D.S. 2000, ApJ, 534, 265
Fixsen, D.J. et al. 1998, ApJ, 508, 123

Flynn, C. & Fuchs, B. 1995, MNRAS, 270, 471
Freese, K., Fields, B. and Graff, D. 2000, in The First Stars - MPA/ESO Workshop (astro-ph/0002058)
Fukugita, M., Hogan, C.J. & Peebles, P.J.E. 1998, ApJ, 503. 518
Gates, E.I. & Gyuk, G. 2000, preprint
Gibson, B.K. & Mould, J.R. 1997, ApJ, 482, 98
Gilmore, G. 1999, in The Identification of Dark Matter, ed. N.J.C. Spooner & V. Kudryavtsev (World Scientific), p 121
Gilmore, G. & Unavane, M. 1998, MNRAS, 301, 813
Gott, J.R. 1981, ApJ, 243, 140
Gould, A., Flynn, C. & Bahcall, J. 1998, ApJ, 503, 798
Graff, D.S., Freese, K., Walker, T.P. & Pinsonneault, M.H. 1999, ApJ, 523, L77
Green, A. 1999, ApJ, 537, 000 (astro-ph/9912424)
Hansen, B.M.S. 1998, Nature, 394, 860
Hansen, B.M.S. 1999, ApJ, 517, L39
Hawkins, M.R.S. 1993, Nature, 366, 242
Hawkins, M.R.S. 1996, MNRAS, 278, 787
Hawkins, M.R.S. 1999, in The Identification of Dark Matter, ed. N.J.C. Spooner & V. Kudryavtsev (World Scientific), p. 206
Henstock, D.R. 1996, PhD thesis (Manchester University)
Hewitt, J.N. 1986, PhD thesis (MIT)
Hills, J.G. 1986, AJ, 92, 595
Hodgkin, S.T. et al. 2000, Nature, 403, 57
Hogan, C.J. 1999, in Inner Space/Outer Space II (astro-ph/9912107)
Honma, M. & Kan-ya, Y. 1998, ApJ, 503, L139
Hut, P. & Rees, M.J. 1992, MNRAS, 259, 27P
Ibata, R.A., Richer, H.B., Gilliland, R.L. & Scott, D. 1999, ApJ, 524, L95
Ibata, R.A. et al. 2000, ApJ, in press (astro-ph/0002138)
Irwin, M.J. et al. 1989, AJ, 98, 1989
James, P. & Casali, M.M. 1996, Spectrum 9, 14
Jedamzik, K. 1997, Phys.Rev.D., 55, R5871
Kassiola, A., Kovner, I. & Blandford, R.D. 1991, ApJ, 381, 6
Kerins, E. 1997, A&A, 328, 5
Kerins, E. & Carr, B.J. 1994, MNRAS, 266, 775
Kerins, E. et al. 2000, MNRAS, in press (astro-ph/000256)
Koopmans, L.V.E. & de Bruyn, A.G. 2000, A&A, 356, 391
Kuijken, P. & Gilmore, G. 1989a, MNRAS, 239, 571
Kuijken, P. & Gilmore, G. 1989b, MNRAS, 605, 651
Kroupa, P., Tout, C. & Gilmore, G. 1993, MNRAS, 262, 545
Lacey, C.G. & Ostriker, J.P. 1985, ApJ, 299, 633
Lequeux, J. et al. 1996, A&A, 312, L1

Lequeux, J. et al. 1998, A&A, 334, L9
Lewis, G.F. et al. 1998, MNRAS, 295, 573
Loveday, J. et al. 1997, ApJ, 489, L29
Madau, P. & Pozzetti, L. 2000, MNRAS, 312, L9
Mathews, G.J., Schramm, D.N. & Meyer, B.S. 1993, ApJ, 404, 476
Melchior, A-L. et al. 1999, A&A, 339, 658
Metcalf, R.B. & Silk, J. 1999, ApJ, 519, L1
Moore, B. & Silk, J. 1995, ApJ, 442, L5
Navarro, J.F., Frenk, C.S. & White, S.D.M. 1997, ApJ, 490, 493
Paczyński, B. 1986a, ApJ, 304, 1
Paczyński, B. 1986b, ApJ, 308, L43
Paczyński, B. 1996, ARA&A, 34, 419
Persic, M. & Salucci, P. 1992, MNRAS, 258, 14P
Pfenniger, D., Combes, F. & Martinet, L. 1994, A&A, 285, 79
Press, W.H. & Gunn, J.E. 1973, ApJ, 185, 397
Rudy, R.P. et al. 1997, Nature 387, 159
Rugers, H. & Hogan, C.J. 1996, ApJ, 459, L1
Ryu, D., Olive, K.A. & Silk, J. 1990, ApJ, 353, 81
Sackett, P.D. et al. 1994, Nature, 370, 441
Samurovic, S., Cirkovic, M.M. & Milosevic-Zdjelar, V. 2000, MNRAS, in press
Sciama, D.W. 2000, MNRAS, 312, 33
Schmidt , R. & Wambsganss, J. 1998, A&A, 335, 379
Smecker, T.A. & Wyse, R.F.G. 1991, ApJ, 372, 448
Songaila, A. et al. 1994, Nature, 368, 599
Surdej, J. et al. 1993, AJ, 105, 2064
Tadros, H., Warren, S. & Hewett, P. 1998, New Astron. Rev., 42, 115
Tamanaha, C.M., Silk, J., Wood, M.A. & Winget, D.E. 1990, ApJ, 358, 164
Thomas, P. & Fabian, A.C. 1990, MNRAS, 246, 156
Turner, M.S. 1999, in Physics in Collision (Ann Arbor, MI, 24 - 26 June 1999), ed. M. Campbell and T.M. Wells (World Scientific, NJ) (astro-ph/9912211)
Tytler, D., Fan, X.M. & Burles, S. 1996, Nature, 381, 207
Walker, M.A. 1999, MNRAS, 306, 504; 308, 551
Walker, M.A. & Wardle, M. 1998, ApJ, 498, L125
Webb, J.K. et al. 1997, Nature 388, 250
Webster, R.L. et al. 1991, AJ, 102, 1939
Weinberg, D.H. et al. 1997, ApJ, 490, 564
Yost, S.A. et al. 1999, ApJ, in press (astro-ph/9908364)
Zaritsky, D. et al. 1993, ApJ, 405, 464
Zepf, S.E. et al., 2000, AJ, 119, 1701
Zheng, Z. et al. 1999, AJ, 117, 2757

Discussion

Bradbury: The matrix of BDM candidates fails to consider the possibility of astroengineered supercomputers. Why assume the universe is dead?

Carr: It is true that "Dyson-sphere-like" objects were not considered.

Bradbury: The possible discovery of planets makes this an omission worth considering.

Graff: Dyson spheres may be ruled out by microlensing signatures, due to mass distribution.

Bradbury: I would comment that classical Dyson Spheres would not give the same signatures as possible Matrioshka Brain architectures.

Carr: Penny, would you please comment on red halos?

Sackett: The presence of faint extended light about NGC 5907 seems fairly firmly established now. The argument has been about the source of this light. The results of Zepf et al (2000) place new constraints on the source, since only ≤1 giant was found in the HST NICMOS field. I wouldn't say that the steep IMF was the favored solution, but the other two possibilities conflict with other measurements, either independent distance or color estimates. It is fair to say that it remains a puzzle.

Gould: How severe is the constraint on Ω_B between (0.019-0.017) and the present WD density (if attributed to microlensing)?

Carr: If one accepts both the cosmological nucleosynthesis bound on Ω_B of $0.019h^{-2}$ and the Lyman-α cloud estimate of $0.017h^{-2}$ at $z \approx 3$, then one has an upper limit $0.002h^{-2}$ on the density of baryons that can have gone into pregalactic white dwarfs (i.e. at $z > 3$). The density required for the microlensing white dwarfs depends on the extent to which they are preferentially concentrated within 50 kpc (compared to the WIMPs), but even in the most extreme case (e.g. the extended protodisc model of Gates) it would exceed $0.002h^{-2}$. This suggests that one needs protogalactic white dwarfs forming at $z < 3$. However, it should be cautioned that the Lyman-α limit depends on other cosmological parameters (e.g. the UV flux) so it is probably not as firm as the nucleosynthesis limit.

Are there enough MACHOs to fill the Galactic Halo?

Thierry Lasserre (EROS collaboration)
CEA/Saclay, DSM/DAPNIA/SPP, F-91191 Gif-sur-Yvette, France

Abstract. EROS2 (Expérience de Recherche d'Objets Sombres) is a second generation microlensing experiment operating since mid-1996 at the European Southern Observatory (ESO) at La Silla (Chile). We present the two year analysis from our microlensing search towards the Small Magellanic Cloud (SMC). We give results from our search towards the Large Magellanic Cloud (LMC); 17.5 million stars spread over 43 square degrees have been analyzed and two candidates have been found. We combine these new results from the search for microlensing towards the LMC with limits previously reported by EROS1 and EROS2 towards both the LMC and SMC. The derived upper limit on the abundance of stellar mass MACHOs rules out such objects as an important component of the Galactic halo if their mass is in the range $[10^{-7} - 1]$ $M_\odot$.

1. Introduction

Following the proposal of (Paczyński 1986) to use gravitational microlensing to probe the dark matter content of the Galactic halo, several groups have been observing both Magellanic Clouds for a decade, and the first candidates in these directions were reported in 1993 by the EROS and MACHO collaborations (Aubourg et al. 1993; Alcock et al. 1993). The main result from the first generation experiments is the strong limit derived by both EROS1 and MACHO groups on the fraction of Galactic dark matter in planet-sized objects, from the absence of candidates with duration shorter than 10 days (Aubourg et al. 1995; Alcock et al. 1996; Renault et al. 1998; Alcock et al. 1998).

In 1996, based on six candidates discovered towards the LMC (with an average duration of about 40 days), the MACHO group estimated the optical depth at half that required to account for the dynamical mass of the standard spherical dark halo (Alcock et al. 1997a). From similar observations, EROS1 found two candidates in the same range of timescales and decided to set an upper limit on the halo mass fraction (Ansari et al. 1996).

Since the results for the mass region above 0.01 $M_\odot$ were controversial, EROS started a second phase (EROS2) in order to increase the number of monitored stars in the Magellanic Clouds by a factor ten. In 1998, the analysis of the first year of data towards the SMC allowed the detection of one microlensing event (Palanque-Delabrouille et al. 1998; see also Alcock et al. 1997b). This single detection allowed EROS2 to further constrain the halo composition, excluding in particular the MACHO central value (Afonso et al. 1999). First results from

Machos and the Galactic Halo? 55

EROS2 towards the LMC came out in 1999 (Lasserre 1999) and are reported in this paper.
From all the independent EROS data sets towards both the LMC and SMC, there is growing evidence that MACHOs are not a substantial part of the Galactic dark halo. To be more quantitative, we combine all the EROS limits discussed above in order to set a strong constraint on the amount of compact halo objects with masses less than 4 $M_\odot$ (Lasserre 2000a). To conclude, we compare this limit with the new MACHO result derived from 5.7 years of LMC observation (Alcock et al. 2000).

2. Experimental setup and observations

The telescope, camera, telescope operation and data reduction are as described in Bauer et al. (1997) and Palanque-Delabrouille et al. (1998). Towards the SMC, 8.6 square degrees have been analysed, covering the period from July 1996 to March 1998, with an average sampling of once every three days. Towards the LMC, 25 square degrees spread over 43 one square degree fields have been analysed from August 1996 to May 1998, with a sampling of one measurement every six days on average.

3. Back to EROS1 results

Two candidates were found in the analysis of data from the EROS1 photographic-plate experiment (Ansari et al. 1996). Because of technical reasons, we could not yet recover the light curve of the candidate EROS1-LMC-1 in the EROS2 data; here, we are only dealing with EROS1-LMC-2. The source star involved in the event was found to be variable (Ansari et al. 1995), but microlensing fits gave a better description of the measurements, and it was accepted as a candidate in order to set an upper limit on the Galactic halo mass fraction (Ansari et al. 1996). Its follow-up by EROS2 revealed a new variation in March 1999, eight years after the first one; this variation was also reported by the MACHO collaboration. It is worth noting that the periodic modulation is found to be the same, within the error bars, in both EROS1 and EROS2 data sets ($P = 2.8169 \pm 0.0005$ days). In conclusion, since the probability to have two microlensing events on the same star is extremely low, the two bumps on the light curve are probably due to intrinsic variations of the source star. Consequently, we will not consider EROS1-LMC-2 as a candidate in what follows.

4. SMC 2 year data analysis

One candidate, MACHO-SMC-97-1/EROS2-SMC-97-1, was found in this analysis among 5.3 million monitored stars (Palanque-Delabrouille et al. 1998; Alcock et al. 1997b). The result of a microlensing fit leads to an Einstein radius crossing time $t_E = 129$ days, significantly longer than the average duration of LMC events $< t_E >_{\text{LMC}} \sim 40$ days. The χ^2 is 261 for 279 degrees of freedom including the fact that the source is intrinsically variable ($P = 5.124$ days). Light curves and residuals of the fit are shown in Figure 1. As discussed in Palanque-Delabrouille

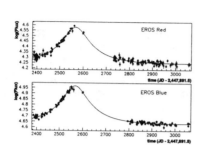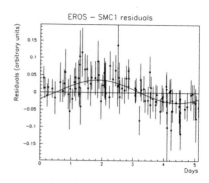

Figure 1. Left: light curve of the microlensing event EROS2-SMC-97-1. The fit shown does not include the periodic modulation. Right: light curve of residuals of the simple microlensing fit folded to one period $P = 5.124$ days. This amplitude of variation is about 5%.

et al. (1998), we consider that the long duration of the SMC candidate together with the absence of any detectable parallax, in our data as well as in that of the MACHO group (Alcock et al. 1997b), indicates that the lens is likely to be located in the SMC. As an example, we expected about four events for one solar mass deflectors (with a standard halo fully comprised of MACHOs.) To be conservative, we can assume that this single lens observed belongs to the Galactic dark halo to further constrain the halo composition. In that way we excluded in particular that more than 50% of the dark halo is made up of $0.01 - 0.5$ $M_\odot$ compact objects (Afonso et al. 1999).

5. LMC 2 year data analysis

In this section we report the discovery of (only) 2 candidates among 17.5 million light curves analysed. We discuss the problem of the background events, singling out the case of the blue bumpers than can mimic a simple microlensing signal. Results of the present analysis are as reported in (Lasserre 1999) and (Lasserre et al. 2000a).

5.1. Selection of the light curves consistent with a microlensing shape

To analyse the LMC data set we used a program independent from that used in the SMC study presented in the above section, with largely different selection criteria. Details of the cuts will be provided in (Lasserre et al. 2000b) and (Lasserre 2000c).

We first select the 8% "most variable" light curves, in order to build "enriched" star catalogues; working on this subset of stars, we apply a first set of cuts to preserve, in each colour separately, the light curves that exhibit a significant variation.

To begin, we calculate the baseline flux in the light curve, basically the most probable flux. We then search for a sequence of consecutive measurements that are all on the same side of the baseline flux (hereafter "run"). The light curves selected either have an abnormally low number of runs over the whole light curve, or show one long run of at least 5 valid measurements that is statistically significant. We then ask for a minimum signal-to-noise ratio by requiring that the group of 5 most luminous consecutive measurements be significantly further from the baseline than the average spread of the measurements. We also ask for a smooth time variation inside the most significant run.

After the selection of significant smooth bumps we compare the measurements with the best fit point-lens point-source microlensing light curve (hereafter "simple microlensing"). This allows us to reject variable stars whose light curve differs too much from simple microlensing, and it is sufficiently loose not to reject light curves affected by blending, parallax or the finite size of the source, and most cases of multiple lenses or sources.

After this second set of cuts, stars selected in at least one passband represent less than 0.01% of the initial sample. The third set of cuts are dedicated to identify and reject variable star background.

5.2. Removing the intrinsic variable star background

Almost all remaining stars are located in two thinly populated zones of the colour-magnitude diagram. The first zone contains stars brighter and much redder than those of the red clump; variable stars in this zone are rejected if they vary by less than a factor two or have a very poor fit to simple microlensing. The second zone is the top of the main sequence. Here we find that selected stars, known as blue bumpers (Alcock et al. 1997a), can mimic a simple microlensing signal; but they display variations that are very often smaller than 60% or lower in the visible passband than in the red one. They cannot correspond to microlensing plus blending with another unmagnified star either, as it would imply blending by even bluer stars, which is very unlikely. We thus reject all candidates from the second zone exhibiting these two features.

5.3. Candidates

The fourth set of cuts tests for compatibility between the light curves in both passbands. We retain candidates selected in only one passband if they have no conflicting data in the other passband. For candidates selected independently in the two passbands, we require that their largest variations coincide in time. Only two stars pass all cuts; these events cannot be labeled as "golden" candidates as the agreement with simple microlensing is not excellent. Microlensing fit parameters are given in Table 1, and light curves of the candidates are shown in Figure 2, in the blue passband.

Table 1. Results of microlensing fits to the two new LMC candidates; t_E is the Einstein radius crossing time in days, u_0 the impact parameter, and $c_{bl}^{R(V)}$ the $R(V)$ blending coefficients. V_J and R_C correspond respectively to Johnson and Cousins passbands. Coordinates (α, δ) are given with equinox reference J2000.

	α	δ	u_0	t_E	c_{bl}^R	c_{bl}^V	χ^2/dof	V_J	R_C
LMC-3	$5^h30'50.1''$	$-67°36'10.4''$	0.23	41	0.76	1	208/145	22.4	21.8
LMC-4	$5^h09'01.2''$	$-69°01'01.4''$	0.20	106	1	1	406/150	19.7	19.4

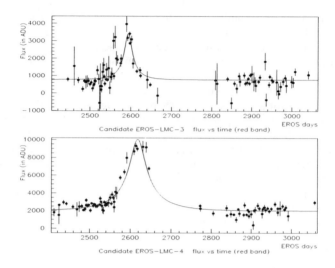

Figure 2. Light curves of candidates EROS2-LMC-3 and EROS2-LMC-4 in the blue passband; time is given in days since Jan. 1, 1990 (JD 2,447,892.5). We superimpose the best point-lens point-source fit.

5.4. Efficiency

The tuning of each cut and the calculation of the microlensing detection efficiency are done with simulated simple microlensing light curves, as described in (Palanque-Delabrouille et al. 1998). For the efficiency calculation, microlensing parameters are drawn uniformly in the following intervals: time of maximum magnification t_0 within the observing period ±150 days, impact parameter normalised to the Einstein radius $u_0 \in [0, 2]$ and timescale $t_E \in [5, 300]$ days. All cuts on the data were also applied to the simulated light curves. The effect of blending, which is the main source of systematic error, was studied with synthetic images by Palanque-Delabrouille (1997). The agreement between the method used in the present analysis and the method that takes blending into account is within 10%; this is because two effects lead to a partial compensation: efficiency is decreased as blending lowers the observed magnification and event duration, but efficiency normalised to the brightest star of the blend is increased as it allows to monitor at least twice the initial numbers of stars. The detection efficiency for the present analysis is given in Table 2.

Table 2. Detection efficiency in % as a function of the Einstein radius crossing time t_E in days, normalised to events generated with $u_0 < 1$, and to $T_{\text{obs}} = 2$ yrs. We analysed $N_* = 17.5 \times 10^6$ stars.

t_E	5	11	18	28	45	71	112	180	225	280
ϵ	2	5	11	15	19	23	26	25	18	2.5

6. Combined limits on Galactic halo MACHOs

Up to now, a total of four microlensing candidates have been observed by EROS towards the Magellanic Clouds, one from EROS1 and two from EROS2 towards the LMC, and one towards the SMC. As discussed previously, the lens of event EROS2-SMC-97-1 is likely to be located in the SMC. For that reason, the limit derived below uses the three LMC candidates; for completeness, we also give the limit corresponding to all four candidates.

The limits on the contribution of MACHOs to the Galactic halo are obtained by comparing the number and durations of microlensing candidates with those expected from the so-called "standard" halo model described in Palanque-Delabrouille et al. (1998) as model 1. The model predictions are computed for each EROS data set in turn, taking into account the corresponding detection efficiencies (Ansari et al. 1996; Renault et al. 1998; Afonso et al. 1999; Table 2 above); the four predictions are then summed. In this model, all dark objects have the same mass M; the model predictions have been computed for many trial masses M in turn, in the range $[10^{-8} - 10^2]$ M$_\odot$.

The method used to compute the limit is described in (Ansari et al. 1996). Two ranges of timescale t_E are considered, smaller or larger than $t_E^{lim} = 10$ days. We can then compute, for each mass M and any halo fraction f, the combined Poisson probability for obtaining, in the four different EROS data sets taken as a whole, zero candidates with $t_E < t_E^{lim}$ and three or less (alt. four or less) with $t_E > t_E^{lim}$. For any value of M, the limit f_{max} is the value of f for which this probability is 5%. Given the timescale of the candidates, we consider $t_E^{lim} = 10$ days to be a conservative choice. We checked the influence of small variations of t_E^{lim} (in between 5 to 20 days); the difference, noticeable only for masses around 0.02 M$_\odot$, is always less than 5%. The 95% C.L. exclusion limit derived from this analysis on the halo mass fraction, f, for any given dark object mass, M, is given in Figure 3. The solid line, that constitutes our main result, corresponds to the three LMC candidates; the dashed line includes the SMC candidate in addition. The standard spherical halo model fully comprised of objects with any mass function inside the range $[10^{-7} - 4]$ $M_\odot$ is ruled out at 95% C.L.

7. Comparison with the MACHO results towards the LMC

The sensitivity of a microlensing experiment is proportional to $N_* T_{obs} \epsilon(t_E)$. The total sensitivity associated to our combined limit is given by the sum of the sensitivity of all EROS data sets. In the region of stellar mass objects, corresponding to a timescale of about 50 days, the new LMC data contributes about 60% to our total sensitivity, the SMC and EROS1 LMC data contribute 15% and 25% respectively. Our total sensitivity can be compared with that of MACHO which is about twice greater (Alcock et al. 2000). The combined limit presented in this paper gives strong constraints for halo lenses with masses less than 1 M$_\odot$. Nevertheless, the new MACHO result derived from the analysis of 10.7 million stars monitored for 5.7 years towards the LMC is now consistent with present as well as previous EROS results (Alcock et al. 2000). It is important to point out that, although the EROS and MACHO results are consistent, the interpretation of the microlensing events remains different. Our position is the following:

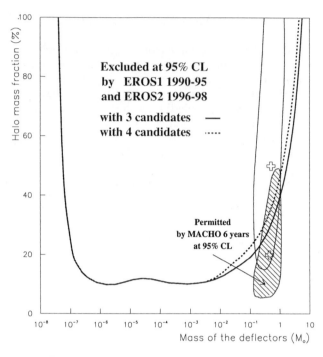

Figure 3. 95% C.L. exclusion diagram on the halo mass fraction in the form of compact objects of mass M, for the standard halo model (4×10^{11} M$_\odot$ inside 50 kpc), from all LMC and SMC EROS data 1990-98. The solid line is the limit inferred from the three LMC microlensing candidates; the dashed line includes in addition the SMC candidate. The MACHO 95% C.L. accepted region is the hatched area, with the preferred value indicated by the cross (Alcock et al. 2000). The area defined by the thin line around the cross at 50% and 0.5 M$_\odot$ represent the region previously allowed by the MACHO 2-year LMC analysis (Alcock 1997a).

we cannot exclude some variable star background in our sample of candidates. Consequently, we only set upper limits on the fraction of the Galactic halo comprised of dark compact objects with masses up to a few solar masses; we do not want to quote a non-zero lower limit.

8. Discussion

There is now strong evidence, from both EROS and MACHO data sets, that MACHOs with masses less than 1 M$_\odot$ are not a substantial component of the Galactic dark halo (Lasserre et al. 2000a; Alcock et al. 2000). It is important to point out two things: first, the EROS combined limits include both LMC and SMC observations; and second, stars used in the analysis are spread over a large solid angle in both Magellanic Clouds, 43 square degrees towards the LMC and 10 towards the SMC. Whereas EROS and MACHO results are consistent, our interpretations remain different, mainly on two points: the variable star background discussed above and the localization of the lenses. From a study of the spatial spread of their events on the LMC, the MACHO group argues for the discovery of a halo lens population rather than Magellanic Cloud lenses (Alcock et al. 2000). Nevertheless, since the LMC model used for this analysis (Gyuk et al. 2000) is not unique (cf Aubourg et al. 1999 and references therein) we do not think that this interpretation can be considered conclusive.

While the comparison of LMC and SMC events is often suggested as a way to measure the flattening of the dark halo (Alcock et al. 1997b), we think that the comparison of both numbers and durations of candidates should mainly be used as a test of the halo hypothesis. At present, with the EROS data alone, such a comparison is not yet significant; consequently, continued monitoring of the SMC is very important. In addition, the longest LMC and SMC events will show some signs of parallax if lenses lie in the halo (Gould 1998).

References

Afonso C. et al. (EROS), 1999, A&A, 344, L63
Alcock C. et al. (MACHO), 1993, Nature, 365, 621
Alcock C. et al. (MACHO), 1996, AJ, 471, 774
Alcock C. et al. (MACHO), 1997a, AJ, 486, 697
Alcock C. et al. (MACHO), 1997b, AJ, 491, L11
Alcock C. et al. (MACHO), 1998, AJ, 499, L9
Alcock C. et al. (MACHO), 2000, ApJ, submitted (astro-ph/0001272)
Ansari R. et al. (EROS), 1995, A&A, 299, L21
Ansari R. et al. (EROS), 1996, A&A, 314, 94
Aubourg É. et al. (EROS), 1993, Nature, 365, 623
Aubourg É. et al. (EROS), 1995, A&A, 301, 1
Aubourg É. et al., 1999, A&A, 347, 850
Bauer F. et al. (EROS), 1997, in Proceedings of the "Optical Detectors for Astronomy" workshop (ESO, Garching)
Gould A., 1998, AJ, 506, 000
Gyuk et al., 2000, ApJ, 535, 000 (astro-ph/9907338)
Lasserre T. (EROS), 1999, in Proceedings of "Gravitational lensing : Recent Progress and Future Goals", Boston (astro-ph/9909505)
Lasserre T. et al. (EROS), 2000a, A&A, 355, L39
Lasserre T. et al. (EROS), 2000b, in preparation
Lasserre T., 2000c, PhD thesis, Université de Paris 6
Paczyński B., 1986, AJ, 304, 1
Palanque-Delabrouille N., 1997, PhD thesis, University of Chicago and Université de Paris 7
Palanque-Delabrouille N. et al. (EROS), 1998, A&A, 332, 1

Renault C. et al. (EROS), 1998, A&A, 329, 522

Discussion

Bennett: What are the coordinates of your two new microlens candidates in the LMC?

Lasserre: I will send you the coordinates by e-mail.

Gould: Did you check the double bump events for binary-source and binary-lens events?

Lasserre: No, we didn't.

Kerins: Does your detection efficiency vary much across your fields?

Lasserre: Yes, The efficiency varies a lot, field to field, principally because the sampling is different, especially for short duration events (less than 20 days) – one point every 3 days towards the centre, one point every 5-6 days in the outermost regions. NB: We use all the fields to calculate the efficiency.

Binney: I found your discovery of a second bump in what had seemed a good microlensing event very disturbing. Is it possible that further years of observation will reveal secondary bumps in the light curves of other "microlensed" stars and thus gradually bring down the optical depth?

Lasserre: It wasn't a "Golden" event, because soon after its discovery in 1990, we found that it (the source star) is an intrinsic variable star (Ansari et al., AA, 299, L21-24 (1995)), but we couldn't interpret the first bump by another interpretation than microlensing. So it was accepted, but only to set an *upper limit*. We never claimed that it was a *real* microlensing event.

Bennett: The EROS #2 microlensing candidate has been identified as a variable star since shortly after it was announced. So, it is not the case that a "good" microlensing candidate was removed due to a recent light curve bump.

Lasserre: I agree with that. It was *not* a good microlensing candidate, but it was kept only to be conservative, in order to put an upper limit on the halo mass fraction. We couldn't explain the 1.1 magnitude bump (peak on 29 Dec 1990) with an eclipsing binary model. See also Ansari et al. (1995).

MACHO Results from 5.7 Years of LMC Observations

K. Cook and the MACHO Collaboration
Lawrence Livermore National Laboratory,MS L-413, P.O.Box 808, Livermore, CA 94550, USA

Abstract. I will summarize the MACHO Collaboration's results from an analysis of 5.7 years of data monitoring 12 million stars toward the Large Magellanic Cloud for microlensing. We have utilized improved efficiency determinations, an improved likelihood analysis and a more thorough treatment of systematic errors. The events we have detected (13 to 17 depending upon analysis criteria) represent a clear excess of microlensing events over that expected from known populations. If this excess results from compact halo objects, it would represent objects of about 0.5 solar mass comprising about 20% of the halo dark matter.

LMC Self-Lensing Constraints

G. Gyuk, N. Dalal, and K. Griest

University of California, San Diego, Department of Physics, 9500 Gilman Dr., La Jolla, CA 92093, U.S.A.

Abstract. LMC self-lensing, which invokes ordinary LMC stars as the long sought-after lenses, has recently gained considerable popularity as a possible solution to the microlensing conundrum. We carefully examine the full range of LMC self-lensing models, including for the first time the contribution of the LMC bar in both sources and lenses. We review the pertinent observations made of the LMC, and show how these observations place limits on such self-lensing models. We find that, given current observational constraints, *no* purely LMC disk models are capable of producing optical depths as large as that reported in the MACHO collaboration 5-year analysis. We also introduce a new quantitative measure of the central concentration of the microlensing events and show that it discriminates well between disk/bar self-lensing and halo microlensing. We find that an LMC halo geometry may be able to explain the observed events. However, since *all* known LMC tracer stellar populations exhibit disk-like kinematics, such models will have difficulty being reconciled with observations. We find a self-lensing optical depth contribution for the LMC between $0.47 \cdot 10^{-8}$ and $7.84 \cdot 10^{-8}$, with $2.44 \cdot 10^{-8}$ being the value for the set of LMC parameters most consistent with current observations.

1. Introduction

If the handful of events discovered towards the Magellanic Clouds are due to a population of objects in an extended Milky Way halo, they can be interpreted to represent between 20% and 100% of the dark matter in our Galaxy (Alcock et al. 1997; Gates, Gyuk & Turner 1996). However, the most probable masses of these objects lie in the 0.1 to $1 M_\odot$ mass range (Alcock et al. 1997). Such a large number of objects in this mass range is quite problematic (e.g. Fields, et al. 1998). Therefore alternatives to MW halo lensing have been sought to explain the LMC microlensing events.

One alternative, proposed by Sahu (1994a) and Wu (1994), suggests that stars within the LMC itself, lensing other LMC stars, could produce the observed optical depth. This claim has been disputed by several other groups (Gould 1995; Alcock et al. 1997), who claim that the rate of LMC self-lensing is far too low to account for the observed rate.

The main reason that previous work has produced such discordant results is that different papers have treated the LMC differently. For example, Gould

(1995) and Alcock et al. (1997) treated the LMC as a thin exponential disk, while Sahu (1994a) and Aubourg et al. (1999) modeled the LMC as being much more extended along the line of sight. These two qualitatively different prescriptions give wildly different predictions for the optical depth and rate of self-lensing.

We have provided a set of calculations of LMC microlensing that treats LMC self-lensing in a systematic, thorough and realistic fashion including contributions of both disk and bar. We relate the known LMC observations to microlensing predictions, and provide a framework within which future observations will easily translate into microlensing predictions. We hope this will serve as a general basis for comparison between observation and theory in the future.

2. Observations and Models of the LMC

To answer the question of whether LMC self-lensing is significant, we must understand the distribution of stars within the LMC. If the LMC is a thin disk, then the small rates and optical depths derived by Gould and others will be valid. Conversely, if the LMC is puffy, then the large rates and optical depths claimed by Sahu and others will be correct. The basis for any description of the LMC is the set of observations that have been made of the LMC. We therefore turn to the current state of observations of the LMC.

2.1. LMC disk

Since the pioneering work of de Vaucouleurs (1957), it has been well accepted that the stellar component of the LMC has an exponential profile. The value de Vaucouleurs measured for the exponential scale length, R_d, continues to agree with the current value of 1.8° (Alves et al. 1999), which corresponds to a physical scale length of 1.6 kpc for a distance to the LMC of 50 kpc. In addition to this stellar population, the LMC possesses significant quantities of HI gas, which has recently been mapped out by Kim et al. (1998). Their images show clear spiral structure in the gas, supporting the notion that the LMC is a typical dwarf spiral galaxy. The gas is confined to a thin disk, inclined at roughly 30°, with a position angle $\sim 170°$. Based on these observations, in this paper we describe the stellar disk by a double exponential profile, given by

$$\rho_d = \frac{M_{disk}}{4\pi z_d R_d^2} e^{-\frac{R}{R_d} - |\frac{z}{z_d}|},$$

where R_d is the radial scale length, z_d is the vertical scale height, and M_{disk} is the disk mass. Note that R_d is well constrained by observation, but we have some leeway in the scale height and in the mass of the disk. The disk is inclined at angle i to our line of sight and has position angle PA.

2.2. LMC Bar

The LMC hosts a prominent bar, of size roughly $3° \times 1°$. The bar has the unusual property of being offset from the dynamical center of the HI gas. The offset is $\approx 1.2°$ (Westerlund 1997), corresponding to a physical offset of ~ 1 kpc. The kinematics of the LMC bar are consistent with solid body rotation (Odewahn

1996), as is seen in numerous barred galaxies. The distribution of matter within the bar is not well known. With little guidance from observations, we have treated the bar simply as a triaxial Gaussian, with axis ratios chosen to match the observed ratios. We let

$$\rho_b = \frac{M_{bar}}{(2\pi)^{3/2} x_b y_b z_b} e^{-\frac{1}{2}[(\frac{x}{x_b})^2+(\frac{y}{y_b})^2+(\frac{z}{z_b})^2]},$$

where x, y, z are coordinates along the principal axes of the bar, and x_b, y_b, z_b are the scale lengths along the three axes. M_{bar} is the total mass of the bar. This form is somewhat similar to models used to describe other galactic bars, e.g. Dwek et al. (1995). We place the bar in the same plane as the disk; however, we place the bar center at the position of the observed bar centroid, at ($\alpha = 5h24m, \delta = -69°48'$) (de Vaucouleurs 1957). We use a position angle for the bar of 120°.

The HI maps of Kim et al. show that the LMC bar does not dominate the central dynamics. We thus apply the following restriction: $M_b < 25\% M_d$ to avoid bar domination. This agrees nicely with the estimates of Sahu (1994b), who suggested a bar to disk mass ratio in the range 15-20% based on luminosity considerations.

Table 1. Observed velocity dispersions for various populations.

Population	Study	Velocity Dispersion	Age
supergiants	Prevot et al. (1989)	6	young
HII	"	6	young
HI	Hughes et al. (1991)	5.4	young
VRC	Zaritsky & Lin (1997)	18.4	young?
PNe	Meatheringham et al. (1998)	19.1	intermediate
OLPV	Hughes et al. (1991)	33	old
ILPV	"	25	intermediate
YLPV	"	12-15	young
OLPV	Bessel et al. (1986)	30	old
metalpoor giants	Olszewski (1993)	23-29	old
metalrich giants	"	16.0	intermediate?
new clusters	Schommer et al. (1992)	20	intermediate
old clusters	"	30	old
carbon stars	Kunkel et al. (1997)	15	young
CH stars (disk)	Cowley & Hartwick (1991)	10	yng/intermed?
CH stars (halo)	"	20-25	old

2.3. LMC Velocity Distribution and Vertical Scale Height

Perhaps the best determination of the inner velocity curve of the LMC is in the work of Kim et al. (1998). We supplement this by outer rotation curves derived from carbon stars (Kunkel et al. 1997) and clusters and planetary nebulae (PNe) (Schommer et al. 1992) to obtain a basic outline: the circular velocity rises rapidly in the first two kpc and then levels off and is flat at about 70 km/s out to at least 8 kpc. We approximate the rotation curve by solid body rotation out to a radius $r_{\text{solid}} = 2$ kpc, followed by flat rotation at $v_c = 70$ km/s.

These studies help to define the bulk motions of gas and stars in the LMC. However, it is velocity dispersions (and the implied scale heights) that are crucial in determining the optical depth and rate of microlensing. So let us consider measurements of the velocity dispersion of LMC populations. Table 1 lists some of the more recent kinematic studies of the LMC by population type, velocity dispersion and probable age. The general trend among the many kinematic studies of the LMC seems to be clear: tracers have velocity dispersions ranging from ~ 5 km/s for very young ages to ~ 30 km/s for the most ancient populations. Note that a spheroidal distribution would require velocity dispersions of $\sigma \approx v_c/\sqrt{2} \approx 50$ km/s. Thus *all* LMC populations studied to date have disk-like kinematics regardless of age (Olszewski et al. 1996).

From the velocity dispersions, let us now turn to the vertical scale heights. Bessel et al. (1986) estimated the vertical scale height of the oldest population to be roughly 0.3 kpc, while Hughes et al. (1991) estimated the scale height to be $\lesssim 0.8$ kpc. They emphasized that this was the oldest population, accounting for at most 2% of the mass of the LMC. The majority of the LMC disk, they contend, should possess a more compact vertical distribution and smaller vertical velocity dispersions. This is supported by RR Lyrae and cluster studies which suggest that the ancient extended populations make up considerably less than 10% of the LMC stars (Olszewski et al. 1996; Kinman et al. 1991). We thus allow our scale height (which should characterize the bulk of the LMC population) to range up to 0.5 kpc and adopt velocity dispersions in a corresponding range of 10-30 km/s. In theory these parameters should be tied together by the vertical Jeans equation (however see Weinberg 1999). In practice, our knowledge of the total mass and mass distribution of the LMC is poor enough that we simply note that the opposite extremes of these ranges (i.e. 10 km/s with 0.5 kpc and 30 km/s with < 0.2 kpc) are likely inconsistent.

Table 2. Estimates of the LMC mass. Note that some entries refer to specific components, such as the disk or halo, while other entries correspond to the LMC as a whole.

Study	Mass Estimate $M_\odot$	Radius kpc	Component
Hughes et al. (1991)	$6.0 \cdot 10^9$	4.5	Total
Kim et al. (1998)	$2.5 \cdot 10^9$		Disk
"	$3.4 \cdot 10^9$	8	Halo
Schommer et al. (1992)	$\sim 2.0 \cdot 10^{10}$	5	Total
"	$1.0 \cdot 10^{10}$	8	Total
Meatheringham et al. (1998)	$3.2 \cdot 10^9$		Disk
"	$6.0 \cdot 10^9$	5	Total
Kunkel et al. (1997)	$6.2 \cdot 10^9$	5	Total
"	$< 1.0 \cdot 10^{10}$		Total

2.4. LMC Mass

Estimates of the LMC mass (Table 2) range from only a few times 10^9 $M_\odot$ to $\sim 2 \cdot 10^{10} M_\odot$. Close inspection of these authors' methodologies, however, reveals a few regularities. The highest estimates are based on the spheroidal estimator of Bahcall & Tremaine (1981), which assumes both velocity isotropy and a spherical mass distribution. Since both of these conditions are likely to be violated, the spheroidal estimators should be taken as upper limits. A similar argument can be made for the point mass estimation of Kunkel et al (1997). With these caveats in mind the data seem consistent with a disk of perhaps $3 \cdot 10^9 M_\odot$ and a halo whose mass within 8 kpc is roughly $6 \cdot 10^9 M_\odot$. While the extremely high quality HI data of Kim et al. (1998) would appear to rule out a disk mass much in excess to this, the halo component is much more uncertain. We thus take $M_{disk} + M_{bar} \leq 5.0 \cdot 10^9 M_\odot$ and allow the total LMC halo mass within 8 kpc to range up to $1.5 \cdot 10^{10} M_\odot$.

Table 3. Model parameters. $M_{d+b} = M_{disk} + M_{bar}$. All masses are for an LMC distance of 50 kpc.

Parameter	Preferred Value	Allowed Range
inclination	30°	20 − 45°
R_d	1.8°	1.8°
z_d	0.3 kpc	0.1-0.5 kpc
v_c	70 km/s	60-80 km/s
L	50 kpc	45-55 kpc
σ_v	20 km/s	10-30 km/s
a_h	2 kpc	1-5 kpc
M_{d+b}(8kpc)	$3 \cdot 10^9 M_\odot$	$< 5 \cdot 10^9 M_\odot$
M_{dark}(8kpc)	$6 \cdot 10^9 M_\odot$	$< 1.5 \cdot 10^{10} M_\odot$
M_{bar}/M_{d+b}	0.15	0.05-0.25
$M_{stellar\ halo}$	$0 \cdot 10^8 M_\odot$	$0 - 5.0 \cdot 10^8 M_\odot$

3. LMC Optical Depths and Concentration

The measured total optical depth towards the LMC from the 5-year MACHO collaboration analysis is $\approx 1.3 \cdot 10^{-7}$ (Alcock et al. 2000). How does this compare with the predicted range of optical depths of the above models? The total optical depth for a particular set of model parameters is somewhat involved to calculate. Since the spatial distribution of the various source populations is important one cannot simply calculate a single optical depth. Instead, one needs to average the optical depths on an observed field by field basis. We choose to use the MACHO 5 yr fields to allow direct comparison with the observations. We find no combination of parameters within our ranges allows an optical depth greater than $8.0 \cdot 10^{-8}$. For the preferred parameter values the optical depth is $2.4 \cdot 10^{-8}$. Note that this value is *five times* smaller than the observed optical depth.

We have also calculated the optical depth due to a possible LMC MACHO halo. For a $6 \cdot 10^9 M_\odot$ LMC halo 100% composed of MACHOs, we find values of the optical depth between $\sim 7.5 \cdot 10^{-8}$ and $\sim 8.5 \cdot 10^{-8}$ depending on the other parameters. It is clear that a halo type configuration is much more effective at producing optical depth than the disk/bar. However, as discussed above, the

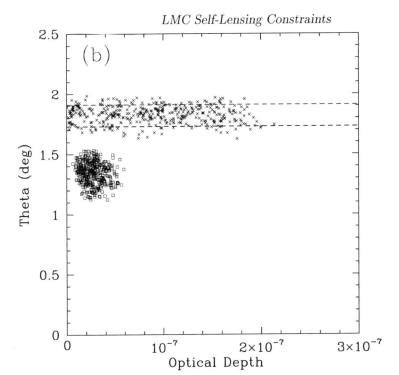

Figure 1. τ vs. $\tilde{\theta}$ for the 30 field set appropriate for the MACHO 5 year results. The open boxes are for the disk/bar, and the x's are for the LMC halo. The points were randomly selected uniformly in parameter space, within the allowed ranges. The dashed lines show the range for MW halo lensing.

mass of such a halo is tightly constrained by numerous observations unless it is non-luminous (i.e. a dark matter signal!).

One potentially powerful way of distinguishing Milky Way (MW) halo microlensing from LMC self-lensing is to compare the spatial distribution of the observed microlensing events with the predictions of LMC and halo models (Alcock et al. 1997). For microlensing events due to a MW halo population of lenses, the lens population is uniform across the source distribution, so one expects the events to be distributed in proportion to the LMC source density times the experimental efficiency. For LMC/LMC self lensing, both the sources and lenses are distributed like the LMC stars, so there should be a more rapid drop off of measured optical depth at large distances from the LMC bar. We quantify this observation using a simple measure. We write $\tilde{\theta} \equiv \langle \theta_{ij} \rangle$, where θ_{ij} is the angle on the sky between the location of events i and j, and the average is over all pairs. $\tilde{\theta}$ is a statistic that measures the average separation of events, and therefore the extent of the spatial distribution of events.

We have computed $\tilde{\theta}$ and τ for our models using the MACHO 5-year field sample. Figure 1 shows a scatterplot of τ versus $\tilde{\theta}$ for LMC disk-, LMC halo-, and MW halo models. The LMC disk/bar (open boxes) and the LMC halo (x's) models are resolved in both $\tilde{\theta}$ and τ. Unfortunately, the paucity of actual events limits our ability to rule out models based upon the observed event distribution. In addition, to be useful the monitored sources must span a sufficiently wide area, since obviously we will be unable to discern any intrinsic central concentration if the data sample only a small swath of the sky. Future data, however, with increased sky coverage and more events should be a strong discriminant between disk models and halo models. Note, however, that the plots also indicate that the LMC halo and MW halo models will probably remain degenerate.

4. Discussion

Overall, we find that self lensing models typically suffer two major defects. First, it is quite difficult for such models to produce enough lensing to account for the observed optical depth, while remaining within the bounds set by observation. Second, the optical depth due to disk or bar self-lensing is strongly concentrated on the sky, in contrast to the relatively uniform distribution of events seen to date. These two statements have a major caveat: if the LMC lenses are distributed in an extended or halo-like geometry, it is possible to produce the required optical depth, and the central concentration of the predicted events is significantly diminished. Such an extended or halo-like distribution, however, requires either a hitherto undetected stellar population, or a dark MACHO component to the LMC halo. If a dark LMC halo is invoked, then one might expect it to have a similar fraction of dark MACHOs as the Milky Way Halo. Otherwise, the presence of such a component in the LMC but not in the Galactic halo would be puzzling. On the other hand, if a stellar LMC halo with a luminosity function similar to the disk is invoked, direct observation of these LMC halo stars should be presently possible. Indeed, as discussed above, several stellar populations which correspond to stars that do trace the spheroid in our Galaxy have been observed in the LMC, and *all* of them fail to exhibit a halo geometry. Therefore, current observations suggest that the number of stars in any such stellar halo is small, and that an LMC stellar halo probably does not greatly contribute to microlensing.

Acknowledgments. It is a pleasure to thank Thor Vandehei for numerous enlightening and stimulating discussions on this and related matters. We acknowledge support from the U.S. Department of Energy, under grant DEFG-0390ER 40546, and from Research Corporation under a Cottrell Scholar award.

References

Alcock, C., et al. 1997, ApJ, 486, 697
Alcock, C. et al. 2000, ApJ, in press (astro-ph/0001272)
Alves, D. 1998, PhD thesis, University of California, Davis
Aubourg, E., et al. 1999, A&A, submitted (astro-ph/9901372)
Bahcall, J. & Tremaine, S. 1981, ApJ, 244, 805
Bennett, D., et al. 1996, Nucl. Phys. Proc. Suppl., 51B, 152-156
Bessel, M.S., Freeman, K.C., & Wood, P.R., 1986, ApJ, 310, 710
Boden, A., Shao, M. & Van Buren, D. 1998, ApJ, 502, 538

Cowley & Hartwick, 1991, ApJ, 373, 80
de Vaucouleurs, G. 1957, AJ, 62, 69
Dwek, et al. 1995, ApJ, 445, 716
Fields, B.D., Freese, K. & Graff, D.S. 1998, New Astronomy, 3, 347
Gates, E., Gyuk, G., & Turner, M. 1996, PRD, 53, 4138
Gould, A. 1995, ApJ, 441, 77
Hughes, S.M.G., Wood, P.R. & Reid, N. 1991, AJ, 101, 1304
Kim, S. et al., 1998, ApJ, 503, 674
Kinman et al. 1991, PASP, 103, 1279
Kunkel, W.E., Demers, S., Irwin, M.J., & Albert, L. 1997, ApJ, 488, L129
Meatheringham, S.J., Dopita, M.A., Ford, H.C., & Webster, B.L. 1988, ApJ, 327, 651
Odewahn, S. C. 1996, in "Barred Galaxies", ed. Buta, R., Crocker, D. A., & Elmegreen, B. G., ASP Conf. Ser. 91, p. 30
Olszewski, E. W. 1992, in Proceedings 11th Santa Cruz Summer Workshop in Astronomy and Astrophysics, Smith, G. & Brodie, J., ASP Conf. Ser. 48, p. 351
Olszewski, E.W., Suntzeff, N.B., & Mateo, M., 1996, ARA&A, 34, 511
Prevot, L., Rousseau, J. & Martin, N., 1989, A&A, 225, 303
Sahu, K.C. 1994a, Nature, 370, 275
Sahu, K.C. 1994b, PASP, 106, 942
Schommer, R.A., Olszewski, E.W, Suntzeff, N.B. & Harris, H.C. 1992, AJ, 103, 447
Weinberg, M. 1999, ApJ, submitted (astro-ph/9905305)
Westerlund, B. 1997, "The Magellanic Clouds". Cambridge University Press: Cambridge, UK.
Wu, X. 1994, ApJ, 435, 66
Zaritsky, D. & Lin, D.N.C. 1997, AJ, 114, 2545

Discussion

Graff: Why, if the velocity dispersion is roughly constant across the LMC, do you find a higher concentration of events in self-lensing models than in halo models?

Gyuk: Most of our models took a constant scale height instead of a constant velocity dispersion. We have indeed looked at the constant velocity dispersion models of Alves et al. These models do have a smaller central concentration. However, the central concentration is still higher than in the corresponding halo models. This is partly because the LMC bar has a somewhat different velocity dispersion/ optical depth relation. Further, typically in these models the velocity dispersion is small enough that the self-lensing optical depth is still quite small.

Sackett: You mentioned that there was considerable uncertainly in the inclination of the LMC. I wondered if this full range was reflected in your self-lensing simulations self-consistently, since inclination also affects the interpretation of the rotation curve.

Gyuk: Yes, it was included.

Zinnecker: You concluded that LMC self-lensing would be unlikely, yet three out of a dozen LMC/SMC microlensing events are likely to be due to self-lensing in the LMC/SMC as you also pointed out. So I am somewhat confused. Please explain again how to make these two seemingly conflicting statements consistent.

Gyuk: The three events in question are LMC 9 and the two SMC events. The interpretation of LMC 9 is uncertain. If the source is a binary then the solution is consistent with a halo lens. At least one of the SMC events is certainly SMC self-lensing. However the SMC is a very different galaxy from the LMC. Almost all models of the SMC predict substantial self-lensing, so seeing SMC self-lensing is not surprising. In any event the structure of the SMC is not highly relevant to the LMC self-lensing.

What are MACHOs? Interpreting LMC Microlensing

David S. Graff
The Ohio State University
Departments of Physics and Astronomy
Columbus, OH 43210 USA

Abstract. I discuss two hypotheses that might explain LMC microlensing: the halo stellar remnant lensing hypothesis and the unvirialized LMC lensing hypothesis. I show that white dwarfs cannot contribute substantially to the cosmic baryon budget; they are strongly constrained by chemical evolution and background light measurements. Although there have been some claims of direct optical detections of white dwarfs in the halo, I show how the full sample of direct optical searches for halo lenses do not support the halo lens hypothesis.

N-body simulations suggest that the LMC may be naturally excited out of virial equilibrium by tidal forcing from the Milky Way. New measurements of LMC kinematics not only do not rule out the unvirialized LMC lensing hypothesis, but even moderately favor it (at 95% confidence).

1. Introduction

Microlensing today has a wide variety of scientific applications ranging from stellar atmospheres (Sackett, this volume) to Galactic structure (Binney, this volume) to searching for planets (Gaudi, this volume), but the initial prime mover of microlensing, which seduced so many particle astrophysicists and cosmologists such as myself into dirtying our hands with grungy astronomy, has been the search for dark matter in the Milky Way halo. Specifically, we wish to determine what fraction of the halo is composed of lumps of dark matter that could lens background stars in other galaxies, such as the Magellanic Clouds. These lumps of matter are known as MACHOs (MAssive Compact Halo Objects) by which I refer to any halo objects massive enough to cause detectable gravitational microlensing. The MACHOs stand in contrast to WIMPs (Weakly Interactive Massive Particles), putative particle-dark matter candidates that are not massive enough to microlens. MACHOs are also distinct from halo stars, sometimes known as the stellar halo. Halo stars are massive enough to lens, but are not dark. They make up less than 1% of the mass of the halo (Graff & Freese 1996) and thus cannot be responsible for the microlensing events seen towards the LMC.

There are several candidate objects that in principle could be MACHOs; they are reviewed by Carr in this volume.

As dark matter experiments, microlensing searches have been phenomenally successful, among the most successful of all dark matter searches. The MACHO and EROS experiments have published a combined limit in which they have ruled out enormous regions in parameter space (Afonso et al. 2000), limits that have been tightened by results presented at this meeting (Lasserre, this volume; Cook, this volume). These results exclude objects with mass in the range $10^{-7} - 1 M_\odot$ from making up all the dark matter in the halo.

However, microlensing results still present a mysterious signal; the MACHO experiment has reported an excess of microlensing events towards the Magellanic Clouds with time scales of tens of days. These events are difficult to explain. If one interprets them as being due to a halo of lenses, then their total mass is $9^{+4}_{-3} \times 10^{10} M_\odot$, which, depending on the halo model chosen, makes up some $\sim 20\%$ of the mass of the halo. They have long time scales which, if interpreted as being due to MACHOs in the halo, suggest that the lenses have masses in the range $0.1 - 1 M_\odot$ (Cook, this volume). The only known dark astrophysical objects that have masses in this range are white dwarfs. Taken at face value, this result could be interpreted to mean that white dwarfs are common in the universe, perhaps being a significant fraction of all baryons.

The white dwarf hypothesis gained strength when Ibata et al. (1999) claimed to have detected them in the Hubble Deep Field and Ibata et al. (2000) claimed further detection of one additional halo white dwarf in a photographic proper motion survey.

In this paper, I will show that white dwarfs make poor dark matter candidates, and contribute negligibly to the cosmological baryon budget. Thus, microlensing experiments have not identified a significant baryonic dark matter candidate. I will show that the detections of Ibata et al. do not require a large population of halo white dwarfs.

I have not been able to completely rule out white dwarfs as being responsible for microlensing; in fact, Gates (this volume) has suggested that the MACHO experiment may have detected a new component of the Galaxy composed of white dwarfs. Still, I will establish that white dwarfs are strongly constrained, exotic lensing candidates requiring several conditions.

Before one accepts any potentially controversial interpretation of an experimental result, one must be sure that the result is not due to an unaccounted for background signal in the experiment. I will discuss a possible background for the microlensing experiments, lensing by ordinary stars near the LMC. I will show that this background has not been ruled out observationally, is suggested by some theoretical calculations, and has even been detected observationally, though the statistical significance of these detections is not compelling.

2. Can MACHOs be White Dwarfs?

2.1. Cosmology and MACHOs

We begin this section with a discussion of the total cosmological density of MACHOs in units of the critical density, Ω_{macho}. Most of this work is taken from Fields, Freese, & Graff (1998). Since microlensing has as its ancestry the search for dark matter, the cosmological density of MACHOs is of critical importance – if anything, more important than the mass density of MACHOs around the Milky Way.

Fields et al. (1998) placed upper and lower limits on the Cosmological densities of MACHOs, which I here update to include the new MACHO results presented by Cook in this volume. If we assume that MACHOs trace dark matter, then $\Omega_{\text{macho}} = \Omega_m f_{\text{halo}} \sim 0.06$. Here, Ω_m is the density of cosmological density of collisionless matter and f_{halo} is the fraction of the mass of the halo within the LMC radius of 50 kpc composed of MACHOs. Making the minimal assumption that only Milky Way type spiral galaxy halos contain MACHOs, the lower bound on the cosmic density of MACHOs is $\Omega_{macho} > 0.001 - 0.01$

The above discussion is independent of what the MACHOs actually are. What if the MACHOs were stellar remnants, white dwarfs or neutron stars? Stellar remnants are in some respects attractive MACHO candidates; they have the right mass, and are dark. However, as I will show, stellar remnants make poor dark matter candidates.

The constraints against stellar remnants are ultimately due to limits on the quantity of nuclear fusion. Large numbers of white dwarfs mean that lots of nuclear fusion had to take place. While a traditional stellar population will convert perhaps $\sim 5\%$ of its mass from hydrogen to helium, most of the hydrogen being locked up in dwarf stars, a stellar remnant will contain no hydrogen. Pound for pound, a white dwarf MACHO population implies thus that ~ 20 times more fusion will have taken place than a similar mass of stars.

Since a large white dwarf MACHO population implies that there has been copious fusion, we can place limits on this population by placing limits on the two byproducts of fusion, energy and heavy elements. A cosmologically significant population of white dwarfs severely modifies the chemical enrichment of its surroundings. Under a typical scenario, a 3 $M_\odot$ progenitor star will eventually die into a 0.6 $M_\odot$ white dwarf, but the remaining 80% of its mass is released as chemically enriched gas. The chemical evolution implied by this gas is analyzed in Fields, Freese & Graff (2000, FFG00).

The strongest limits come from the abundances of carbon and nitrogen. Standard chemical yields of low metallicity stars suggest that the ejected gas of white dwarfs is enriched with a solar abundance of either carbon or nitrogen (depending on the extent of hot bottom burning). Since Galactic halo stars have a carbon abundance of $\sim 10^{-2}$ solar, only $\sim 10^{-2}$ of halo material can have been processed through stars at the time of the formation of the halo. Similarly, the mean carbon enrichment of the universe at $z \approx 3$ is $\sim 10^{-2}$ solar as measured in Lyman-α absorption systems. Thus we see that only $\sim 10^{-2}$ of all baryons can have passed through stellar progenitors of white dwarfs by $z \approx 3$. (See FFG00 for a more detailed calculation).

Chabrier (1999) has suggested that zero metallicity white dwarfs may not emit carbon, finessing these limits. In that case, more robust, but less restrictive chemical evolution limits can be placed with helium. FFG00 showed that the helium evolution due to a cosmological remnant population limits Pop. III remnants to $\Omega_{WD} < 0.002h^{-1}$, still an insignificant component of baryons. Thus, remnants cannot contribute significantly to baryonic dark matter.

A parallel limit can be placed using the light emitted by the stellar progenitors. This limit is slightly weaker than the chemical evolution limits, but far more robust. Zero-metallicity stars may not emit carbon, but they certainly emit light! Using limits on the infrared background, Graff et al. (1999) showed that $\Omega_{wd} < 0.004h$. Again, we conclude that stellar remnants cannot contribute to baryonic dark matter.

The original scientific goal of microlensing was to identify baryonic dark matter. As we have seen, we can place robust limits on the cosmological density of stellar remnants, the only astrophysical objects that could be responsible for the signal seen by the MACHO experiment. Thus, no stellar object can contribute significantly to baryonic dark matter.

2.2. White Dwarfs in the Milky Way Halo

Even though white dwarfs cannot be significant components of cosmological dark matter, might they not still be present in the Milky Way halo? Could not the Milky Way halo be an exceptional place in which much of the mass was composed of stellar remnants? Gates (this volume) suggested such a model. In her system, the lensing detected by the MACHO experiment is due to a new stellar remnant component of the Milky Way, not associated with the dark matter halo.

It is difficult to constrain the possibility that the Milky Way halo alone could contain large numbers of white dwarfs. Obviously, limits on background light do not apply to our own halo, but only to distant galaxies. Gibson & Mould (1997) noted that such a halo of white dwarfs was inconsistent with measured low carbon abundances of halo stars. However, FFG00 noted that a galactic wind driven by SNe Ia from these white dwarfs could blow the carbon-enriched gas out of the halo.

A possible way to confirm the existence of a spheroid population of white dwarfs is to try to directly detect them. If such a spheroid exists, then its white dwarfs are the most common type of star in the Milky Way. Their detection is not trivial however, because they can be quite dim, $M_V \sim 18$, and it is difficult to separate them from the overwhelming background of brighter Milky Way stars and distant galaxies.

Before I begin discussion of direct detection of white dwarfs, I should mention that there has been a recent revolution in the theoretical study of white dwarf luminosities, colors, and cooling curves. Before 1997, no one had calculated what white dwarf atmospheres cooler than 4000 K, roughly the temperature of the then coolest observed white dwarf, were like. However, a MACHO population of white dwarfs would be older and thus likely cooler than disk white dwarfs. The new theory of white dwarfs is discussed by Hansen in this volume.

Two results from this analysis pertinent to optical searches for white dwarfs are as follows: (1) old white dwarfs with helium-dominated atmospheres would have cooled to invisibility, and could never be directly detected, and (2) cool hydrogen-atmosphere white dwarfs emit most of their light in the V and R bands, and have spectral energy distributions very far from black body.

Point (1) means that no direct search for white dwarfs could ever rule out the existence of a MACHO population of white dwarfs: the MACHOs could be all helium-atmosphere white dwarfs. However, point (2) above makes optical searches for hydrogen-atmosphere white dwarfs relatively more powerful and model-independent, since the white dwarfs are emitting their light in optical frequencies.

There have been two broad strategies to search for a halo population of white dwarfs. In one, deep images are taken at high galactic latitude, thus eliminating the background of disk stars. Here, the strongest background is distant galaxies. The high spatial resolution of the HST can be used to separate galaxies from stars. Flynn, Gould & Bahcall (1996) used this method with the Hubble Deep Field, and did not find evidence of white dwarfs.

The other method is to look for high proper motion objects. Halo white dwarfs should have high proper motions due to their high velocity, and low intrinsic magnitude. Proper motion searches have ranged from shallow photographic searches over wide solid angles (Luyten 1979; Ibata et al. 2000) to deep narrow searches of the Hubble Deep Field (Ibata et al. 1999).

Interpreting the proper motion surveys is controversial, since they are not consistent. Flynn et al. (2000) compared the various proper motion surveys. We found that the LHS survey (Luyten 1979) was by far the most powerful, tens of times more powerful than the other published surveys. Yet, the LHS survey does not find evidence of a MACHO population of white dwarfs while two other surveys, Ibata et al. (1999) and Ibata et al. (2000) find a handful of objects suggesting that the halo might be full of white dwarfs.

There are two different possible interpretations of these conflicting results: either the halo is full of white dwarfs and the LHS survey does not see them, or the halo does not contain many white dwarfs, and the handful of objects seen by the two Ibata et al. surveys are a mix of Poisson fluctuations and background objects. I consider these two possibilities below.

The LHS survey was done some thirty years ago, and its high proper motion stars were often detected by hand. Thus, there is no modern artificial star–Monte Carlo estimate of its efficiency. However, in Flynn et al. (2000), we estimated the efficiency of the LHS survey by the numbers of bright and dim stars, and estimated that it was at least 60% efficient down to magnitude $R_L = 18.5$. This estimate can only be based on dim, low proper motion stars since there are very few dim high proper motion stars in the sample. Possibly, the survey misses the dim high proper motion stars, although there are some reasons to think otherwise, which are discussed in Flynn et al (2000).

The Ibata et al. (1999) survey looked for proper motion of Hubble Deep Field objects with the philosophy that anything that moved must be a star and not a background galaxy. However, in Graff & Conti (in preparation, 2000), we show that Ibata et al. underestimated their proper motion uncertainties. Thus, the objects they found are most likely distant galaxies.

The Ibata et al. (2000) survey used scanned Schmidt plates to search for high proper motion stars. One advantage of this technique is that the objects found are close enough to be observed spectroscopically. They found two dim, low temperature, spectroscopically confirmed white dwarfs, one of which was also in the LHS survey.

In their paper, Ibata et al. (2000) suggest that these objects imply that white dwarfs make up 10% of the local mass density of the Galactic halo. This number is highly uncertain. Just from Poisson statistics of 2 objects, the 90% confidence lower limit is 4 times lower than the detected value, 2% of the local mass density of the halo. Ibata et al. also made poorly constrained assumptions about the mass of their objects and the absolute magnitude of the new white dwarf in their sample, the one that is not in the LHS survey. Thus the real lower limit based on their observations is below 1% of the local mass density of the halo.

This limit is low, comparable to the local density of halo stars. Thus, these two white dwarfs could simply be representatives of the local Pop II halo white dwarf population, which likely make up a bit less than 1% of the local mass density of the halo. It is still premature to say whether this survey has found evidence of a MACHO population of white dwarfs.

2.3. Conclusion

White dwarfs cannot make a significant component of Baryonic Dark Matter. They would cause too much enrichment of carbon, nitrogen, and helium. They would also create an infrared background above current limits.

Even though white dwarfs cannot greatly contribute to the cosmic baryon census, they could be over-represented in the Milky Way halo. Direct searches can never rule out this possibility since a high fraction of white dwarfs could have helium atmospheres and be too dim to see. There have been some claimed detections of white dwarfs, but, in my opinion, these detections are not yet compelling.

3. Other lensing candidates, especially LMC self lensing

Various authors have proposed several exotic, non-baryonic lensing candidates that are reviewed by Carr in this volume. However, before we are driven to these candidates, we must first verify that there are no other viable baryonic lensing candidates.

One obvious baryonic candidate is lensing by ordinary stars. Ordinary stars have the right mass, typically $0.1 - 1 M_\odot$, but they are not usually thought of as making good dark matter candidates, since they are not dark! Ordinary stars are known to make up a negligible fraction of the local halo mass density (Graff & Freese 1996). However, various schemes have been proposed in which stars do cause a large optical depth towards the LMC.

Wu (1994) and Sahu (1994) proposed that the LMC could generate self-lensing, lensing of LMC stars by other LMC stars. However, Gould (1995) examined the self lensing optical depth of a virialized disk galaxy and showed it was proportional to the velocity dispersion, $\tau = 2\sigma^2/c^2 \sec^2 i$, where σ is the line of sight velocity dispersion and i is the inclination angle. The velocity dispersion of the LMC is measured to be low in a variety of stellar populations representing a wide variety of ages and metallicities (reviewed by Gyuk, Dalal & Greist 1999). Thus, the bulk of stars in the LMC are in a thin, face-on disk with a low optical depth, lower than that observed by the MACHO group.

Zhao (1998) suggested that the Milky Way halo could contain small recently accreted objects such as the Sagittarius Dwarf, and tidal tails ripped off these objects and the Magellanic Clouds. He proposed that if one of these lumps of matter were interposed along the line of sight to the LMC, it could perhaps cause sufficient microlensing to account for the MACHO observations.

If such systems of objects were distributed randomly in the halo, as we would expect if they arose from Sagittarius-Dwarf-like systems, then the probability that an individual one would by chance be aligned with the LMC is quite low, $\sim 10^{-4}$ (Gould 1999). Yet they cannot be numerous, since if such objects were composed of ordinary luminous stars, in order to have a density sufficient to cause the measured microlensing, they would have a surface brightness high enough to be visible to the naked eye. The only way they could have escaped detection is if they lie exactly in front of the LMC, or, like the Sagittarius dwarf, behind the Galactic center. Thus, the only likely way that there could be an object along the line of sight to the LMC is if it was somehow associated with the LMC, perhaps a tidal tail lifted off the LMC by interactions with the Milky Way or SMC, or perhaps a part of the LMC itself, a thick "shroud" (Evans & Kerins 2000).

Several papers have been written about whether or not there is an object along the line of sight to the LMC, too many to review here. They are discussed in detail in Zaritsky et al. (1999), whose basic conclusion is that such a population *cannot be ruled out* if it is close enough to the LMC in both positions and velocities, within approximately 10 kpc and 30 kms^{-1}. If the object is in this range, it is swamped by stars within the LMC proper, and is difficult to detect.

In Graff et al. 2000, we searched for a kinematically separate population from the LMC using carbon stars. The idea behind this search was that any separate population should have a velocity that is different from that of the LMC. We found, at 95% statistical confidence, evidence of a population displaced 30 kms^{-1} from the LMC. If this population exists, it would likely be at a distance from the LMC sufficient to cause a significant microlensing signal.

Numerical simulations of the LMC-Milky Way system support the idea that a chunk of the LMC was ripped off by the Milky Way. If such a chunk lay along the line of sight, it would cause enough microlensing to explain the MACHO-group results. In unpublished work, Mark Galpin and I have analyzed the N-body simulation of Weinberg (2000). We found that the tidal features of this simulation are large enough to cause a microlensing optical depth of 1×10^{-7}, consistent with the MACHO-group results, along some lines of sight.

3.1. Conclusion

A yet undetected population of ordinary stars could cconstitute the measured microlensing population. This population must be close to the LMC in both distance and radial velocity, otherwise it would have already been detected. N-body simulations suggest that such a population could have been ripped off the LMC by tidal interactions with the Milky Way. Graff et al. (2000) have found, at only the 95% confidence level, evidence of a non-LMC disk population, the Kinematically Distinct Population, or KDP. If this population is confirmed, it could explain the microlensing detected by the MACHO collaboration.

4. Editorial Musings

I have discussed two leading explanations for LMC microlensing events: that the halo is full of white dwarfs, or that the LMC is sufficiently far from virial equilibrium to generate a large optical depth with a small velocity dispersion. To some extent, both explanations have similarly weak observational support; neither can be ruled out, and each has statistically weak observational evidence.

However, the white dwarf hypothesis requires a chain of unlikely events: massive amounts of baryons must have undergone star formation with a radically different, extremely narrow mass function, and these stars must not have emitted any carbon. Further, this star formation mechanism can only have occurred in the inner halos of Milky Way-type galaxies, from which a galactic wind must remove ejected gas. Occam's razor suggests that before we believe such a long chain of new cosmology, we should adopt the simpler notion that microlensing experiments have simply found a background of lensing by ordinary stars.

References

Afonso, C. et al. 2000, ApJ, 532, 340
Chabrier, G. 1999, ApJ, 513, 103
Evans, N. W., & Kerins, E. J. 2000, ApJ, in press (astro-ph/9909254)
Fields, B. D., Freese, K., & Graff, D. S. 1998, New Astron. 3, 347
Fields, B. D., Freese, K., & Graff, D. S. 2000, ApJ in press (astro-ph 9904291)
Flynn, C., Gould, A. & Bahcall, J. N. 1996, ApJ, 466, 55
Flynn, C., Sommer-Larsen, J., Fuchs, B., Graff, D. S., & Salim, S. 2000, MNRAS, in press (astro-ph/9912264)
Gibson, B. K. & Mould, J. R. 1997, ApJ, 482, 98
Gould, A. P. 1995, ApJ, 441, 77
Gould, A. P. 1999, ApJ, 525, 734
Graff, D. S. & Freese, K. 1996, ApJ, 456, 49
Graff, D. S., Freese, K., Walker, T. P. & Pinsonneault, M. H. 1999, 523, 77
Graff, D. S., Gould, A. P., Suntzeff, N., Scommer, R. A. & Hardy, E. 2000, ApJ, in press (astro-ph/9910360)
Gyuk, G., Dalal, N., & Greist, K. 1999, ApJ, submitted (astro-ph 9907338)
Ibata, R., Richer, H. B., Gilliland, R. L., & Scott, D. 1999, ApJ, 524, 95
Ibata, R., Irwin, M., Bienaymé, O., Scholtz, R., & Guibert, J. 2000, ApJ, 532, 41
Luyten, W. J. 1979, LHS Catalogue: a catalogue of stars with proper motions exceeding $0''\!.5$ annually (Minneapolis: University of Minnesota)
Sahu, K. 1994, Nature, 370, 275
Weinberg, M. 2000, ApJ, 532, 922
Wu, X.-P. 1994, ApJ, 435, 66
Zaritsky, D., Schectman, S. A., Thompson, I., Harris, J. & Lin, D. N. C. 1999, AJ, 117, 2268
Zhao, H. S. 1998, MNRAS, 294, 139

Discussion

Carr: Is the γ-ray limit for Mkn 501 not weakened if the WDs form at high redshift? Could Hansen's "beige dwarfs" not also explain the Ibata et al. observations, in which case one would avoid most of your limits?

Graff: The limit constrains the co-moving photon number density and so is largely redshift-independent. There is an additional constraint that the product of the energies of the IR photon and the TeV γ-ray must be greater than the square of the electron mass. Thus we constrain starlight emitted since redshift 30.

Hansen: Are HST objects beige dwarfs or Ibata objects ?

Gyuk: Given the time scales, how do you keep the shroud from virializing?

Graff: The shroud is present in the simulations of M.Weinberg. I have not determined what the virialization time scale is in these simulations, but I believe that the shroud is kept in an unvirialized state by continual tidal forcing from the Milky Way. H S Zhao has suggested that the virialization time scale is longer than the dynamical time scale.

Evans: There are halo objects in Gliese's catalogue. Are there any halo white dwarfs? What are the ratios of halo to disk population in Luyten's catalogue?

Graff: As far as I know, there are no halo white dwarfs in the Gliese Catalogue that are not Luyten objects. Dimmer than $M_R = 13$ there are 29 disk white dwarfs and 4 halo white dwarfs in the Luyten sample. I caution that the proper motion limits of the Luyten survey must be taken into account when turning this ratio into a ratio of disk/halo white dwarf densities.

Gyuk: Do you expect to see the receding limb of the polar ring?

Graff: Our KDP, if it is real, is the velocity signature of a tidal structure such as a polar ring.

Old White Dwarfs as a Microlensing Population

Brad M. S. Hansen

Department of Astrophysical Sciences, Peyton Hall, Princeton University, Princeton, NJ, 08544, USA

Abstract. A popular interpretation of recent microlensing studies of the line of sight towards the Large Magellanic Cloud invokes a population of old white dwarf stars in the Galactic halo. Below I review the basic properties of old white dwarf stars and the ongoing efforts to detect this population directly.

1. Introduction

Other authors in this volume cover the microlensing motivations much better than I can, so I shall suffice to remind you that one possible explanation of the microlensing events towards the LMC invokes a population of objects in the range $0.3 - 0.8 M_\odot$. Potentially these could be either normal hydrogen-burning stars or white dwarfs, the burnt-out remnants of stellar evolution. To distinguish these populations, we turn to direct searches at optical wavelengths, since the latter population is $10^{-3} - 10^{-4}$ as bright as the former. The number counts of faint red stars suggest that hydrogen burning stars cannot account for the microlensing population (Bahcall et al. 1994; Graff & Freese 1996). The question I wish to address is how well one can constrain the white dwarf hypothesis by similar means.

2. White Dwarf Cooling

To derive a constraint from direct optical searches, we need to know how to recognise white dwarfs. White dwarfs of moderate age (< 10 Gyr) emit approximately as black bodies. However, once the white dwarf cools to effective temperatures below ~ 5000 K, the atmospheric hydrogen resides in molecular form. This has dramatic consequences for the appearance of the white dwarf (Hansen 1998, 1999a; Saumon & Jacobsen 1999), because the dominant opacity source in such an atmosphere is collisionally induced absorption by H_2, which absorbs primarily in the near infra-red. For black bodies of effective temperatures $\sim 3000 - 5000$ K the peak of the spectrum lies in the same wavelength region, so that the increased absorption leads to dramatic deviations from the traditional assumption of black body colours. The general trend of the colour evolution is that the flux is forced to emerge preferentially blueward of $\sim 0.8 \mu m$. With the evolution of the black body flux redward, the peak of the spectrum for old white dwarfs lies around $0.6 \mu m$.

A correct quantitative analysis of the atmospheric conditions is not only important for determining the colours of the white dwarf, but is also of critical importance for the cooling evolution. For white dwarfs cooler than $T_{\text{eff}} \sim 6000$ K, the internal structure has no radiative zones. The bulk of the white dwarf mass resides in a degenerate core which is approximately isothermal due to the efficient thermal conduction by the electrons near the surface of the Fermi sea. This isothermal core is joined to a thin convective envelope which extends all the way to the photosphere. Thus, the entire white dwarf structure is critically dependant on the surface boundary condition and hence the atmospheric conditions. It must be noted that, while sophisticated analyses of white dwarf atmospheres have been around for several years (e.g. Bergeron, Saumon & Wesemael 1995), the white dwarf cooling calculations have lagged significantly behind, employing grey atmosphere boundary conditions. Thus, while one may trust the masses and gravities inferred in recent atmospheric analyses, one cannot completely trust the ages inferred for the older white dwarfs in most recent papers. To my knowledge, the only cooling calculations using boundary conditions based on proper atmosphere models are those of Hansen (1999a) and Salaris et al. (2000). For isochrones and luminosity functions using the Hansen (1999a) models, see Richer et al. (2000).

The models described above concern white dwarfs with pure hydrogen atmospheres. Empirically, a significant fraction of known white dwarfs appear to have no hydrogen in their atmospheres (Liebert et al. 1979; Bergeron, Ruiz & Leggett 1997). When pure helium atmospheres reach effective temperatures < 5000 K, they have no molecular component to provide significant opacity and, as a result, the photosphere lies at much higher densities, where the beginnings of pressure ionization start to provide a small ionized component. This higher density and the concomitant difference in the boundary condition results in much faster cooling for old white dwarfs with helium atmospheres. Furthermore, these objects are expected to appear approximately as black bodies (contrary to the hydrogen atmosphere case) but will be hard to detect because they cool much faster.

The matter of a component of the white dwarf population sporting helium atmospheres is one that is often not mentioned in the recent studies attempting to tie direct searches to microlensing populations. It is important to note that this represents an essentially unobservable (due to their much more rapid cooling) component and, as such, will always introduce some uncertainty into the comparison.

3. Observational Progress

3.1. The story thus far

The last couple of years have been fun for those interested in the question of halo white dwarfs. The realisation that old white dwarfs would appear somewhat bluer than black bodies meant that previous constraints on the local number density based on Hubble Deep Field point source counts were somewhat overstated. Much excitement was generated by the detection of proper motions of several faint blue point sources in the HDF North (Ibata et al. 1999). The colours and magnitudes of these objects are consistent with old white dwarfs

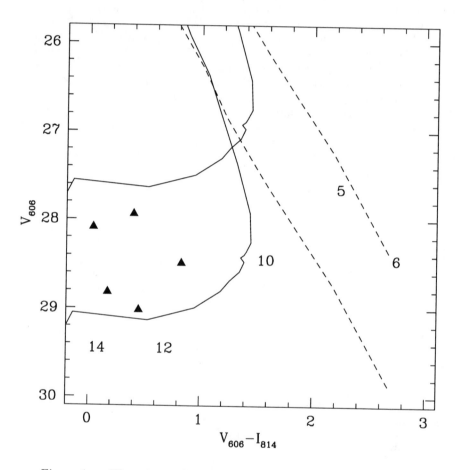

Figure 1. The source of all the fuss. The five points are the magnitudes & colours of the 5 proper motion objects of Ibata et al. (1999). The solid curves describe the cooling of a 0.6 solar mass white dwarf with hydrogen atmosphere located at 1 and 2 kpc. Representative ages are indicated along the lower curve. The dashed curves are the same but for a pure helium atmosphere. The much more rapid cooling is evident.

~ 12 Gyr old at distances ~ 1 kpc (Fig. 1). Given the radical consequences of this observation, some kind of confirmation would be nice. A third epoch, to confirm the measurement, has been approved but was postponed due to the HST gyro problems in December 1999. Nevertheless, the veracity of the HDF proper motions should be addressed within the next year or so.

In the interim, several other interesting results have come to light. Most importantly, Hodgkin et al. (2000) presented a near-infrared spectrum of a nearby cool white dwarf, demonstrating the reality of the flux-suppression due to molecular hydrogen absorption. This provides a welcome check that the theory behind the predicted colour change is at least qualitatively correct. The high proper motion of this object also suggests membership of the Galactic halo. Another object showing such a long wavelength depression is LHS 3250 (Harris et al. 1999)

A further interesting indirect argument supporting the white dwarf hypothesis comes from Mendez & Minniti (2000), who assert that the number counts of faint blue point sources (the class amongst which Ibata et al. discovered proper motions) in the Hubble Deep Field South is approximately twice that in the HDF North. This is consistent with the objects (whatever they are) being Galactic in origin, since the southern field looks in a direction closer to the Galactic centre and consequently the stellar density is expected to be higher.

A note of caution, however, is sounded by Flynn et al. (2000), who find little evidence for a population equivalent to that of Ibata et al. in the Luyten proper motion survey. They conclude that hydrogen atmosphere white dwarfs cannot comprise a significant fraction of the Galactic halo. On the other hand, a new search of wide field photographic plates by Ibata et al. (2000) has uncovered several high proper motion objects, at least two of which are spectroscopically confirmed to be white dwarfs and with some modicum of IR flux suppression.

In my not-completely-unbiased interpretation of the above, the evidence in favour of the white dwarf hypothesis is encouraging, although not conclusive. The recent results of Hodgkin et al. (2000) and Ibata et al. (2000) have demonstrated at least the existence of a population of high proper motion (and thus probably halo) white dwarfs. The question is now a matter of number density. It is on this point that we must await further data.

3.2. Future prospects

While the above results indicate that some white dwarfs do show halo kinematics, the fact that there is little evidence for a large halo white dwarf population in existing large-scale proper motion surveys casts some doubt that this population may be responsible for the microlensing results. This last statement relies heavily on the data set of Luyten (1979), so it would be nice to check this with another large scale survey.

Towards this end, a program led by Peter Stetson has begun at CFHT to image almost 20 square degrees to $V \sim 25$. Repeated with an interval of 2-3 years, this program should provide a conclusive test of the white dwarf halo hypothesis. Using the new MACHO mass fraction estimate ($\sim 20\%$) and assuming only half of all white dwarfs have hydrogen atmospheres, this program is still expected to find about 8 white dwarfs per square degree.

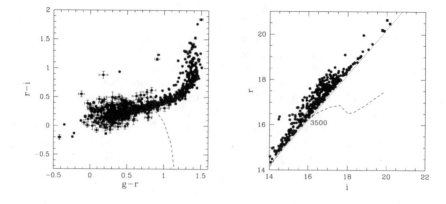

Figure 2. The left hand panel shows the SDSS stellar locus (York et al. 2000) in the g-r/r-i plane, with the evolution of a representative white dwarf shown by the dashed line. Thus, it is possible to select very cool white dwarfs by colour alone. However, objects that cool are also faint. The right hand panel shows the apparent magnitudes of the stars with g-r>0.8 as well as the dashed line for the cooling of a 0.6 $M_\odot$ white dwarf at 10 pc (i.e. the absolute magnitude). The dotted line shows the apparent magnitude of a 3500 K white dwarf at various distances.

Another fascinating prospect is the hope of finding such old white dwarfs in the Sloan Digital Sky Survey (SDSS). Although this comprehensive large scale survey will not provide proper motions as a matter of course, some proper motion information is potentially available from comparison with Palomar Sky Survey plates. Furthermore, there will be a strip of sky in the southern hemisphere that will be multiply imaged. Even without proper motions, very old white dwarfs are potentially detectable by colours alone in the SDSS five-band photometric system. Photometric selection alone is unlikely to provide a complete census as only the oldest white dwarfs will stand out from the locus of other stars (the problem is that even the reddest bandpass doesn't extend significantly beyond 1μm). Nevertheless, very old white dwarfs are interesting in their own right.

Figure 2 shows a representative stellar locus and how one might select old white dwarfs based on colours. The colour selection suggests that only white dwarfs cooler than 3500 K can be selected in this manner. Using an i-band detection limit of 18, we find such objects can be detected to distances $\sim$ 30 pc. If the entire MACHO fraction (taken to be 20%) were in 0.5$M_\odot$ white dwarfs, 50% of whom had hydrogen atmospheres, and all had 3500 K temperatures, we would expect approximately 43 white dwarfs in the Sloan survey, or 1 every 480 square degrees. This is probably an overestimate. Hotter white dwarfs cannot be distinguished from the stellar locus (it is worth noting that the Hodgkin et al. and Ibata et al. detections all correspond to $T_{\rm eff} \sim 3500 - 4000$ K, so in the marginal region) and cooler objects are fainter and thus the effective volume is smaller. Nevertheless, this simple estimate offers encouragement that at least some white dwarfs are potentially detectable in this manner. More detailed calculations are underway to provide a more robust estimate.

Another large scale proper motion survey that should offer interesting constraints is that being undertaken by the EROS project (Goldman 1998). With a survey area of 350 square degrees and an I limit of 20.5, we can use the same approximate scenario as above to estimate ~ 23 white dwarfs in their sample. This is a somewhat more realistic ballpark number in this case as the proper motion selection will allow detection of hotter objects.

The bottom line of the above is that, although uncertainty about mass distributions and chemical compositions make prognostication difficult, the fact that naive estimates lead to predictions of 10-100 detections in ongoing surveys suggests that we should see something if the hypothesis of a significant white dwarf halo is correct.

4. Beige Dwarfs

Although it was not my intention to address this subject at the conference, it was raised a couple of times, so I will conclude with a few remarks about "Beige" dwarfs (Hansen 1999b). Given that brown dwarfs are strongly ruled out by the microlensing timescales, red dwarfs by direct observations and white dwarfs (as well as neutron stars) are an uncomfortable fit due to issues of chemical pollution, it appears that there are no surviving baryonic candidates for halo MACHOs. However, there is one remaining possibility. Lenzuni, Chernoff & Salpeter (1992) demonstrated that one could circumvent the traditional hydrogen burning limit by starting with a brown dwarf and accreting material slowly enough that it could cool. In this fashion one could construct a degenerate hydrogen/helium object with mass $> 0.1 M_\odot$. If one could build such objects to masses $> 0.3 M_\odot$, they would make ideal MACHO candidates as they would be faint and would have no nuclear burning history and thus no chemical pollution problem. The term "Beige" dwarf comes from the superposition of brown dwarf and white dwarf characteristics. The primary problem for this scenario is that there is little evidence that this mode of "star" formation was ever important.

The existence of Beige dwarfs is observationally testable. They have similar radii to white dwarfs but can only exist at effective temperatures $T_\mathrm{eff} < 2000$ K, so the detection of hotter objects would suggest white dwarfs. Indeed, the spectroscopically confirmed objects of Hodgkin et al. (2000) and Ibata et al. (2000) indicate temperatures $\sim 3500 - 4000$ K. Therefore these are unlikely to be Beige dwarfs.

Acknowledgments. I would like to thank the conference organisers for a thoroughly enjoyable (and efficiently run) meeting. Support for this work was provided by NASA through Hubble Fellowship grant #HF-01120.01-99A, from the Space Telescope Science Institute, which is operated by the Association of Universities for Research in Astronomy, Inc., under NASA contract NAS5-26555.

References

Bahcall, J. N., Flynn, C., Gould, A. & Kirhakos, S. 1994, ApJ, 435, L51

Bergeron, P., Ruiz, M.-T. & Leggett, S. K. 1997, ApJS, 108, 339

Bergeron, P., Saumon, D. & Wesemael, F. 1995, ApJ, 443, 764

Flynn, C., Sommer-Larsen, J., Fuchs, B., Graff, D. S. & Salim, S. 2000 (astro-ph/9912264)

Goldman, B., in The Third Stromlo Symposium: The Galactic Halo, eds. Gibson, B. K., Axelrod, T. S. & Putman, M. E., ASP Conference Series Vol 165, p 413; San Francisco

Graff, D.S. & Freese, K. 1996, ApJ, 435, L516

Hansen, B. M. S. 1998, Nature, 394, 860

Hansen, B. M. S., 1999a, ApJ, 520, 680

Hansen, B. M. S., 1999b, ApJ, 517, L39

Harris, H. C., Dahn, C. C., Vrba, F. J., Henden, A. A., Liebert, J., Schmidt, G. D. & Reid, I. N. 1999, ApJ, 524, 1000

Hodgkin, S. T., Oppenheimer, B. R., Hambly, N. C., Jameson, R. F., Smartt, S. J. & Steele, I. A. 2000, Nature, 403, 57

Ibata, R. A., Richer, H. B., Gilliland, R. L. & Scott, D. 1999, ApJ, 524, L95

Ibata, R. A., Irwin, M., Bienayme, O., Scholz, R. & Guibert, J. 2000, ApJ, 532, L41

Lenzuni, P., Chernoff, D. F. & Salpeter, E. E., 1992, ApJ, 393, 232

Liebert, J., Dahn, C. C., Gresham, M. & Strittmatter, P. A. 1979, ApJ, 233, 226

Luyten, W. J. 1979, *LHS Catalogue: A catalogue of stars with proper motions exceeding 0.5" annually*, University of Minnesota, Minneapolis

Mendez, R. A. & Minniti, D. 2000, ApJ, 529, 911

Richer, H. B., Hansen, B., Limongi, M., Chieffi, A., Straniero, O. & Fahlman, G. G. 2000, ApJ, 529, 318

Salaris, M., Garcia-Berro, E., Hernanz, M., Isern, J. & Saumon, D. 2000, ApJ, submitted

Saumon, D. & Jacobsen, S. B. 1999, ApJ, 511, L107

York, D. et al. 2000, AJ, submitted

Discussion

Reid: A comment regarding the completeness of Luyten's proper motion surveys: one of the principal criteria used in identifying candidates was detection on both POSS-I E and O plates. This tends to exclude red dwarfs, which disappear from the O plates even if visible on E plates. Hence an estimate of the completeness of Luyten's survey as a function of magnitude will tend to underestimate the completeness for blue objects, i.e. possible halo WDs.

Hansen: I agree.

Part 2: Extra-Solar Planets

From Low-Mass Star Binaries down to Planetary Systems

S. Udry[1], M. Mayor[1], J.-L. Halbwachs[2], F. Arenou[3]

[1] *Observatoire de Genève, CH-1290 Sauverny, Switzerland*
[2] *Observatoire de Strasbourg UMR 7550, F-67000 Strasbourg, France*
[3] *DASGAL, Observatoire de Meudon, F-92195 Meudon, France*

Abstract. In the context of possible progress arising from large microlensing surveys in the field of binaries with low-mass stars, long-term radial-velocity surveys of G, K and M dwarfs of the solar neighbourhood are presented. The inferred orbital elements are discussed, focusing on the $(e, \log P)$ diagram, the mass-ratio and secondary-mass distributions and on the binary frequency of the studied samples. For G and K dwarfs, most of the characteristics pointed out in previous studies are confirmed. The proportion of companions to M dwarfs is not found to be significantly different from the binary frequency among G- and K-dwarf primaries.

The orbital-element distributions and mass function of stellar companions to solar-type stars compared to the equivalent distributions for extra-solar giant planets strongly suggest different formation and evolution mechanisms for the two populations. In particular, the secondary minimum-mass function shows the paucity of companions in the brown-dwarf domain whereas there is a large peak of giant-planet candidates. Other characteristics of the period and eccentricity distributions also point out the different origins of stellar+brown-dwarf and planetary companions to solar-type stars.

1. Introduction

Binary stars are so widespread that we can expect a significant fraction of microlensing events to be affected by stellar duplicity. In the long term, if the number of microlensings of multiple stars is large enough, we have a chance to get new insights into the statistical properties of multiple systems. However, it represents a quite complex effort to derive distribution functions depending on orbital eccentricities, separations, mass ratios, percentages of multiple systems, primary masses, and so on. Probably at the beginning it will be easier to use existing orbital distributions $f(m_1, m_2, e, P, ...)$ derived from classical approaches like spectroscopy, astrometry, imagery, common proper motion or interferometry to help understand the data provided by the microlensing events.

Multiple-star lensing events will most frequently appear when the system components are not too far apart. The statistical properties of binaries with semi-major axes less than a few AU ($P \leq 10$ years) are most relevant for microlensing. Moreover, most of the lensing events will result from stars at the lower end of the main sequence or from substellar objects. This paper will thus mainly discuss the orbital distributions derived from spectroscopic binaries at the bottom of the main sequence, and below.

For about 20 years, systematic searches have been made to determine the distribution of orbital elements of companions to dwarf stars between F7 and M5. The working horses of these RV searches have been the CORAVEL spectrometers (Baranne et al. 1979) in both hemispheres monitoring at a moderate precision ($\sim 250\,\text{ms}^{-1}$) the radial velocities of more than 1000 stars in the solar vicinity. In particular, they have provided an almost complete census of spectroscopic binaries (SB) part of a volume-limited sample of some 570 F7 to K9 dwarf stars (Duquennoy & Mayor 1991; Halbwachs et al. 2000a). The M part of the survey is also bringing new results (Udry et al. 2000a) to be compared with what is obtained for the G-K stars. Improvements of our knowledge on M-dwarf binaries are expected from on-going high-precision RV surveys with ELODIE and FEROS of M dwarfs in the solar neighbourhood (Delfosse et al. 1999a).

The surveys will be briefly presented in Section 2 and their comparative results discussed in Section 3. Binaries detected in systematic long-term searches among the members of open clusters can also provide useful additional information for the mass-ratio distribution, improving the available statistics by combining the data accumulated in clusters and from the field. They are included in the discussion as well. Section 4 discusses the mass function of substellar companions. Giant planets more massive than 0.3 Jupiter masses (M_{Jup}) and with periods shorter than 3 years have been found around $\sim 5\,\%$ of solar-type stars. In the same domain of periods, $P \leq 1000\,\text{d}$, spectroscopic binaries are detected only around 8–9 % of low-mass stars. We will present in Section 5 the present status of distributions of orbital elements for giant planets. Finally, Section 6 will be devoted to the future developments expected in the field of multiple stars.

2. Large programmes to monitor radial velocities of well-defined samples of dwarfs

The definition of the survey samples is of prime importance for obtaining unbiased estimates of the binary frequency and the distribution of mass ratios. For field stars, a volume-limited sample is a mandatory condition to avoid the well known over-representation in the mass-ratio distribution of binaries with Δm close to zero. Precise parallaxes provided by the Hipparcos astrometric mission (ESA 1997) have been of fundamental importance in the last years to more carefully redefine the stellar samples.

In the early eighties, systematic RV monitorings were initiated with the two CORAVEL spectrographs (Baranne et al. 1979) for different populations of nearby stars in both hemispheres. The two CORAVEL spectrographs were mounted on the 1-m Swiss telescope at Haute-Provence Observatory (France) and on the 1.54-m Danish telescope at La Silla Observatory (Chile). The modest precision of these instruments of about $0.25\,\text{kms}^{-1}$ has already permitted a binarity study of close dwarf stars, namely the G, K and M dwarfs of the solar neighbourhood. Originally the stars were taken from the *Catalogue of Nearby Stars* (CNS2, Gliese 1969) and from its supplement (Gliese & Jahreiss 1979). They were selected using trigonometric parallaxes derived from ground-based observations, expected to give unbiased volume-limited samples. The results are summarized in the following subsections.

2.1. The CORAVEL G- and K-dwarf surveys

The G-dwarf sample. The statistical study of the G-dwarf part of the sample was initially presented in Duquennoy & Mayor (1991, DM91). Thanks to the rich data collected for the nearby stars (visual and astrometric companions, common proper motion stars) this survey is rather complete, at least for the detection of stellar companions. Part of the 269 stars of that sample, a smaller volume-limited subset of 164 stars has allowed DM91 to derive:

1) The distribution of periods on a broad domain. The mode of the distribution $f(\log P)$ appears around 180 years (figure 7 of DM91).

2) The distribution of orbital eccentricities as a function of the orbital periods (figure 5 of DM91). The main characteristics of the $(e, \log P)$ distribution still hold with additional recent data (see Section 3).

3) The distribution of mass ratios $f(q) \equiv f(m_2/m_1)$ for the complete sample (all periods; figure 10 of DM91) whose most important characteristics were: a maximum value of the distribution for $q \sim 0.2$ and a declining distribution up to $m_2/m_1 = 1$ (no peak close to $q = 1$). A rediscussion of that material but restricted to SB with periods $P \leq 3000$ days revealed a flat distribution ($f(q) = $ cst; Mazeh et al. 1992). Note that in 1991 the mass distribution for substellar companions was completely unknown.

4) The global percentage of binaries (or multiple stars) in the solar vicinity is close to 2/3.

The combined G+K-dwarf sample. Significant steps towards improved orbital-element distributions for companions to solar-type stars have been brought by the extension of the sample to K dwarfs (Halbwachs et al. 1998), a better definition of the volume-limited sample thanks to the precise Hipparcos parallaxes, the determination of the companion real masses by using Hipparcos astrometric data (Halbwachs et al. 2000b) and the enlargement of the survey span. The final analysis of the combined G-K sample is about to be published (Halbwachs et al. in prepaaration). Some aspects are already presented in Halbwachs et al. (2000a) and the distributions of masses and periods will be further discussed in Section 3.

2.2. Spectroscopic binary surveys of M-dwarf stars

Despite the quoted importance of information on very low-mass star duplicity, systematic spectroscopic studies of the binarity of M dwarfs have been very sparse. Few orbital parameters have been derived so far and the binarity frequency was generally estimated from a statistical treatment of the RV variations (Young, Sadjadi & Harlan 1987; Marcy & Benitz 1989). The existing studies suffer from small number statistics and large, uncertain, incompleteness corrections. A broad discussion of binaries with M-dwarf primaries within 20 pc may also be found in Fischer & Marcy (1992).

The CORAVEL M-dwarf surveys. Preliminary results of a long-term RV search for companions of nearby M dwarfs in both hemispheres, with the two CORAVELs, have been reported in Udry et al. (2000a). The CORAVEL M-dwarf samples consist of about 450 stars of the solar vicinity. Although not complete yet these surveys are already drastically improving the amount of data available for M-dwarf SB by providing a first set of orbital elements for 31 binaries (among which 23 were previously unknown).

The high-precision M-dwarf surveys. The situation is also improving with the results of a high-precision RV survey initiated by Delfosse et al. (1998a, 1999a) in 1995 with the ELODIE spectrograph (Baranne et al. 1996) at the Haute-Provence Observatory. A systematic RV monitoring of all northern M dwarfs closer than $\sim 10\,\mathrm{pc}$ is in progress. The first unbiased view of the low-mass star duplicity is expected soon. For the moment only a first estimate of the occurrence frequency of short-period M-dwarf binaries is available and can be compared with more massive primaries (see Section 3). This survey has nevertheless already been very fruitful by revealing previous unknown companions to our closest neighbours (Delfosse et al. 1999a) and a first planet orbiting the low-mass M4 dwarf Gl 876 (Delfosse et al. 1998b; also found by Marcy et al. 1998). Adding new SB orbits from the high-precision RV surveys (Delfosse et al. 1999a, b, c; Forveille et al. 1999) and from the literature to the CORAVEL data, we have 48 M-dwarf orbits that permit a first qualitative comparison of their orbital properties ($(e, \log P)$ diagram) with solar-type star characteristics (see Section 3).

A high-precision planet search around M dwarfs in the south has recently been started with FEROS on the 1.5-m ESO telescope at La Silla by the same team, under the leadership of Thierry Forveille. This programme will also provide, as a by-product, useful information on M-dwarf binary characteristics.

2.3. Other on-going large programmes

We can anticipate further significant progresses in the distributions of orbital elements for companions to solar-type stars thanks to other large programmes. For example, a volume-limited survey of ~ 1650 late F to early M dwarfs is on-going at La Silla with the CORALIE spectrograph on the 1.2-m Euler Swiss telescope, at a 5–7 ms^{-1} precision level (Udry et al. 2000b). Already now, a few long-term drifts of RV in the programme have revealed SB of long periods and small amplitudes not detected by previous surveys.

Another very large RV monitoring programme of about 3400 G dwarfs was also initiated a few years ago in order to obtain an unequaled set of orbital parameters for solar-type stars and thus provide a significant improvement on the classical systematic study by DM91. The observational part of the survey has been carried out in the context of a collaboration between the Geneva Observatory, CfA and Tel Aviv University. Preliminary results, mainly on the $(e, \log P)$ diagram, are presented in Udry et al. (1998).

Systematic adaptive optics searches for distant companions are also promising progress in the definition of the long-period part of the period distribution.

From Low-Mass Star Binaries down to Planetary Systems   95

3. Orbital-elements distributions of companions to solar-type stars

3.1. The $(e, \log P)$ diagram

The distribution of orbital eccentricities as a function of period brings information on formation mechanisms of binaries and their dynamical evolution. The comparison of $(e, \log P)$ diagrams of spectroscopic binaries for G-K primaries in the field (DM91, Halbwachs et al. 2000a) and in open clusters (Mermilliod et al. 1992; Stefanik & Latham 1992; Mermilliod & Mayor 1999) with the equivalent distribution for M dwarfs (Udry et al. 2000a) is given in Figure 1. Qualitatively, no fundamental differences are found between the different star populations. The $(e, \log P)$ diagram seems independent from the primary masses. It presents essentially the usual features already pointed out in DM91.

– The short-period SB with $P \leq P_{\text{circ}} \simeq 10\,\text{d}$ are circularized by tidal interactions. For stars with convective envelopes, turbulent viscosity retarding the equilibrium tides is the most efficient mechanism for spin-orbit synchronism and circularization through tidal friction (Zahn 1992). The limit is well marked for M dwarfs: the highest period with $e \leq 0.05$ is 9.5 d whereas the smallest period for eccentric orbits is 10.3 d. P_{circ} is equivalent for G and K dwarfs but with a few "eccentric" stars at smaller periods possibly experiencing partial circularization (Duquennoy, Mayor & Mermilliod 1992).

– "Longer"-period systems with small eccentricities are missing. Above P_{circ} imposed by the tidal circularization, all SB are eccentric. A lower envelope at about $e = 0.1$ is observed. Explanations invoke eccentricity pumping mechanism acting at an early phase when stars have disks (e.g. Lubow & Artymowicz 1992).

– The eccentricity scatter increases with period attesting for a correlation between e and P. An upper envelope to the distribution is easily recognized. It probably results from the Roche lobe limitation at the pre-main sequence time.

3.2. The mass-ratio distribution: $f(q)$

Systematic RV surveys carried out over the last 20 years have allowed the detection of numerous stellar companions for the stars in the solar vicinity. Distributions of orbital elements for these SB are reasonably well determined, at least down to the substellar regime ($m_2 \sin i \geq 0.075\,\text{M}_\odot$). However, the mass function of secondaries is much more sensitive to biases of "badly"-defined samples: non volume-limited samples, volume-limited samples incomplete or based on imprecise parallaxes will all be affected by biases. Unfortunately, the price to pay for the strict definition of a volume-limited sample, complete and based on precise parallaxes, is a drastic reduction of the number of useful SB.

Among our ~ 570 stars of the G+K-dwarf sample, 57 SB have been found with periods less than 10 years. This group of SB is called our *extended sample* and will serve to derive the lower part of $f(q \leq 0.5)$, not affected by the possible excess of binaries with comparable magnitudes. The CNS2 + complement are almost complete for $\pi \geq 65\,\text{mas}$ ($\delta \geq -15°$) and are almost unbiased for $\pi \geq 46\,\text{mas}$ (Halbwachs et al. 2000a). This latter *unbiased sample* consists of 249 stars among which only 26 SB have periods smaller than 10 years. That sample has to be used to define the frequency of binary stars with $q > 0.5$.

For double-line spectroscopic binaries (SB2), the measured ratio of RV amplitudes directly provides the mass ratio: $q = m_2/m_1 = K_1/K_2$. For single-line

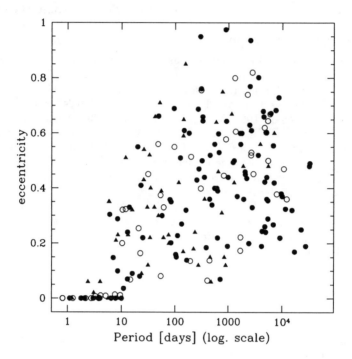

Figure 1. Comparative $(e, \log P)$ diagrams between M dwarfs (open symbols; Udry et al. 2000a) and G+K dwarfs in the field (filled circles; Halbwachs et al. 2000a) and in open clusters (triangles; Mermilliod et al. 1992; Stefanik & Latham 1992; Mermilliod & Mayor 1999).

binaries (SB1), from the radial velocities, we only have access to

$$Y \equiv f(m)/m_1 = \frac{q^3 \sin^3 i}{(1+q)^2}.$$

For limited-size samples the deconvolution of the distribution from the unknown $\sin i$ is hazardous. However, if the periods are not too short, the orbital-plane inclination can be derived from the Hipparcos astrometric data taking into account the precise orbital elements derived from the RV curve (Halbwachs et al. 2000a, b; Arenou et al. 2000). In our 57-SB sample we have 25 SB2. For 22 among the remaining 32 SB1, the astrometric orbit has been solved. Only 10 SB1 are left without additional constraints on their mass ratios. Finally, an iterative inversion technique is applied (Boffin, Paulus & Cerf 1992; Mazeh & Goldberg 1992).

Open clusters also provide stellar samples convenient for studying the mass-ratio distribution. Complete samples of late F to G dwarfs members of the Pleiades and Praesepe have been monitored in RV with CORAVEL at the Haute-Provence Observatory during more than 10 years. A set of 11 SB (9 SB1 +

From Low-Mass Star Binaries down to Planetary Systems

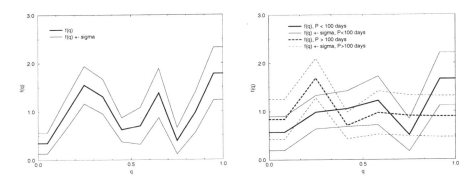

Figure 2. Mass-ratio distributions for: *Left:* the whole sample (field + open clusters); *Right:* 2 domains of periods, $P \leq 100$ d and $P > 100$ d

2 SB2) have been found in the Pleiades with periods shorter than 10 years (Mermilliod et al. 1992) and 16 (13 SB1 + 3 SB2) in Praesepe[1] (Mermilliod & Mayor 1999). However, for open clusters, photometry is sufficient to set in most cases a fair estimate of the companion masses or at least to set an upper bound (the RV curve gives a lower limit for the secondary mass).

Comparing the binaries in the field and in these clusters, we have a 26% probability that these two samples are issued from the same parent population. We thus merge the two samples to derive the mass-ratio distribution (Halbwachs et al. in preparation). The global sample is then composed of 84 SB among which only 10 are SB1 without additional constraints. The *unbiased part*, limited to 46 mas for the field stars and used to analyse the $f(q > 0.5)$ side of the distribution, restricts the sample to 50 SB of which only 6 SB1 without constraints.

Figure 2 (left) presents the mass-ratio distribution for the whole sample ($P \leq 10$ y). We see 3 peaks in the distribution, at $q = 0.25$, 0.6 and 1. We also observe the significant decrease of the distribution for $q < 0.2$. The peak at $q = 0.6$ possibly results from binaries with white-dwarf companions. By estimating the statistical significance of the peak at $q = 1$, we only can reject a constant distribution at an 11% level, a percentage not really significant. If we now separately analyse SB with periods smaller and larger than 100 days (but with $P \leq 10$ y), we see that the possible peak of the mass-ratio distribution at $q = 1$, if real, comes from the "shorter-period" binaries ($P \leq 100$ d; Fig. 2 right).

The preliminary mass-ratio distribution derived by Fischer & Marcy (1992) for M dwarfs is in agreement with a flat distribution above $q = 0.4$. This distribution has to be improved in the near future based on the extended M-dwarf samples currently observed.

---

[1] The SB2 percentage is smaller than for the field due to the relative faintness of cluster stars

98   Udry et al.

3.3. The period distribution and binary frequency

From the enlarged G+K sample we observe a rise of $f(\log P)$ up to $P \sim 10$ years. The full analysis including visual binaries and common proper-motion stars has still to be finalized to improve the $f(\log P)$ distribution of DM91.

We have found 26 SB in the *unbiased sample* ($\pi \leq 46$ mas) composed of 249 F7-K9 main-sequence primaries. A statistical correction of 0.8 ± 0.5 "star" has been estimated for the undetected-binary selection effect. In conclusion, for $q > 0.02$ and $P \leq 10$ years, we observe $11 \pm 2\%$ of double stars among G-K stars. The rate is $9 \pm 2\%$ for $P < 1000$ days.

For comparison, the percentage of SB detected among close M dwarfs is $8\pm 3\%$ for $P \leq 1000$ days (Delfosse & Forveille 2000). **We thus do not observe a significant smaller percentage of SB among M stars compared to solar-type stars**, at least for the unbiased domain of periods ($P \leq 1000$ d).

4. The mass function below the main sequence

The moderate precision programmes have already revealed a few companions with $m_2 \sin i$ small enough to be brown-dwarf candidates (Latham et al. 1989; DM91; Mayor et al. 1997). At the present time, the spectrographs with higher-precision capabilities have brought much further insight into orbital-element distributions and mass function of secondaries with minimum masses below $0.1\,M_\odot$. In particular, more than 50 companions have been detected with $m_2 \sin i$ smaller than $20\,M_{\rm Jup}$, most of them having minimum masses lower than $5\,M_{\rm Jup}$ (Marcy, Cochran & Mayor 2000; Vogt et al. 2000; ESO 2000; Fischer et al. 2000; Udry, Mayor & Queloz 2000c) . The information collected so far on SB, combined with the most recent detection of very low-mass companions (planets and brown-dwarf candidates), give us a synthetic view on companions to solar-type stars from $q = m_2/m_1 = 1$ to $q < 0.001$ (Fig. 3).

The histogram of minimum masses for the companions of solar-type stars with $m_2 \sin i$ less than $0.1\,M_\odot$ is drawn in Figure 4. The impressive peak for $m_2 \sin i \leq 8\,M_{\rm Jup}$ is certainly not part of the very flat tail of stellar and brown-dwarf secondaries. Despite the huge observational bias against the detection of small mass companions we observe an increasing number of low mass gaseous planets. In the same time, brown-dwarf candidates easier to detect are rare. If we consider the still limited span of precise RV surveys as well as the sometimes still limited number of measurements per star, the correction of observational biases is premature. In this sense the dotted line in Figure 3 in only indicative of a potential estimate of the bias-corrected distribution.

Radial velocities only give access to the minimum mass $m_2 \sin i$ of low-mass companions. In addition to RV data, we have collected information constraining the orbital-plane inclinations of brown-dwarf and planetary candidates from different sources (see Mayor & Udry 2000 for a review): Hipparcos astrometric data (Halbwachs et al. 2000a, Zucker & Mazeh 2000), planetary transit (Charbonneau et al. 2000, Henry et al. 2000, Mazeh et al. 2000), dust disks around stars (Trilling, Brown & Rivkin 2000) or synchronization properties for short-period systems (Fuhrmann, Pfeiffer & Bernkopf 1998). For these companions we obtain real masses. It is very impressive to see that most of the $\sin i$ of the brown-dwarf candidates are found to be small leading to masses of the secondaries above the

From Low-Mass Star Binaries down to Planetary Systems 99

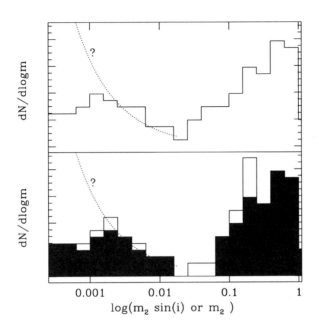

Figure 3. Mass function of companions to solar-type stars. Log scale. The dotted line is indicative of a potential bias-corrected distribution. Top: $m_2 \sin i$. Bottom: composite histogram of m_2 (open part) and $m_2 \sin i$ (if $\sin i$ not known; colored part)

H-burning limit, at the bottom of the main sequence. They possibly represent the tail of the binary $\sin i$ distribution of the peak at $f(q = 0.2)$. Only 4 objects with $\sin i$ determination have real-mass estimates in the brown-dwarf domain (three from Halbwachs et al. (2000b) and one from Zucker & Mazeh (2000)).

Taking into account available orbital inclinations a composite histogram is built up for masses m_2 or $m_2 \sin i$ (if $\sin i$ not known) (Figs. 3 and 4, bottom). The gap between SB and planets is remarkable, emphasizing the paucity of brown-dwarf companions. It clearly suggests different formation and evolution processes for the two populations. Further indications for such differences have been pointed out in the period and eccentricity distributions (see Sect. 5).

The sharp drop of the mass function observed below $\sim 8\,M_{Jup}$ strongly suggests a maximum mass for giant planets close to $10\,M_{Jup}$, without any relation with the D-burning limit around $13\,M_{Jup}$ (Fig. 4). The latter has very probably nothing to do with the maximum mass of objects initially formed from dust/ice particles in an accretion disk. At the opposite extreme, present samples cannot provide any real estimate of the minimum mass for brown dwarfs orbiting solar-type stars. The D-burning limit seems as well to be unrelated to the min-

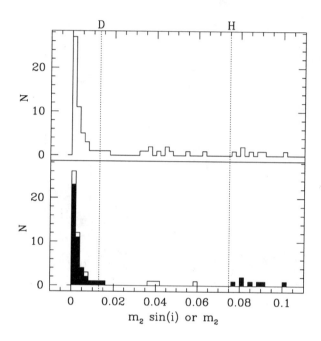

Figure 4. Mass function of companions to solar-type stars. Linear scale. The dotted lines indicate the H- ($75\,M_{Jup}$) and the D- ($13\,M_{Jup}$) burning limits. Top: $m_2 \sin i$. Bottom: composite histogram of m_2 (open part) and $m_2 \sin i$ (if $\sin i$ not known; filled part).

imum mass of brown dwarfs formed by fragmentation in a cloud. Free-floating "planets" (i.e. brown dwarfs) with masses probably below $10\,M_{Jup}$ have been reported in σ-Orionis (Zapatero Osorio et al. 2000). **The D-burning mass is clearly of no help to understand the observed mass distribution.**

5. Orbital-element distribution of giant planets

The observed gap in the mass distribution between giant planets and stellar secondaries strongly suggests the existence of two distinct companion populations for solar-type stars. Giant planets and stellar binaries are believed to have different formation mechanisms whose fossil traces should be revealed by a comparison of their orbital properties.

5.1. The distribution of periods

Clearly, due to the still strong bias affecting the detection of long-period planets, a significant comparison of the period distributions of planetary candidates

From Low-Mass Star Binaries down to Planetary Systems 101

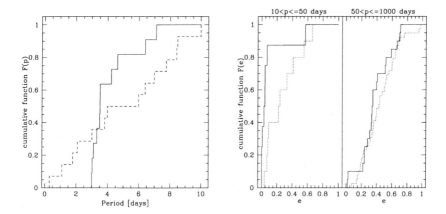

Figure 5. *Left:* Cumulative distributions of periods smaller than 10 days for planetary (solid line) and stellar companions (dashed line) to solar-type stars. *Right:* Cumulative distributions of eccentricities for planets (solid line) and stellar binaries (dotted line) for 2 domains of periods: $10 \leq P \leq 50$ d and $50 \leq P \leq 1000$ d.

($m_2 \sin i \leq 10\,\mathrm{M_{Jup}}$) and spectroscopic binaries ($m_2 \sin i \geq 10\,\mathrm{M_{Jup}}$) is only possible for relatively short-period systems ($P \leq 100$ d), for which the detection bias is vanishing. In the domain of very-short periods ($P \leq 10$ d), the distribution of periods of giant planets is steeply rising for decreasing periods down to about 3 days. A cut-off is observed around that period. The shortest period observed to date is $P = 2.986$ days for HD 83443 b (ESO 2000, Mayor et al. 2000). It is clearly emphasized in Figure 5 (left) comparing the cumulative distributions of periods smaller than 10 days for giant planets and spectroscopic binaries. Several reasons may be invoked to set this limit, like e.g. magnetospheric central cavity of the accretion disk, tidal interaction (Lin et al. 1996), Roche lobe overflow (Trilling et al. 1998) or evaporation (Mayor & Udry 2000).

5.2. The distribution of eccentricities

The comparison of orbital eccentricities of spectroscopic binaries for G, K and M primaries (DM91; Halbwachs et al. 2000a; Udry et al. 2000a) with the equivalent distribution for giant planets is remarkable. In Figure 6 we have plotted the orbital eccentricities as a function of $\log P$ for both double stars and giant planets. At first glance we do not observe significant differences of orbital elements between planets and SB. If the formation mechanisms for planets and double star are different, why do we observe such similar ($e, \log P$) distributions? In fact some differences may be emphasized in particular domains of periods.

For planets with very short periods, most of the orbits are quasi-circular as for double stars. Planets having suffered a strong orbital migration are probably circularized by the tidal interaction with the accretion disk. We observe quite a number of quasi-circular planetary orbits with $P \leq 40$ days. It is however also true that two short-period planets present eccentric orbits: around HD 108147 ($P = 10.9$ d, $e = 0.56$) and HD 217107 ($P = 7.11$ d, $e = 0.14$). Such configurations

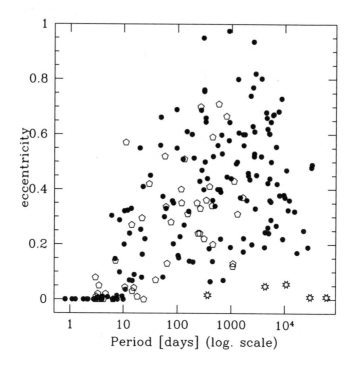

Figure 6. Comparative $(e, \log p)$ diagram for planetary (open pentagons) and stellar companions to G+K+M solar-type stars of the field (filled circles). Starred symbols indicate planets of the solar system

could be explained by the presence of an additional companion revealed by a drift observed in the systemic velocity of those stars or by gravitational interaction between giant planets formed in the outer regions of the system (Weidenschilling & Marzari 1996; Lin & Ida 1997).

We have seen that most of the orbits of double stars with long periods ($P \geq 50$ d) are fairly eccentric. Quasi-circular ($e < 0.01$) long-period orbits are very rare. A similar situation exists for extra-solar giant planets (with $e = 0.13$, 47 UMa is just at the border), although they are still strongly observationally biased in that period range. We of course know one planetary system with long-period giant planets on quasi-circular orbits, namely our own Solar system.

The origin of the eccentricity of extra-solar giant planets has been sought in the gravitational interaction between multiple giant planets (Weidenschilling & Marzari 1996; Rasio & Ford 1996; Lin & Ida 1997) or between the planets and the planetesimals in the early stages of the system formation (Levison, Lissauer & Duncan 1998). Among the giant planet candidates several eccentric orbits show a drift of their mean velocity indicating the presence of a long-period companion (probably stellar) whose gravitational perturbation can also be suspected to be responsible for the observed (high) planetary eccentricity as e.g. for the planet orbiting 16 Cyg B (Mazeh, Krymolowski & Rosenfeld 1997).

In summary, for periods larger than about 50 days, the comparison of double star and planet eccentricities is really surprising. With the limited-size samples presently available, we cannot see any significant distinctions between both populations. This is shown in Figure 5 (right) which gives the cumulative function of eccentricities for periods between 50 and 1000 days, for the two populations. If the orbital eccentricity of binaries finds its origin in the disruption of small N-body systems, maybe the generated distribution of eccentricities is close to the distribution generated by the gravitational interactions between giant planets.

Nevertheless, a clear difference may be noticed between the two populations for periods between 10 and 50 days. This is illustrated by the corresponding cumulative functions given in Figure 5 (right). In this range of periods, outside the circularization domain, planetary systems have smaller eccentricities than stellar binaries indicating different formations or evolutions.

In conclusion, the differences observed between planetary systems and double stars in the shape of the secondary mass functions and in the period and eccentricity distributions argue for the two populations being formed by distinct processes. More observations are however still needed to bring clear constraints on the possible formation and evolution scenarios.

6. Future prospects in the domain of multiple stars

Most of our knowledge of the orbital-element distributions of short-period binaries ($P \leq 10$ y) is derived from spectroscopic surveys done in the 80's-90's with a moderate precision of about 0.25 kms^{-1}. New precise RV surveys ($\varepsilon \simeq 10$ ms^{-1}) carried out on F-K stars (e.g. the CORALIE survey, Udry et al. 2000b) or for M dwarfs (Delfosse et al. 1999a) will have an almost 100 % completeness from jovian-type objects to normal spectroscopic binaries.

If we want to explore the domain of longer periods, drastic improvements are expected both from imagery and astrometry. Follow-up studies with adaptive optics of stars with giant planets or brown-dwarf companions quite frequently reveal distant visual additional companions, not previously identified, and clearly show the incompleteness still existing in surveys searching for distant, very faint, stellar secondaries: e.g. the companions to τ Boo and HD 114762 (Lloyd et al. 2000) or the companion to HD 174457 (Naef et al. 2000). Systematic adaptive optics surveys of all stars in a volume-limited sample have to be done.

Ground-based interferometry will prove to be very fruitful for binary star research (Davis 1996). In dual-beam mode (Shao & Colavita 1992), foreseen for the VLT and Keck telescopes, a higher precision is achieved. For the VLTI the astrometric reflex motion of a primary star will be detected with a precision of 10-50 μas. This precision will be sufficient to determine the orbital orientations (and real masses) of most of the detected giant planets and brown-dwarf candidates. Space missions like SIM, FAME or GAIA will still further improve our knowledge of orbital distributions with unprecedented astrometric precision (down to 4 μas for GAIA for a huge stellar sample and maybe 1 μas for SIM for some specific high-priority targets). Figure 7 illustrates the domain in the ($\log a$, $\log m_2$) plane where astrometry and Doppler spectroscopy are sensitive.

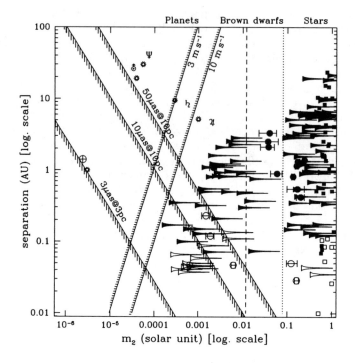

Figure 7. Separation-mass diagram for companions to solar-type stars from binaries to planetary systems. Elongated symbols represent the $\sin i$ probability. Open symbols are for low-e orbits. Limits of the Doppler spectroscopic and astrometric techniques are indicated.

We can now ask ourselves what are the domains where the microlensing technique can provide unique results. Several brown dwarfs have been found to be double (PPL 15, Basri & Martin 1999; DENIS-PJ1228.2-1547, Delfosse et al. 1997, Martin, Brandner & Basri 1999). Some results on the duplicity of young brown dwarfs could be expected from adaptive optics and in a few cases from spectroscopy. However, most old brown dwarfs in the field will only be revealed (as well as their duplicity) through microlensing surveys.

At the present time our knowledge of giant planets is restricted to objects with masses from just below the mass of Saturn to a few times the mass of Jupiter. The minimum mass detected so far is for HD 83443 c with $m_2 \sin i = 0.5\,M_{\text{Saturn}}$ (Mayor et al. 2000). An increase in the number of measurements per star and in the precision of spectrographs will perhaps allow us to detect planets just at the border of the Neptune mass, but only for very short orbital periods. Before the arrival of space interferometers, microlensing surveys should have the possibility to explore the still unknown domain of very low-mass planets.

References

Arenou F., Halbwachs J.-L., Udry S., Mayor M. 2000, in *Birth and Evolution of Binary Stars*, IAU Symp. 200P, eds. B. Reipurth & H. Zinnecker, p. 135
Baranne A., Mayor M., Poncet J.-L. 1979, Vistas in Astron., 23, 279
Baranne A., Queloz D., Mayor M., et al. 1996, A&AS, 119, 373
Basri G., Martin E. 1999, AJ, 118, 2460
Boffin H., Paulus G, Cerf N. 1992, in *Binaries as Tracers of Stellar Formation*, eds. Duquennoy A. & Mayor M., Cambridge University Press, p.26
Charbonneau D., Brown T., Latham D., Mayor M. 2000, ApJ, 529, L45
Davis J. 1996, in "Science with the VLT Interferometer", ESO Astrophys. Symp., ed. Paresce F., p.143
Delfosse X., Forveille T. 2000, in "Very-low mass stars and brown dwarfs in stellar clusters and associations", eds. Rebolo R., Zapatero Osorio M.R.& Béjar V., Cambridge University Press
Delfosse X., Tinney C., Forveille T. 1997, A&A, 327, L25
Delfosse X., Forveille T., Perrier C., Mayor M. 1998a, A&A, 331, 581
Delfosse X., Forveille T., Mayor M., et al. 1998b, A&A, 338, L67
Delfosse X., Forveille T., Beuzit J.-L., et al. 1999a, A&A, 344, 897
Delfosse X., Forveille T., Udry S., et al. 1999b, A&A, 350, L39
Delfosse X., Forveille T., Mayor M., Burnet M., Perrier C. 1999c, A&A, 341, L63
Duquennoy A., Mayor M. 1991, A&A, 248, 485
Duquennoy A., Mayor M., Mermilliod J.-C. 1992, in *Binaries as Tracers of Stellar Formation*, eds. Duquennoy A. & Mayor M., Cambridge University Press, p.52
ESA 1997, *The Hipparcos and Tycho Catalogue*, ESA-SP 1200
ESO 2000, http://www.eso.org/outreach/press-rel/pr-2000/pr-13-00.html
Fischer D., Marcy G. 1992, ApJ, 396, 178
Fischer D., Marcy G., Butler P., et al. 2000, ApJ, submitted
Forveille T., Beuzit J.-L., Delfosse X., et al. 1999, A&A, 351, 619
Fuhrmann K., Pfeiffer M.J., Bernkopf J. 1998, A&A, 336, 942
Gliese W. 1969, Ver Astron Rechen Inst Heidelberg, 22
Gliese W., Jahreiss H. 1979, A&AS, 38, 423
Halbwachs J.-L., Mayor M., Udry S. 1998, in "Brown Dwarfs and Extrasolar Planets", eds. Rebolo R., Martin E. & Zapatero Osorio M.R., ASP Conf. Ser. 134, 308
Halbwachs J.-L., Arenou F., Mayor M., Udry S. 2000a, in "Birth and Evolution of Binary Stars", IAU Symp. 200P, eds. Reipurth B. & Zinnecker H., p.132
Halbwachs J.-L., Arenou F., Mayor M., Udry S., Queloz D. 2000b, A&A, 355, 581
Henry G., Marcy G., Butler P., Vogt S. 2000, ApJ, 529, L41

Latham D., Mazeh T., Stefanik R., Mayor M., Burki G. 1989, Nature, 339, 38
Levison H.F., Lissauer J., Duncan M.J. 1998, AJ, 116, 1998
Lin D., Ida S. 1997, ApJ, 477, 781
Lin D., Bodenheimer P., Richardson D.C. 1996, Nature, 380, 606
Lloyd J., Patience J., Liu M., et al. 2000, in "Planetary Systems in the Universe", IAU Symp. 202, eds. Penny A., et al., PASP Conf. Ser., in press
Lubow S.H., Artymowicz P. 1992, in *Binaries as Tracers of Stellar Formation*, eds. Duquennoy A. & Mayor M., Cambridge University Press, p.145
Marcy G.W., Benitz K.J. 1989, ApJ, 344, 441
Marcy G., Butler P., Vogt S., Fischer D., Lissauer J. 1998, ApJ, 505, L147
Marcy G., Cochran W., Mayor M. 2000, in "Protostars and Planets IV", eds. Mannings V., Boss A., Russel S., The University of Arizona Press, p.1285
Martin E., Brandner W., Basri G. 1999, Science, 283, 1718
Mayor M., Udry S. 2000, in "Disks, Planetesimals and Planets", eds. Garzón F., Eiroa C., de Winter D. & Mahoney T., ASP Conf. Ser., in press
Mayor M., Queloz D., Udry S., Halbwachs J.-L. 1997, in "Astronomical and Biochemical Origins of Life in the Universe", eds. Cosmovici C., Browyer S. & Werthimer D., IAU Coll. 161, Editrice Compositori, p.313
Mayor M., Naef D., Pepe F., et al. 2000, in "Planetary Systems in the Universe", IAU Symp. 202, eds. Penny A., et al., PASP Conf. Ser., in press
Mazeh T., Goldberg D. 1992, ApJ, 394, 592
Mazeh T., Goldberg D., Duquennoy A., Mayor M. 1992, ApJ, 401, 265
Mazeh T., Krymolowski Y., Rosenfeld G. 1997, ApJ, 477, 403
Mazeh T., Naef D., Torres G. et al. 2000, ApJ, 532, L55
Mermilliod J.-C., Rosvick J., Duquennoy A., Mayor M. 1992, A&A, 265, 513
Mermilliod J.-C., Mayor M. 1999, A&A, 352, 479
Naef D., Mayor M., Arenou F., et al. 2000, in "Planetary Systems in the Universe", IAU Symp. 202, eds. Penny A., et al., PASP Conf. Ser., in press
Rasio F., Ford E. 1996, Science, 274, 954
Shao M., Colavita M. 1992, A&A, 262, 353
Stefanik R., Latham D. 1992, in "Complementary Approaches to Double and Multiple Star Research", IAU Coll. 135, eds. McAlister H.A. & Hartkopf W.I., ASP Conf. Ser. 173, 32
Trilling D.E., Benz W., Guillot T. et al. 1998, ApJ, 500, 428
Trilling D.E., Brown R.H., Rivkin A.S. 2000, ApJ, 529, 499
Udry S., Mayor M., Latham D., et al. 1998, in "Cools Stars, Stellar Systems and the Sun", eds. Donahue R. & Bookbinder J., ASP Conf. Ser. 154
Udry S., Mayor M., Delfosse X., et al. 2000a, in "Birth and Evolution of Binary Stars", IAU Symp. 200P, eds. Reipurth B. & Zinnecker H., p.158
Udry S., Mayor M., Naef D., et al. 2000b, A&A, 356, 590
Udry S., Mayor M., Queloz D. 2000c, in "Planetary Systems in the Universe", IAU Symp. 202, eds. Penny A., et al., PASP Conf. Ser., in press
Vogt S., Marcy G., Butler P., Apps K. 2000, ApJ, 536, 902

From Low-Mass Star Binaries down to Planetary Systems 107

Weidenschilling S.J., Marzari F. 1996, Nature, 384, 619
Young A., Sadjadi S., Harlan E. 1987, ApJ, 314, 272
Zahn J.-P. 1992, in *Binaries as Tracers of Stellar Formation*, eds. Duquennoy A. & Mayor M., Cambridge University Press, p.253
Zapatero Osorio M.-R., Béjar V., Martin E., et al. 2000, Science, in press
Zucker S., Mazeh T. 2000, ApJ, 531, L67

Discussion

Gould: Can it be a selection effect that the intermediate $m \sin i$ objects are moved to higher masses, but no low-mass objects are moved to intermediate masses?

Mayor: No, the "desert" between giant planets and spectroscopic binaries cannot be explained by selection effects. The numerous SJ at the bottom of the main sequence can easily give a few $m_2 \sin i$ in the range of brown dwarf masses. But the statistics on the orbital planes of objects with $m_2 \sin i$ less than 5–10 M_J cannot create a significant number of objects at about 50 M_J.

Reid: A comment: A fair amount of work has been conducted on imaging surveys of M dwarfs by Henry, Gizis and myself. Those observations tend to suggest a preference for equal mass systems and a scarcity of brown dwarf companions (i.e. extending the brown dwarf deficit to wide binaries). As for L dwarfs, we have a large programme of HST imaging which aims to identify wide (> 3–5 AU) binaries in that sample. Since the sample is magnitude limited, we expect a bias to equal-mass systems.

Mayor: A lot of systematic surveys have been carried out in the last 10 to 15 years to search for visual binaries among the dwarfs. As mentioned in the introduction to my talk I have not included the discussion of the wide binaries as being less relevant to microlensing searches.

Fischer: What is the error bar on mass determination for HD10476 from Hipparcos astrometry? Given the 2 mas separation and Hipparcos precision ($\sim$1 mas) I wonder if this uncertainty is generous enough?

Mayor: I believe that the quoted uncertainly on HD10476 given by Zucker & Mazeh (2000) is correct as the separation is derived from a large number of ($\sim$100) astrometric measurements.

Gould: Are 1% M dwarf mass measurements really possible? Also, why do you mention SIM and GAIA but not FAME?

Mayor: Yes, already a few masses of M dwarfs have been determined at the level of about 1% precision, combining precise velocities and interferometry and/or adaptive optics measurements. I only mention SIM and GAIA as an example of ambitious space missions supposed to greatly contribute to that field but it's true that we also have to mention the FAME and DIVA missions.

Reid: There is evidence for a lower binary fraction amongst M dwarfs (than amongst G dwarfs) if one includes wider systems. As you say, this may be less interesting for microlensing.

Mayor: We know from the spectroscopic surveys of M dwarfs carried out with CORAVEL, and with ELODIE with a much higher precision, that a significant fraction of short period binaries exists, but the full discussion and comparison of these samples with G stars is still in progress.

Gould: Where does information about eccentricity for P = 3000 years come from?

Mayor: Above P = 20 years the orbital elements are derived from visual orbits. For a few of the closest stars some estimates exist for orbits in excess of 1000 years, but with increasing uncertainties.

PLANET Observations of Anomalous Microlensing Events

J. Menzies[1], M. D. Albrow[2,8], J.-P. Beaulieu[3], J. A. R. Caldwell[1], D. L. DePoy[4], B. S. Gaudi[5], A. Gould[5], J. Greenhill[5], K. Hill[5], S. Kane[6], R. Martin[6], M. Dominik[7], R. M. Naber[6], R. W. Pogge[4], K. R. Pollard[2], P. D. Sackett[6], K. C. Sahu[8], P. Vermaak[1], R. Watson[5], A. Williams[6] (The PLANET collaboration)

[1] *South African Astronomical Observatory, P.O. Box 9, Observatory 7935, South Africa*

[2] *University of Canterbury, Dept. of Physics & Astronomy, Private Bag 4800, Christchurch, New Zealand*

[3] *Institut d'Astrophysique de Paris, INSU CNRS, 98bis Boulevard Arago, 75014 Paris, France*

[4] *Ohio State University, Dept. of Astronomy, Columbus, OH 43210, USA*

[5] *University of Tasmania, Physics Dept., G.P.O. 252C, Hobart, Tasmania 7001, Australia*

[6] *Perth Observatory, Walnut Road, Bickley, Perth 6076, Australia*

[7] *Kapteyn Astronomical Institute, Postbus 800, 9700 AV Groningen, The Netherlands*

[8] *Space Telescope Science Institute, 3700 San Martin Drive, Baltimore, MD 21218, USA*

Abstract. In the last five years, PLANET (Probing Lensing Anomalies NETwork) has detected several anomalies in ongoing microlensing events in real-time. The frequent sampling and the high photometric accuracy of the PLANET data allowed and required accurate modelling of some anomalous events. As a consequence, we were able to measure (1) relative proper motions between lens and source star, (2) limb-darkening coefficients for distant source stars, (3) the motion of binary lens components around their center of mass. Results for MACHO 97-BLG-28, MACHO 98-SMC-1 and MACHO 97-BLG-41 are discussed. The last of these was successfully modelled as a rotating binary. Using the parameters found for the binary motion, we deduce that the lens responsible for this event is likely to be an M dwarf binary in the Galactic disk with a period $P \sim 1.5$ yr.

1. Introduction

The PLANET collaboration was formed some five years ago with the aim of determining the proportion of stars in the direction of the Galactic Bulge with Jupiter-mass planetary companions. The procedure adopted was to monitor closely Bulge stars undergoing gravitational microlensing. A single star at the distance of the Bulge being lensed by a star at a fraction of that distance would be seen to brighten and fade in a predictable way. If the lensing star had a planetary companion, the latter's presence would be betrayed by an anomalous feature, lasting about one day, in the light curve. By making sufficiently precise, frequent photometric measurements from observatories at different longitudes around the globe, PLANET hoped to detect planets, if they accompanied lensing stars. Alerts are issued for ongoing microlensing events in the Bulge by the MACHO (up to the end of 1999), EROS and OGLE survey groups, and a subset of these events has been monitored by the PLANET group. Most of the light curves are well described by what is expected for a point-source point-lens event. However, there are some events showing anomalous behaviour, often involving very high amplifications and rapid changes of brightness on a short timescale. This is the case when the lens is a binary, and its influence on the light from the background source star is strongly influenced by caustics whose structure depends on the mass ratio and separation of the binary components. The caustics are linear features in the plane of the source star where the amplification is formally infinite.

The caustic structure acts a surface probe as it sweeps across the face of the background source star. At the simplest level, a model fit to the light curve together with knowledge of the diameter of the source leads to a measure of the relative lens-source proper motion. This helps to locate the lens with respect to the source and in principle it would be possible to determine the mass of the lens, the distances of the lens and source, as well as their relative velocities. Modelling the fine details of the light curve can lead to a measurement of the limb darkening of the source star, a quantity that is not practically obtainable by any other method at present. If the magnification is high enough it should be possible to obtain detailed spectroscopic data for comparison with stellar atmosphere models (see Sackett, this volume).

Once a sufficiently large number of binary events has been observed, it should be possible to obtain useful information on the distribution of mass ratios and component separations for a population of stars at a considerable distance from the solar neighbourhood.

The PLANET group has observed many anomalous events and this presentation is intended to give some idea of what has been learned and of what perhaps can be done in future. More detail can be obtained from the published papers listed at the end.

2. MACHO 97-BLG-28

Following an alert from MACHO, regular monitoring was commenced by PLANET. A sudden anomalous brightening suggested that a caustic crossing event was in progress. Intensive observations were carried out at all PLANET sites

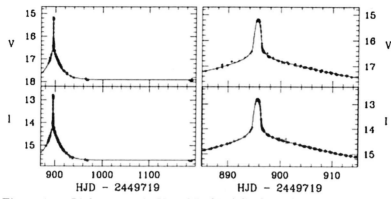

Figure 1. Light curves in V and I of MACHO BLG-97-28. The right hand panels show in more detail the part of the curve around the cusp crossing.

for the approximately 30 hour duration of the anomaly. Sample times of 3 to 30 minutes between data points were achieved. A total of 686 data points were obtained in the 1997 season, while 10 baseline points were obtained at SAAO in 1998. In combining data from the different sites, it is generally our experience that, in spite of our use of a network of stars near the lensing event to obtain relative photometry, it is necessary to apply small corrections to bring all sites onto the same zero point. It is also usually important to include seeing corrections to reduce scatter in the photometric data. This was particularly so for 97-BLG-28, and it was the case that points for which I > 14.7 mag and the seeing > 2.2 arcsec FWHM had to be excluded from the modelling.

To reproduce all details of the light curve shown in Figure 1, it was necessary to include effects due to lens binarity, source blending, extended source size and source limb darkening giving 19 parameters to be fitted. The final model implies that the event was produced by the passage of an isolated cusp of the central caustic of the lensing binary over the face of the source star (see Sackett, this volume). While a uniformly illuminated disc was ruled out for the source, the derived square root limb darkening coefficients agree well with theoretical expectations for a K2III star. This is the first time that limb darkening coefficients have been measured by microlensing, one of the few measurements for a normal giant, and the first for any star at the distance of the Galactic bulge. A full discussion of the result can be found in Albrow et al (1999b).

3. MACHO 98-SMC-1

Although the PLANET collaboration concentrates its efforts on events in the Galactic Bulge, considerable time was spent on MACHO 98-SMC-1. One day after the alert was issued by MACHO, all sites started observing this binary lens event. Our data from the period 9 – 17 June, 1998 allowed us to refine the original MACHO prediction for the second caustic crossing of 19.2±1.5 June UT to 18.0 June UT. More intensive observations were begun and the caustic crossing was duly observed at the predicted time with the 1.0-m telescope at Sutherland. A spectrum was obtained around the peak of the crossing with the 1.9m telescope at Sutherland, and showed the source star to be an A-type dwarf.

The significance of this event is that the relative source-lens proper motion can be determined from a knowledge of the angular radius of the source star and the time taken to cross the caustic. This in turn allows the location of the lens to be determined.

A nine-parameter model is required to fit the light curve generated by a static binary lens and a uniform surface-brightness finite source. We found two more or less equally acceptable models (designated I and II in what follows). They give almost the same transit time for the source across the caustic (8.52 and 8.46 hours, respectively), but the angle of crossing is rather different (43.2 and 30.6 degrees, respectively). From our photometry and an estimate of the reddening, the intrinsic colour of the source star can be found. The source's position in the colour-magnitude diagram suggests it is an A6 dwarf, consistent with the appearance of the spectrum. The derived proper motion is 1.26±0.10 $kms^{-1}kpc^{-1}$ for model I and a factor of 1.59 times greater for model II. A comparison of the derived proper motion with the distribution expected from a standard isothermal halo rotating at 220 kms^{-1} with the source in the SMC at 60 kpc, indicates that there is a negligible chance that the lens is in the halo. If, on the other hand, the lens is in the SMC then, taking the SMC to be a one-component system with velocity dispersion $21 kms^{-1}$, we find that some 9% of events could have a proper motion greater than that found for model I, while <0.1% would have proper motions greater than for model II. The SMC is known to have structure on the basis of radial velocity maps, so it is plausible that there is material in the SMC that could produce the observed relative proper motion. If the lens is indeed in the SMC then the durations implied by the models lead to masses characteristic of a typical low-mass binary. The details of the discussion and references can be found in Albrow et al (1999a).

Subsequently, a combined analysis of data from five different groups based on a robust method for finding binary lens solutions proposed by Albrow et al (1999c), confirmed the finding that the lens was in the SMC (Adorf et al 2000). It was not possible to discriminate between a close- and a wide binary solution for the lens, but a measurement of limb darkening was obtained, the first for a metal-poor A star.

4. MACHO 97-BLG-41

Yet another unusual light curve was presented by the microlensing event MACHO 97-BLG-41. The photometric behaviour suggested that the source star had crossed two disjoint caustic regions separated in time by about one month. Five PLANET sites monitored the event for about 90 days. Reductions were done using both the ISIS image subtraction package and the DoPHOT profile-fitting program. Seeing corrections were applied and small zero point corrections were needed to align the data sets from all sites. A total of 46 V and 325 I points were used in the final modelling, and the combined light curve is shown in Figure 2(a).

PLANET Observations of Anomalous Microlensing Events

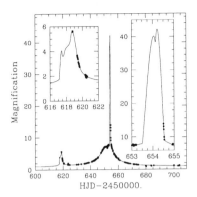 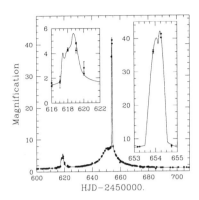

Figure 2. (a) Light curve (points) of MACHO BLG-97-41 obtained by PLANET. (b) Comparison of PLANET model (continuous line) and MACHO/GMAN data points.

Modelling this event proved to be particularly difficult. A preliminary qualitative assessment of the light curve suggested the lens was probably a close binary with three caustics, two outlying triangular ones and a diamond-shaped central one. The first peak was probably due to the passage of the source over one of the outliers while the main peak was associated with the central caustic. However, it was impossible to find a solution involving a static binary. Of course a binary must rotate, but the relative values of event timescale and rotation period determine whether the rotation has a significant effect on the light curve. No evidence of rotation had so far been seen in a binary microlensing event light curve. The complexity of the model would have been increased substantially if a full representation of the binary were included. It was found however that it was only necessary to consider the first order effect of rectilinear relative motion of the two components to get a good fit to the light curve. This requires two extra parameters over the static binary fit. The light curve resulting from the best-fitting model is shown as a solid line in Figure 2(a). In the time interval of 35.17 days between the two caustic crossings the projected binary separation changed by -0.07 ± 0.009, in units of the Einstein radius, and rotated by 5.61 ± 0.36 degrees. Our model leads to a limb-darkening parameter in the I band that agrees well with that expected from stellar atmosphere models for K giants. Using reasonable assumptions for the distributions in space and separation of Galactic binaries, we were able for the first time to derive kinematic probability distributions for the total mass and period for a binary microlens. The most probable lens distance of $\sim$5.5 kpc implies a lens mass of $\sim$1.03$M_\odot$. On the other hand the most probable lens mass, $\sim$0.3$M_\odot$, appropriate to an M dwarf binary, would correspond to a lens distance of $\sim$3.1 kpc, the source being assumed to be at $\sim$8 kpc. The period distribution is sharply peaked wit a maximum at 1.5 yr. Our success in finding a rotating binary solution arose in part because we had such a high quality data set, and partly thanks to strategic choices made in the fitting

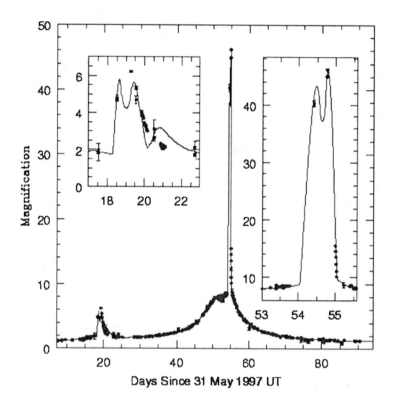

Figure 3. Comparison of PLANET data points with MACHO/GMAN three-body model

procedure: the search was done on a grid using empirical parameters rather the more non-linear canonical ones, and the rotation was treated only to first order.

Completely independently, the MACHO/GMAN collaboration obtained an extensive photometric data set covering this event. They were unable to find either a static or a rotating binary model to fit their data. However, the PLANET model does fit their data rather well as shown in Figure 2(b). MACHO/GMAN resorted to a model involving a triple system comprising a static binary orbited by a Jupiter-mass planet. However, their model does not fit the PLANET data, particularly around the first caustic event, as is seen in Figure 3. It seems that a rotating binary lens provides a natural, physically plausible model for the system, and that it is not necessary to invoke a third body. A detailed discussion of the PLANET data and modelling for this event can be found in Albrow et al (2000).

References

Adorf et al, 2000, ApJ, 532, 340
Albrow et al (The PLANET Group), 1999a, ApJ, 512, 672
Albrow et al (The PLANET Group), 1999b, ApJ, 522, 1011
Albrow et al (The PLANET Group), 1999c, ApJ, 522, 1022
Albrow et al (The PLANET Group), 2000, ApJ, 534, 894

Discussion

Bennett: I think that it is the case that your model for MACHO-97-BLG-41 implies a surprisingly large orbital velocity for the binary lens system. I find that the lens system is probably un-bound if you put it at the most likely distance. I think this must be why your estimated mass is so low.

Menzies: I don't think that is right. This point is discussed in some detail in our paper (Albrow et al. 2000).

Bennett: Here are a couple of comments on MACHO-97-BLG-41: 1) The PLANET fit is inconsistent with the MPS data, but there is a nearby fit that is OK. I suspect that the situation is similar for the MPS triple lens and the PLANET data. 2) Also, I think that the PLANET fit has some difficulty with the rapid implied velocity. At the most likely distance, the implied velocity is too high for a bound system.

Sackett: In answer to Bennett: The data place a constraint on a certain combinations of mass, distance and period, so one must consider self-consistent solutions. The peaks of the individual distributions for d, M and P are *not* a self-consistent solution. A face-on orbit has more than 4 times the potential energy it needs to be bound, so there is a lot of parameter space available for bound orbits.

Popowski: PLANET collects data using several telescopes. How difficult is it to put these data on a consistent photometric system?

Menzies: The degree of difficulty varies from one lens to the next. Even if the same set of reference stars is used at each site, we usually find small offsets ($\lesssim$ 0.05 mag) between data sets – presumably reflecting different plate scales and average seeing and depending on how crowded the field is. Since we know that this does happen we include the zero point for each site as a free parameter in the fitting procedure.

Popowski: If the sampling of a given event is sparse, doesn't your procedure of solving for photometric zero-points push you off the correct solutions or cause ambiguities?

Menzies: Yes, that could be a problem. We demand that there be a good overlap between sites and reject a data set if it is too sparse compared with the others available to us.

Doppler Search for Extrasolar Planets

Debra A. Fischer

UC Berkeley, 601 Campbell Hall, Astronomy Dept., Berkeley, CA 94720

Abstract. Precise Doppler measurements of several hundred FGKM main sequence stars have revealed 34 extrasolar planets. Radial velocity detections are strongly biased toward detection of planets with masses ($M \sin i$) greater than 0.5 times the mass of Jupiter and orbital periods less than a few years. The mass distribution rises toward the low mass bins despite the fact that more massive planets are easier to detect. Planets which orbit more closely than 0.1 AU all reside in circular orbits, plausibly induced by tidal interactions with the host star. In contrast, all planet candidates that orbit farther than 0.2 AU reside in non-circular orbits with eccentricities greater than 0.1.

The information from Doppler techniques must be supplemented with data from other detection methods to build a complete picture of planet formation mechanisms and characteristics. The first transit observation has provided firm evidence that the detected planets are gas giant planets. For planets with less favorably aligned orbits, astrometry will resolve the orbital inclination and thereby determine the absolute masses. Microlensing techniques will provide complementary statistical evidence on the rate of occurrence of planets with lower masses and orbital radii more characteristic of our own solar system.

1. Introduction

In 1995, Michel Mayor and Didier Queloz (Mayor & Queloz 1995) astonished the world with the discovery of the first extrasolar planet. In the last 5 years, 34 planet candidates have been discovered with high-precision radial velocity surveys (Vogt et al. 2000, Fischer et al. 1999, Mayor et al. 1998, Marcy & Butler 1998, Noyes et al. 1997, Cochran et al. 1997) including the first system of multiple planets (Butler et al. 1999), the first detection of a photometric transit by an extrasolar planet (Henry et al. 2000, Charbonneau et al. 2000), and the first sub-Saturn-mass planets (Marcy et al. 2000).

2. The Lick Survey

The original survey at Lick Observatory (Butler et al. 1996) began in 1987 with a sample of 107 main sequence F–M type stars. This magnitude-limited sample excluded binary systems with separations less than two arcseconds and initially operated with a velocity precision of 10 – 15 ms^{-1}. In 1994, refurbishment of

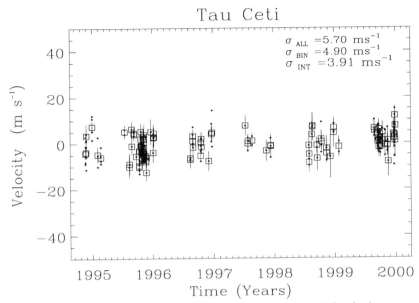

Figure 1. Velocities from Lick Observatory of a radial velocity standard star. The internal error bars are typically 3 ms^{-1} and the rms scatter is under 6 ms^{-1}.

the Hamilton spectrograph optics resulted in an improved instrumental PSF (Valenti et al. 1995) and the velocity precision dropped to 3 – 5 ms^{-1}. The Lick survey provides a well-sampled, long time baseline of observations from which preliminary statistics on the rate of occurrence of extrasolar planets may be extracted. Eight extrasolar planets have been detected from this original sample of 107 stars. All have $M \sin i$ less than 8 M_{JUP} with orbital periods that span from just a few days to over three years. It is notable that there are no brown dwarf candidates in this sample, despite the relative ease with which they would have been detected.

As important as the stars with Keplerian velocity signatures, are the twenty or more chromospherically-quiet standard stars which show no velocity variations. These stars, like τ Cet (Fig. 1), serve as radial velocity standards and have velocities constant to better than 1 ms^{-1} per year.

3. Other High Precision Velocity Surveys

There are several radial velocity programs currently underway. The Doppler survey at Haute-Provence Observatory (Mayor & Queloz 1995) began in 1993 and detected the first extrasolar planet around the star 51 Peg. This survey has since expanded to include more than 320 stars. A large, volume-limited survey with more than 1600 stars has been launched by the Mayor team at La Silla in the Southern Hemisphere. Because this sample is volume-limited it will provide compelling statistical results as the survey matures (i.e., after the stars are well-sampled over a long time baseline). Several extrasolar planet discoveries (Delfosse et al. 1999, Queloz et al. 1999, Santos et al. 2000, Udry et al. 2000) have already emerged from these surveys.

The AFOE program (Noyes et al. 1997) began in 1992. Four planet detections were made here from a sample of about 100 stars. The McDonald Observatory program (Cochran et al. 1997) discovered one planet from a sample of 32 stars. This program will be expanded soon at the Hobby-Eberly Telescope. A new ESO survey has detected one planet (Kurster et al. 1999) from a sample of 40 stars. A Keck survey of 550 stars began in 1996 (Vogt et al. 2000; Marcy et al. 2000) and has discovered a dozen planets. Another Keck survey of 1000 G dwarfs started in 1997 has detected one planet (Mazeh et al. 2000). In addition, some new surveys started up in 1998: a Doppler survey of 150 stars at the AAO (by Paul Butler), and expansion of the Lick planet search project as well as a survey of Hyades stars at Keck by Cochran and Hatzes. There is some overlap in the samples, but the combined statistics from all of these radial velocity surveys suggest that giant planets will be detected closer than a few AU around a few percent of solar type stars.

The precision of Doppler surveys is typically 10 ms^{-1} with some programs obtaining slightly better precision on chromospherically inactive stars. Almost in a class by itself is a Keck survey of 530 F-M type stars with a routine velocity precision of 2 to 3 ms^{-1} (Vogt et al. 2000). With this unprecedented precision, this survey will be in a good position to detect Saturn-mass companions and true analogs of our own solar system: Jupiter-mass companions in decade-long orbits.

In radial velocity work, there are perhaps three rules: large sample sizes, high precision and frequent sampling. It takes a year or two to ramp up and obtain enough observations per star to begin detecting planets. After that, the sample has a finite productive lifetime, closely tied to the velocity precision. Only the highest precision surveys will continue to be productive, finding low amplitude planetary companions with orbital periods of one or two decades.

4. A New Planet: HD 12661

Precision Doppler observations at Lick and Keck have detected a new planetary candidate (Fischer et al., 2000) to the star HD 12661 (Fig. 2). This planet is typical of extrasolar planets discovered to date, with $M \sin i = 2.8$ M$_{Jup}$, orbital period of 265.4 d and eccentricity of 0.32. The non-zero eccentricity observed here is a ubiquitous feature of all extrasolar planets with semimajor axes greater than 0.2 AU.

HD 12661 (=HIP 9683) is classified in most references (Simbad, Hipparcos) as a K0 star; however, the color, B–V = 0.71, is more consistent with a a spectral type G6 V star. The Hipparcos parallax yields a distance of 37.2 pc (Perryman et al. 1997) for this V=7.43 star. Prieto & Lambert (1999) use the Hipparcos distance with evolutionary tracks to derive $\log g = 4.43$, $M_V = 4.58$ Teff=5754K, R=1.096 R$_{sun}$ (all of these parameters are also consistent with a spectral type earlier than the listed type, K0). HD 12661 is a chromospherically quiet, slowly rotating star. We estimate [Fe/H]=+0.32 from $ubvy$ photometry, reinforcing the

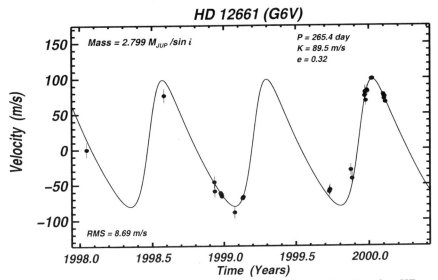

Figure 2. The combined Lick and Keck radial velocities for HD 12661. The solid line is the radial velocity curve from the best–fit orbital solution.

trend originally noted by Gonzalez (1998) of high metallicity in the host stars of gas giant extrasolar planets.

5. Characteristics of Detected Planets

5.1. Doppler Sensitivity

Figure 3 illustrates how the detectability of extrasolar planets with precision Doppler techniques depends on the induced velocity amplitude and the orbital period. The solid lines drawn on Figure 3 trace the velocity amplitudes induced by an 8 M_{Jup} planet (top line) and a 0.5 M_{Jup} mass planet (bottom line) over a range of orbital periods. Jupiter is plotted and labeled for reference.

Until recently, the induced velocity amplitudes for all detected planets were larger than about 30 ms^{-1} (i.e. 3 to 4 sigma detections for most precision Doppler surveys. This velocity amplitude barrier has been broken at Keck where 2 – 3 ms^{-1} precision is now routinely obtained, enabling the detection of companions having $M \sin i \approx 1$ M_{Sat} with orbital periods less than one year. The fact that the detection limit is a constant for all detected orbital periods is a good sign - it suggests that systematic errors in the radial velocity surveys are low, even over long time baselines. This $K_1 = 40$ ms^{-1} detection threshold is set by the velocity precision, typically under 10 ms^{-1}.

One obvious feature in Figure 3 is the truncation of detections at long orbital periods. This is simply a bias in the Doppler technique because measurements are required over at least one full radial velocity orbit. Planets with orbital

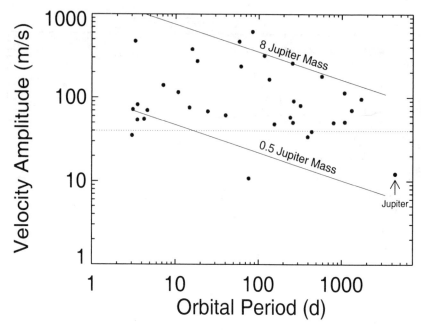

Figure 3. Velocity amplitude vs. orbital periods for 34 extrasolar planets detected to date. The distribution of data shows a bias toward velocity amplitudes greater than 30 m s^{-1} (dotted line) and orbital periods shorter than a few years. The velocity amplitudes induced by 8 M$_{Jup}$ (top) and 0.5 M$_{Jup}$ (bottom) are shown as solid lines and Jupiter is plotted and labeled to serve as a reference point.

periods like HD 12661 (also, the recently discovered v And c, HD 89744 and HD 134897) often require a couple of observing seasons to fill in phase coverage and confirm the orbital parameters. Phase coverage is less of a problem for planets with orbits that are longer than a few years because even a few observations per year (typical for most programs) will provide sufficient sampling. However, at longer orbital periods, the velocity amplitude begins to introduce a challenge to detectability.

Extrapolating the 8 M$_{Jup}$ line in Figure 3 toward the dotted line that defines the threshold for most precision Doppler surveys, the mandate to improve velocity precision below 10 ms^{-1} is clear. If 8 M$_{Jup}$ represents a physical limit to the mass of planets that typically form in a protoplanetary disk, then detections will become more difficult as the parameter search space shrinks to challengingly low velocity amplitudes at long orbital periods. Only the highest precision surveys are likely to detect many planets with orbital periods longer than a few years.

The growth industry for precision Doppler surveys will be the detection of Saturn-mass companions with velocity amplitudes below 30 ms^{-1} and companions with orbital periods longer than a couple of years. One important question

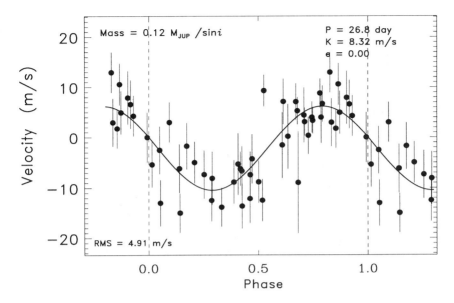

Figure 4. Radial velocities obtained at Lick for the star HR 1729. The low amplitude periodicity is not apparent to the eye in the raw data, however a periodogram shows a strong peak at 26 days which can be nicely phased. More data and analysis are required to strengthen confidence in this result.

to investigate will be how low we can go with good precision and many observations. Work is in progress to use Bayesian methods to extract low amplitude signals from the velocity data (Scargle 2000).

At Lick, the velocity data are routinely run through a periodogram analysis (floating mean, Scargle-Lomb). A few stars show very strong power at particular periods. One example is HR 1729 (Fig. 4). To the eye, the velocities look quite constant. However, there is a strong peak in the periodogram at 26 d that persists despite an intense observing program designed to drive out spurious window functions in the data. This particular star has more than 100 observations since 1995, so the results are intriguing if not compelling. We have added this star to the Keck program where the velocity precision will be about 2 ms^{-1}. If the periodicity persists in those independent observations, then we will have more confidence in this marginally detectable result.

Once the instrumental sources of velocity errors are controlled, the radial velocity precision is proportional to S/N, with one caveat. Every star contributes a characteristic astrophysical noise to the velocities. We expect that in chromospherically quiet stars, convection contributes of order 1 ms^{-1} to velocity scatter. However, chromospherically active stars can show intrinsic velocity flutter of 10 ms^{-1} or more. If a star is rotating faster than 5 kms^{-1}, the velocity precision softens because the broader lines cannot be centroided as precisely. Active stars also have photospheric spots that rotate on the surface of the star and cause the velocities to wander as light from the receding edge of the star and then

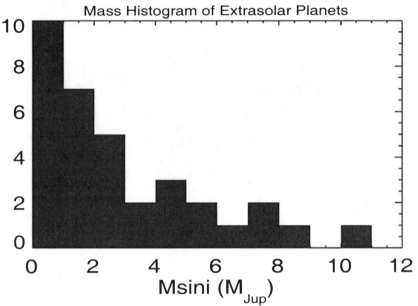

Figure 5. Mass histogram of the 34 extrasolar planets discovered to date. The mass distribution shows a steep rise toward lower mass planets that are intrinsically more difficult to detect.

the approaching edge of the star fades out of the spectrum. Saar et al. (1998) have described how rotation and chromospheric activity correlate with scatter in the radial velocities. In order to use the limited telescope time as efficiently as possible, the target S/N should be well–matched to the best velocity precision attainable for a given star. That is, the target velocity precision should be matched to the astrophysical noise of the star. Our goal is to achieve a velocity precision of about 1 ms^{-1} for the brightest, most inactive stars; we're not quite at that precision, yet. But when we add a new star to our sample, the first step is to assess the chromospheric activity of the star. That assessment tells us what kind of velocity precision to expect and essentially sets our exposure times.

One question now being explored is whether an activity-based correction can be made to the radial velocities. We obtain the CaII H&K lines in the spectral format at Keck, and the Ca infrared triplet (IRT) lines at Lick. Emission in these lines is used to characterize the chromospheric activity (Saar & Fischer 2000; Baliunas et al. 1998; Noyes et al. 1984), providing a contemporaneous activity indicator with every velocity measurement. Saar & Fischer (2000) have used a weighted linear fit between Ca IRT emission and velocity to reduce the rms velocity scatter by up to 40% in about 30 frequently observed Lick stars. The best results were obtained with chromospherically quiet G and K dwarfs. The goal is to eventually derive an activity-based correction for young, active stars, where our efforts to detect planets have been hampered by the large intrinsic photospheric noise.

5.2. Mass Distribution of Extrasolar Planets

Figure 5 is a mass histogram of the 34 detected extrasolar planets around main sequence stars. The mass distribution clearly demonstrates that even in the narrow mass range detectable by radial velocities, lower mass planets are more common than higher mass planets. This is a remarkable result, considering that detectability for this technique is biased toward high masses. The paucity of objects from 10 to 70 M_{JUP} coupled with the slope in the mass function suggests an upper limit to the maximum mass for planets that form in a protoplanetary disk. If brown dwarfs orbit solar-type stars, they must exist in fairly wide orbits.

5.3. Eccentricity Distribution

With a sample of 34 extrasolar planets, it is possible to consider trends in the the orbital characteristics and the statistical properties of the host stars. One of many surprises was the relatively high eccentricity for every planet with an orbital semimajor axis greater than 0.2. There is no reason to expect a bias that would favor detection of eccentric orbits. In fact, since the planet spends a greater fraction near apastron with a slowly changing radial velocity it is easy to miss the short-lived periastron passage if the star is only observed a few times a year. The eccentricity distribution (Fig. 6) shows that with one exception, all of the planets within 0.1 AU of their host star have relatively circular orbits. The one exception, HD 217107, is unusual in that it has a velocity trend of about 40 ms^{-1} per year superimposed on the velocity variation due to the known planet. This trend is most likely caused by an additional, as yet unidentified companion. Another large planet in the system could tidally interact and pump up the eccentricity of the detected planet.

The eccentricity distribution is an important clue to early planet evolution. It is unlikely that the planets formed where we are finding them now; the ambient temperatures are too hot for efficient grain condensation. So the interactions that destabilized these planets from their primordial circular orbits and started them moving inward likely resulted in the observed relatively high eccentricity. For planets that end up parked close to the host star, with orbital period less than about 7 days, tidal interactions can circularize the orbits on timescales of about 1 Gyr. One unanswered question is whether orbital eccentricity will persist for planets in wide, long-period orbits. If eccentricity is a signature of orbital migrations, then will the planets that have not migrated close-in to their host stars maintain their primordial circular orbits? The next decade of data should answer this question.

5.4. Metallicity of Host Stars

Gonzalez (1998) was the first to make the interesting observation that stars with detected gas giants had unusually high metallicities. Relative to a volume-limited sample of G and K dwarfs, even the Sun appears to be slightly metal rich. Figure 7 shows the normalized metallicity distribution for stars with planets. There is a clear skew in the distribution toward high [Fe/H]. The origin of this high metallicity is unclear. There has been speculation that the increased

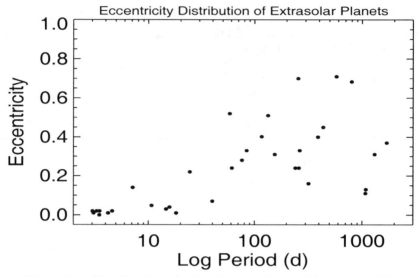

Figure 6. Distribution of orbital eccentricity with period. All extrasolar planets with orbital periods longer than about 20 days have significant, nonzero eccentricity.

metallicity could be a result of planet cannibalism as the convective zone of the star swallows up planets whose inward migration is not halted. Alternatively, the planets might not be the source of the high metallicity, but a by-product of high metallicity. Metal rich clouds would have higher grain content and more efficient cooling that could promote condensation and accretion. Still another possibility is that we are seeing a selection effect. Metal–rich stars are intrinsically brighter and most of the Doppler surveys have magnitude–limited samples. Michel Mayor's southern hemisphere sample will be the largest volume–limited sample and should address the possibility of a selection bias.

6. Stars with Multiple Planets

A year ago the first multiple planet system was discovered around the star v And (Butler et al. 1999). The inner planet in this system was discovered in 1996 (Butler et al. 1996). At that time, an additional trend in the velocities was noted that suggested the presence of a second planet in a more distant orbit. After the expected outer planet had executed a full orbit, Keplerian models were constructed to fit the data. A large rms in the residual velocities to that two-planet fit flagged a problem. On closer inspection, the residuals showed a clear periodicity – the signal of one more planet in an intermediate orbit. Figure 8 shows the residual velocities after theoretical velocities due to the short-period companion are subtracted off.

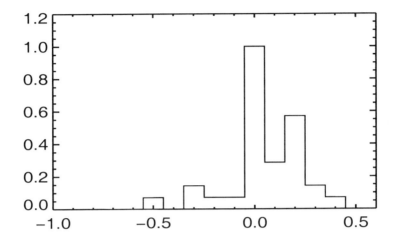

Figure 7. Distribution of [Fe/H] for stars with detected extrasolar planets.

This discovery was important on several fronts. First, the requirement for dynamic stability constrained the masses of the three companions to the planetary regime. Secondly, the dynamical models that were constructed showed islands of stability for the system, but also made it clear that (like our own solar system) this was a dynamically full system. This suggests that the protoplanetary disk started out with an enormous number of planetesimals and ended up with those planets that happened to reside in dynamically stable niches.

In retrospect, if we had been looking, we could have found the middle planet even before the outer planet had completed an orbit. However, at the time, two planets seemed complicated enough. Since v And, we have looked more carefully at the velocities of stars with detected companions. A handful of them show significant trends, suggesting that multiple gas giant planets may not be uncommon. Multiple planets represent a challenge to Doppler analysis, and they will represent a challenge to data analysis for all future planet hunting techniques.

7. Summary

The several radial velocity surveys underway are collectively monitoring nearly 3000 FGKM dwarfs. These surveys provide good constraints on some of the parameter space around stars. First, the radial velocity surveys have not found any unambiguous brown dwarfs. This null result places a 1% upper limit on the rate of occurrence of brown dwarfs, at least within about 10 AU of solar

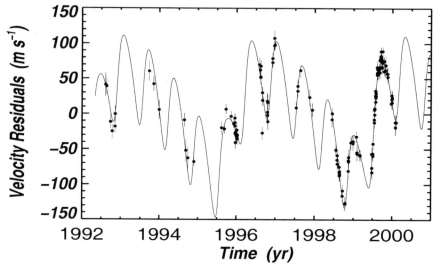

Figure 8. Lick velocities for the star υ And. Theoretical velocities for the inner companion have been subtracted from the data, revealing the 242-day periodicity, modulated by the longer 3.5-year velocity variations. These two cycles in the radial velocity are caused by the middle and outer companions in the system.

type stars. Second, the Doppler surveys are finding giant planets (planets more massive than 0.5 M_{JUP}) with semimajor axes less than 3 AU around about 5% of solar type stars. The mass histogram strongly suggests a physical cutoff in the high-mass end of the distribution at about 10 M_{JUP}. The mass distribution rises steeply toward the low-mass end, suggesting that low-mass planets (like low-mass stars) are more common than high-mass planets. A third interesting result is that all of the detected planets beyond 0.1 AU are in non-circular orbits. The host stars appear to be metal-rich relative to normal field stars.

Another interesting result from Doppler surveys is the detection of a dynamically saturated, multiple planet system. Other stars are showing similar secondary velocity trends, suggesting that multiple planets may be common. Last Fall, the detection of a transiting planet in a 3 day orbit confirmed the nature of the extrasolar planets as gas giants and provided the first absolute mass for an extrasolar planet.

8. Discussion

Several issues would seem to bear on microlensing planet searches. First, the Doppler surveys may be observing quite a different stellar sample than the microlensing surveys. In particular, radial velocity samples are well-pruned and dominated by middle-aged G dwarfs. To date, only one extrasolar planet has been detected around an M dwarf; however, this appears to be a selection effect. There are relatively few M dwarfs in radial velocity samples because they are intrinsically so faint. One niche that microlensing surveys may be able to fill is statistical information regarding the rate of exoplanets around M dwarfs (probably the most common lensing star). It is not known whether planet mass or the rate of formation of planets is correlated with stellar mass, stellar age or metallicity. What are the implications for a typical lensed sample?

Another important issue will be the effect of multiple planets on the microlensed light curve. How difficult will it be to uniquely model a system with two or three planets near an Einstein ring?

Finally, the velocity surveys will almost certainly hit a detectability wall at Saturn- or Neptune-mass planets. Microlensing surveys that could detect lower-mass planets would be extremely important.

References

Baliunas, S. L., Donahue, R. A., Soon, W., & Henry, G. W. 1998, in ASP Conf. Proc 154, The Tenth Cambridge Workshop on Cool Stars, Stellar Systems and the Sun, ed. R. A. Donahue and J. A. Bookbinder (San Francisco: ASP), p. 153

Butler, R. P., Marcy, G. W., Williams, E., McCarthy, C., Dosanjh, P., & Vogt, S. S. 1996, PASP, 108, 500

Butler, R. P. et al. 1999, ApJ, 526, 916

Charbonneau, D., Brown, T.M., Latham, D.W. & Mayor, M. 2000, ApJ, 529, L45

Cochran, W. D., Hatzes, A. P., Butler, R. P., & Marcy, G. W. 1997, ApJ, 483, 457

Fischer, D. A., Marcy, G. W., Butler, R. P., Vogt, S. S. & Apps, K. 1999, PASP, 111, 50

Fischer, D. A., Marcy, G. W., Butler, R. P., Vogt, S. S., Frink, S. & Apps, K. 2000, ApJ, submitted

Gonzalez, G. 1998, A&A, 334, 221

Henry, G.W., Marcy, G.W., Butler, R.P. & Vogt, S.S. 2000, ApJ, 529, L41

Kurster, M. et al. 2000, A&A, 53, L33

Levison, H. F., Lissauer, J. J., & Duncan, M. J. 1998, AJ, 116, 1998.

Lin, D. N. C., Bodenheimer, P., & Richardson, D. C. 1996, Nature, 380, 606

Marcy, G. W. & Butler, R. P. 1998, ARA&A, 36, 57

Marcy, G. W., Butler, R. P., Vogt, S. S. 2000, accepted by ApJ

Marcy, G. W., Cochran, W. D., Mayor, M. 2000, in Protostars and Planets IV, ed V. Mannings, A. P. Boss & S. S. Russell (Tucson: University of Arizona Press), in press.

Mayor, M., Beuzit, J. L., Mariotti, J. M., Naef, D., Perrier, C., Queloz, D. & Sivan, J. P. 1998, ASP Conf Ser., IAU Colloq. 170: Precise Stellar Radial Velocities, ed. J. B. Hearnshaw & C. D. Scarfe (San Francisco: ASP)

Mayor, M. & Queloz, D. 1995, Nature, 378, 355
Mazeh T. et al. 2000, ApJ, 532, L55
Noyes, R. W., Jha, S., Korzennik, S. G., Krockenberger, M., Nisenson, P., Brown, T. M., Kennelly, E. J., Horner, S. D. 1997, ApJ, 483, L111
Noyes, R. W., Hartmann, L., Baliunas, S. L., Duncan, D. K., & Vaughan, A. H. 1984, ApJ, 279, 763
Perryman, M. A. C., et al. 1997, A&A, 323, L49
Prieto, A. & Lambert, D. L. 1999, A&A, 352, 555
Queloz, D., Mayor, M. et al. 2000, A&A, 354, 99
Saar, S. H., Butler, R. P., & Marcy, G. W. 1998, ApJ, 498, L153
Saar, S. H. & Fischer, D. 2000, ApJ, 534, L105
Santos, N., Mayor, M. et al. 2000, A&A, 356, 599
Scargle, J. D. 2000, Proc. AAS
Udry, S., Mayor, M. et al. 2000, A&A, 356, 590
Valenti, J. A., Butler, R. P. & Marcy, G. W. 1995, PASP, 107, 966
Vogt, S. S., Marcy, G. W. & Butler, R. P. 2000, to appear in ApJ

Discussion

Bradbury: Have fits been attempted for non-standard mass distributions, i.e. non-solar system models such as Dyson spheres?

Gaudi: They have not been tried but would probably be inconsistent with the observed curves.

Gaudi: 1) Are the fitting procedures for multiple planets robust? Are you biasing your results by successively subtracting off each planet? Do you do simultaneous multi-parameter fits? 2) Multiple planets and microlensing are discussed in Gaudi, Naber & Sackett (1998).

Fischer: 1) We have done this fitting process both ways and get the same results. We have two software routines - one that simultaneously fits all free parameters for the 2 or 3 putative planets and another that fits the short period companion in a short time baseline and then sequentially fits the additional planet(s). Ultimately, both approaches minimize chi-squared for the 2 or 3 Keplerian orbit solution and both methods agree extraordinarily well.

Gould: I am definitely opposed to your policy of self-censorship. You should publish your 0.12 $M_\odot/\sin i$ planet!

Fischer: Our group prefers to only announce planets for which our confidence is extremely high. Higher precision results will be forthcoming from Keck that will push down the minimum detected mass.

Hansen: When you improve the metallicity measurements do you plan to improve things like C/Fe ratios, since M15-ratio scenarios predict different pollution rates?

Fischer: Yes

Fischer: I'd like to comment on our rejection of chromospherically active stars from radial velocity samples. τ Boo has K$\sim$450m/s so we first look for large amplitude variations like this before dropping active stars from the intensely observed sample.

Mayor: I would like to add two comments: 1) Already now for three young active stars, which are part of the CORALIE survey, giant planets have been detected. 2) Some hints exist that the stars with giant planets closer than 0.1 AU are more metal rich than stars with giant planets with semi-major axes in the range 0.1 to 1 AU (Queloz et al 2000). But evidently such a correlation has to be confirmed with a larger sample.

Photometric Characterization of Stars with Planets

Alvaro Giménez

LAEFF, INTA-CSIC, Apartado 50.727, 28080 Madrid, Spain,
and
Instituto de Astrofísica de Andalucia, CSIC, Apartado 3.004, 18080 Granada, Spain

Abstract. Strömgren *uvby* photometry has been used to estimate temperatures and metallicities in those stars currently known to have planets. A comparison with other sources of information is made pointing out the importance of photometry for massive surveys searching for new extrasolar planets as well as its limitations.

1. Introduction

What I wanted to talk about in this workshop is completely different to microlensing techniques themselves but connected to them through the scientific objectives. Extrasolar planets are becoming the subject of intensive observational and theoretical efforts, and microlensing events may give some clues to the nature of such objects. It is nevertheless important to have a look not only to the planet themselves but also to their host stars. Do they show any special characteristic indicative of the presence of planets around them? If this is the case, the search for new candidates would be greatly facilitated as well as the understanding of their formation process.

It has been shown during the last years by González (1999, and references given therein), using high resolution spectroscopy, that parent stars of planetary systems are systematically more metal abundant than field stars with similar effective temperature, age, and galacto-centric position. The Sun in fact is known since long ago to be metal rich compared to similar stars in its neighborhood. Other proposed characteristics, like the formation of planets at given ages, metallicity with the location or mass of the planet with respect to the host star, etc., could not be confirmed due to the still small number of objects in the sample of known extrasolar planets. Many more discoveries and observations are needed to improve the still poor statistics available.

A new simple method to enlarge the data sample was proposed by Giménez (2000), the publication actually appeared shortly after the workshop in Cape Town, to estimate basic stellar properties of the host stars using Strömgren *uvby* photometry. This is a well-known technique to derive temperatures and metallicities but unfortunately, since the system was essentially defined for B to F stars, calibrations for the spectral range of our interest are not well established. At the time of the mentioned publication, there were 27 stars referenced with confirmed planets around them, though *uvby* photometry was only available

Photometric Characterization of Stars with Planets 131

for 25 of them. Moreover, two stars (55 Cnc and 14 Her) were found to lie outside the range of application. The results presented thus referred to 23 stars while independent values of temperature and metallicity from high-dispersion spectra were only available for half of the sample. For this discussion, planets are defined as companions with minimum mass ($m \sin i$) below the limit of 13 times the value of Jupiter, corresponding to deuterium burning, and obviously a not well defined border line still open to discussion. The latest available list of confirmed extrasolar planets, by August 2000, includes 46 parent stars.

2. Strömgren *uvby* Photometry of Stars with Planets

For the now known 46 stars with planets, we have extracted *uvby* colour indices from the available literature, mainly Hauck & Mermilliod (1998). In addition to the 2 stars previously mentioned with no *uvby* data (namely, GJ 876 and HD 177830), no useful information could be found for HD 83443, 46375, and BD -10 3166). We could thus, in principle, study a total of 41 stars with planets. Unfortunately, calibrations of temperature and metallicity can not be used for colour differences δc_1 larger than 0.15, as it was the case of 55 Cnc and 14 Her, to which we have now to add HD 38259 and HD 12661. A final list of 37 stars could be analyzed as given in Table 1. No reddening correction was applied to the observed colour indices since the stars are all closer than around 50 pc to the Sun.

A plot of the m_1 index versus $(b-y)$ immediately showed that the position of the stars in our sample do not deviate from standard calibrations the same way as those stars, within the same spectral range, showing evidences of stellar activity (Giménez et al. 1991). The relatively calm behaviour of stars with planets may be a consequence of the difficulty to get accurate radial velocities in more active or younger stars.

On the other hand, the plot of the c_1 index versus $(b-y)$, which generally indicates stellar evolution, shows for the stars in our sample that they are slightly more evolved than normal stars without planets in the available photometric catalogues. In this case nevertheless, exceptions very close to the standard calibration representing the zero-age main sequence are also found. In addition, it should be mentioned that in both diagrams, the stars HD 6434 and 114762 show an anomalous position (large δm_1 and negative δc_1) indicating much lower metallicity than the rest of the sample. Moreover the two similar stars 55 Cnc and 14 Her, as well as HD 38259 and 12661, are found far from the expected position for normal stars of the same temperature. At least the two former ones are known to be highly metal-rich stars.

3. Temperatures and Metallicities

Lacking still any new calibration for *uvby* photometry of late-type stars, we have adopted the same described by Giménez (2000), i.e. Olsen (1984) for GK stars and Saxner & Hammarbäck (1985) and Crawford (1975) for F-type stars. The jumps detected in the regions where calibrations change, i.e. $(b-y)$ around 0.37 and 0.51, have been taken into account by averaging the results obtained using the different possible equations. Moreover, with respect to the values given by

Table 1. uvby Photometry and derived parameters

Name	$(b-y)$	m_1	c_1	$[Fe/H]$	T_{ef}
HD 16141	0.422	0.213	0.378	0.02	5710
HD 168746	0.435	0.223	0.342	-0.14	5580
HD 108147	0.346	0.177	0.391	0.14	6180
HD 75289	0.360	0.191	0.405	0.26	6120
$51Peg$	0.416	0.233	0.371	0.18	5770
HD 6434	0.384	0.159	0.274	-0.39	5780
HD 187123	0.405	0.224	0.365	0.15	5820
HD 209458	0.361	0.174	0.362	0.01	6050
υ And	0.346	0.176	0.415	0.13	6180
HD 192263	0.541	0.493	0.275	0.11	4840
ϵ Eri	0.504	0.430	0.263	-0.18	5050
HD 121504	0.381	0.189	0.361	0.01	5920
HD 37124	0.421	0.202	0.28	-0.31	5590
HD 130322	0.475	0.305	0.316	-0.09	5350
ρ CrB	0.394	0.178	0.337	-0.22	5790
HD 52265	0.360	0.190	0.404	0.23	6110
HD 217107	0.456	0.299	0.376	0.31	5560
HD 210277	0.459	0.298	0.353	0.20	5510
$16CygB$	0.416	0.226	0.354	0.08	5740
HD 134987	0.435	0.256	0.374	0.19	5660
HD 19994	0.361	0.185	0.422	0.17	6090
HD 92788	0.433	0.253	0.376	0.19	5680
HD 82943	0.386	0.217	0.390	0.32	5960
HR 810	0.357	0.188	0.364	0.21	6130
$47UMa$	0.391	0.202	0.343	-0.02	5850
HD 169830	0.328	0.177	0.446	0.25	6310
GJ 3021	0.459	0.289	0.300	-0.01	5440
HD 195019	0.419	0.204	0.362	-0.10	5690
GJ 86	0.484	0.337	0.287	-0.06	5270
τ Boo	0.318	0.177	0.439	0.26	6370
HD 190228	0.482	0.264	0.306	-0.53	5220
HD 168443	0.455	0.233	0.377	-0.17	5490
HD 222582	0.406	0.202	0.345	-0.08	5760
HD 10697	0.440	0.238	0.379	0.04	5610
$70Vir$	0.446	0.232	0.351	-0.15	5530
HD 89744	0.338	0.184	0.451	0.28	6250
HD 114762	0.365	0.125	0.297	-0.64	5850

Giménez (2000), we have adopted here a correction for metallicity in terms of δc_1 going from 0.02 for values between 0.075 and 0.100, to 0.04 between 0.100 and 0.125, and 0.06 between 0.125 and 0.150. Stars with δc_1 larger than 0.15 were simply discarded as mentioned above. This reflects the expected correction to the available calibrations better than the single value used previously of 0.04 though a really new photometric calibration should solve the problem in a more definitive way. Again, this correction only affects stars in the range between 0.37 and 0.51 in $(b-y)$.

Results for our enlarged sample of 37 stars is shown in the mentioned Table 1, where columns denote the name of the star, the $b-y$, m_1 and c_1 colours, together with the obtained metallicities and effective temperatures. It should be mentioned here that a misprint in table 2 of Giménez (2000) was found for the effective temperature of HD 168443, that should read 5490 instead of 6490 K. Moreover, the estimated effective temperatures should be accurate within 100 K, while metallicities show a scatter of 0.15 dex. Nevertheless the metal abundacies derived for two stars, namely HD 192263 and 168443, are to be taken with caution. The first one because it has a very low effective temperature and the calibrations in this range are not reliable. The second, because its δc_1 of 0.14 is too close to the limit to be discarded due to non-applicability of the calibration equations.

4. Discussion

The estimated metallicities for the sample of 37 stars confirms the distribution of values around that for the Sun, though the dispersion of the data is larger than the expected uncertainties. A comparison with metallicities derived by means of the same photometric method for a sample of around 3000 stars of F, G, and K spectral types with Hipparcos distances, clearly showed a distribution of values of lower metal abundances. Some metal-deficient stars are nevertheless found to have planets, and now it is not only the case of the very massive companion of HD 114762 but also the lower mass planet around HD 6434.

A plot of the estimated metallicities as a function of the minimum mass of the lightest planet around each star, did not show any clear correlation though actual masses instead of $m \sin i$ values should be used. When taking the semi-major axis, a, again no clear correlation with metallicity is found but, for stars with planets closer than 0.4 AU the their host star, data show some indication of an increased metallicity for stars with closer companions.

If the reason for the high metallicities is the falling of metal-rich planets into the convective envelope of their parent stars, one may argue that the stellar enrichment would be more significant in stars with smaller convective zones (for the same amount of infalling material). Therefore, an increase of the metallicity with effective temperature should be expected but this is not clearly evident in our data sample.

Coming finally back to the problem of the age of stars with planets, two points have been identified as marginal evidence for the relatively old nature of these stars compared to average field stars in similar galactic position. Their location in the $\delta c_1 - (b-y)$ diagram and their slow rotation, also linked to low levels of activity. Both arguments are nonetheless still not strongly supported observationally and some important biases could be present in the sample of discovered planets through the radial velocity method. A detailed comparison with evolutionary models, using accurately determined stellar parameters, is needed. The case of the Sun, which presents a very low specific angular momentum when compared to similar stars without planets, can be easily explained in terms of stellar age, magnetic braking and coupling with a thick accretion disk in early phases. How much of each of these processes is needed for the formation of planetary systems may be deduced from the observation of parent stars as those described above.

References

Crawford, D. L. 1975, AJ, 80, 955

Giménez, A. 2000, A&A, 356, 213

Giménez, A., Reglero, V., de Castro, E. & Fernández-Figueroa, M. J. 1991, A&A, 248, 563

González, G. 1999, MNRAS, 308, 447

Hauck, B. & Mermilliod, M. 1998, A&AS, 129, 431

Olsen, E. H. 1984, A&AS, 57, 443

Saxner, M. & Hammarbäck, G. 1985, A&A, 151, 372

Microlensing Constraints on the Frequency of Jupiter-Mass Planets

B. Scott Gaudi[1], M.D. Albrow[2], Jin H. An[1], J.-P. Beaulieu[3], J.A.R. Caldwell[4], D.L. Depoy[1], M. Dominik[5], A. Gould[1], J. Greenhill[6], K. Hill[6], S. Kane[6], R. Martin[7], J. Menzies[4], R.W. Pogge[1], K. Pollard[8], P.D. Sackett[5], K.C. Sahu[2], P. Vermaak[4], R. Watson[6], A. Williams[7]
(The PLANET collaboration)

[1] *The Ohio State University*

[2] *Space Telescope Science Institute*

[3] *Institut d'Astrophysique de Paris*

[4] *South African Astronomical Observatory*

[5] *Kapteyn Astronomical Institute*

[6] *University of Tasmania*

[7] *Perth Observatory*

[8] *University of Canterbury*

Abstract. Microlensing is the only technique likely, within the next five years, to constrain the frequency of Jupiter analogs. The PLANET collaboration has monitored nearly 100 microlensing events of which more than 20 have sensitivity to the perturbations that would be caused by a Jovian-mass companion to the primary lens. No clear signatures of such planets have been detected. These null results indicate that Jupiter-mass planets with separations of 1.5–3 AU occur in less than 1/3 of systems. A similar limit applies to planets of 3 Jupiter masses between 1 and 4 AU.

1. Introduction

A Galactic microlensing event occurs when a massive, compact object (the lens) passes near to our line-of-sight to a more distant star (the source). If the lens, observer, and source are perfectly aligned, then the lens images the source into a ring, called the Einstein ring, which has angular radius of[1]

$$\theta_{\rm E} \equiv \left[\frac{4GM}{c^2} \frac{D_{\rm LS}}{D_{\rm L} D_{\rm S}} \right]^{1/2} \sim 480\,\mu{\rm as} \left(\frac{M}{M_\odot} \right)^{1/2}, \qquad (1)$$

---
[1] For the scaling relation on the far right of equations (1), (2), and (3), we have assumed $D_{\rm S} = 8$ kpc and $D_{\rm L} = 6.5$ kpc, typical distances to the lens and source for microlensing events toward the bulge.

where M is the mass of the lens, and D_{LS}, D_S, D_L are the lens-source, observer-source, and observer-lens distances, respectively. This corresponds to a physical distance at the lens plane of

$$r_E = \theta_E D_L \sim 3 \text{ AU} \left(\frac{M}{M_\odot}\right)^{1/2}. \quad (2)$$

If the lens is not perfectly aligned with the line-of-sight, then the lens splits the source into two images. The separation of these images is $\mathcal{O}(\theta_E)$ and hence unresolvable. However, the source is also magnified by the lens, by an amount that depends on the angular separation between the lens and source in units of θ_E. Since the lens, observer, and source are all in relative motion, this magnification is a function of time – a "microlensing event". The time scale for such an event is

$$t_E \equiv \frac{\theta_E}{\mu_{\rm rel}} \sim 40 \text{ days} \left(\frac{M}{M_\odot}\right)^{1/2}, \quad (3)$$

where $\mu_{\rm rel}$ is the relative lens-source proper motion.

If the primary lens has a planetary companion, and the position of this companion happens to be in the path of one of the two images created during the primary event, then the planet will perturb the light from this image, creating a deviation from the primary light curve (see Fig. 1). The duration of this perturbation is $\sim \sqrt{q} t_E$, where q is the mass ratio between the planet and primary. Hence, for a Jupiter/Sun mass ratio ($q \simeq 10^{-3}$), the perturbation time scale is $\mathcal{O}(\text{day})$. These short-duration deviations are the signatures of planets orbiting the primary lenses. Note that since the perturbation time scale is considerably less than t_E, the majority of the light curve will be indistinguishable from a single lens.

Three parameters determine the magnitude of the perturbation, and hence define the observables. These are mass ratio q, the instantaneous angular separation d between the planet and primary in units of r_E, and the angle α between the projected planet/star axis and the path of the source. As q decreases, the perturbation time scale decreases, although the magnitude of the deviation does not necessarily decrease. Thus very small mass ratio planets ($q \lesssim 10^{-5}$) can be detected using microlensing, although the detection probability is small. The lower limit to the detectable q is set practically by the sampling of the primary event, and ultimately by the finite size of the source stars (Bennett & Rhie 1996). A microlensing event is generally alerted only if the minimum angular impact parameter in units of θ_E satisfies $u_0 \leq 1$, which corresponds to image positions between 0.6 and $1.6 \theta_E$. Since the planet must be near one of these images to create a perturbation, microlensing is most sensitive to planets with separations $0.6 \lesssim d \lesssim 1.6$, the "lensing zone". The angle α, which is of no physical interest, is uniformly distributed. Only certain values of α will create detectable deviations. Thus integration over α defines a geometric detection probability.

Microlensing as a method to detect extrasolar planets was first suggested by Mao & Paczyński (1991), and was expanded upon by Gould & Loeb (1992) who demonstrated that if all lenses had a Jupiter analog, then $\sim 20\%$ of all light curves should exhibit $\gtrsim 5\%$ deviations. Since these two seminal papers, many authors have explored the use of microlensing to detect planets. It is

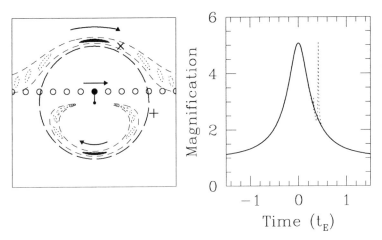

Figure 1. Left: The images (dotted ovals) are shown for several different positions of the source (solid circles), along with the primary lens (dot) and Einstein ring (long dashed circle). If the primary lens has a planet near the path of one of the images, i.e. within the short-dashed lines, then the planet will perturb the light from the source, creating a deviation to the single lens light cure. Right: The magnification as a function of time is shown for the case of a single lens (solid) and accompanying planet (dotted) located at the position of the X in the top panel. If the planet was located at the + instead, then there would be no detectable perturbation, and the resulting light curve would be identical to the solid curve.

not our intention to provide a comprehensive review of this field. However, of particular relevance is the paper by Griest & Safizadeh (1998, GS98) who demonstrated that, for high-magnification events (those with maximum magnification $A_{max} > 10$), the detection probability for planets in the lensing zone is $\sim 100\%$. Thus high-magnification events are an extremely efficient means of detecting extrasolar planets. The results of GS98 also imply that multiple planets in the lensing zone should betray their presence in high-magnification events (Gaudi, Naber & Sackett 1998).

2. Limits on Companions in OGLE 1998-BUL-14

The basic requirements to detect planets with microlensing are good temporal sampling and photometric precision. Since the optical depth to microlensing is low, $\mathcal{O}(10^{-6})$, the survey teams that discover microlensing events toward the Galactic bulge must monitor of order one million stars on a nightly basis in order to detect any events. Therefore, they generally have insufficient sampling to detect the short-duration perturbations to the primary light curve (see, e.g. Fig. 2). However, these survey teams (OGLE (Udalski et al. 1994), MACHO (Alcock et al. 1996), EROS (Glicenstein et al., these proceedings)) issue alerts,

138   Gaudi et al.

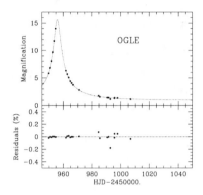
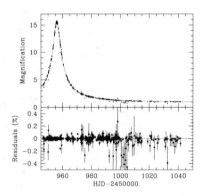

Figure 2. Left: The top panel shows the magnification as a function of time for OGLE data of microlensing event OGLE 1998-BUL-14. The dashed line indicates the best-fit point-source point-lens model (PSPL), which has a time scale $t_E = 40$ days, and a maximum magnification of ~ 16. The bottom panel shows the residuals from the best-fit PSPL model. Right: The top panel shows the magnification as a function of time for PLANET data for microlensing event OGLE 1998-BUL-14. The bottom panel shows the residuals from the best-fit PSPL model (Albrow et al. 2000).

notification of ongoing events. This allows follow-up collaborations (GMAN (Alcock et al. 1997), PLANET (Albrow et al. 1998), MPS (Rhie et al. 1999), MOA (Yock, these proceedings)) to monitor these events frequently with high-quality photometry to search for planetary deviations. In particular, the PLANET collaboration has access to four telescopes located in Chile, South Africa, Western Australia, and Tasmania, and can monitor events nearly round-the-clock, weather permitting.

Figure 2 shows PLANET photometry of an event alerted by the OGLE collaboration, OGLE 1998-BUL-14. This was a high-magnification event ($A_{\max} \sim 16$) with $t_E \simeq 40$ days, making it an excellent candidate to search for planetary deviations. PLANET obtained a total of 600 data points for this event: 461 I-band and 139 V-band. The median sampling interval is about 1 hour, or $10^{-3} t_E$, with very few gaps greater than 1 day. The 1σ scatter in I over the peak of the event (where the sensitivity to planets is the highest) is 1.5%. The dense sampling and excellent photometry means that our efficiency to detect massive companions should be quite high. In fact, examination of the residuals from a single lens model (Fig. 2) reveals no obvious deviations of any kind.

To be more quantitative, we simultaneously search for binary-lens fits and calculate the detection efficiency $\epsilon(d, q)$ of OGLE 1998-BUL-14 as a function of separation and mass ratio using a method proposed by Gaudi & Sackett (2000). For details on the implementation for this event, see Albrow et al. (2000).

We find no binary-lens models in the parameter ranges $0 \leq d \leq 4$ and $10^{-5} \leq q \leq 1$ that provide significantly ($\Delta\chi^2 \gtrsim 10$) better fits to the OGLE 1998-

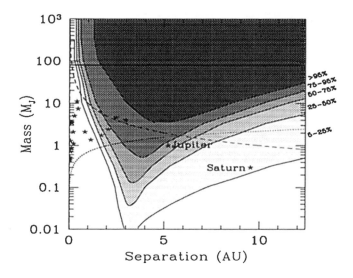

Figure 3. Detection efficiencies for the PLANET data set of OGLE 1998-BUL-14 as a function of the mass and orbital separation of the companion assuming a primary lens mass of $M_\odot$ and Einstein ring radius of $r_E = 3.1$ AU. The contours are (outer to inner) 5%, 25%, 50%, 75% and 95%. In order to convert from mass ratio and projected separation to mass and physical separation, we have averaged over orbital phase and inclination (assuming circular orbits). Jupiter and Saturn are marked with stars, as are the extrasolar planets discovered with radial velocity techniques. The horizontal line marks the hydrogen-burning limit. The dotted line shows the radial-velocity detection limit for an accuracy of 20 ms^{-1} and a primary mass of 1 $M_\odot$. The dashed line is the astrometric detection limit for an accuracy of 1 mas and a primary of mass 1 $M_\odot$ at 10 pc (Albrow et al. 2000).

BUL-14 dataset. We therefore conclude that the light curve of OGLE 1998-BUL-14 is consistent with a single lens.

In Figure 3 we show the detection efficiency ϵ of our OGLE 1998-BUL-14 dataset to companions as a function of the mass M_p and orbital radius a of the companion. Parameter combinations shaded in black are excluded at the 95% significance level. Stellar companions to the primary lens of OGLE 1998-BUL-14 with separations between ~ 2 AU and 11 AU are excluded. Companions with mass ≥ 10 M_J are excluded between 3 AU and 7 AU. Although we cannot exclude a Jupiter-mass companion at any separation, we had an $\sim 80\%$ chance of detecting such a companion at 3 AU. The detection efficiency for OGLE 1998-BUL-14 is $> 25\%$ at $a = 3$ AU for all companion masses $M_p > 0.03$ M_J. We find that we had an $\sim 60\%$ chance of detecting a companion with the mass and separation of Jupiter ($M_p = M_J$ and $a = 5.2$ AU), and an $\sim 5\%$ chance of detecting a companion with the mass and separation of Saturn ($M_p = 0.3\ M_J$ and $a = 9.5$ AU) in the light curve of OGLE 1998-BUL-14.

Thus, although Jupiter analogs cannot be ruled out in OGLE 1998-BUL-14, the detection efficiencies are high enough that non-detections in several events with similar quality will be sufficient to place meaningful constraints on their abundance.

How do the OGLE 1998-BUL-14 efficiencies compare to planet detection via other methods? In Figure 4 we show the radial velocity detection limit on $M_p \sin i$ for a solar mass primary as a function of the semi-major axis for a velocity amplitude of $K = 20$ ms^{-1}, which is the limit found for the majority of the stars in the Lick Planet Search (Cumming, Marcy & Butler 1999). Although we show this limit for the full range of a, in reality the detection sensitivity extends only to $a \lesssim 5$ AU due to the finite duration of radial-velocity planet searches and the fact that one needs to observe a significant fraction of an orbital period. In addition, we plot in Figure 4 the $M_p \sin i$ and a for planetary candidates detected in the Lick survey. Radial velocity searches clearly probe a different region of parameter space than microlensing, in particular, smaller separations. Note, however, that our OGLE 1998-BUL-14 data set gives us a $> 75\%$ chance of detecting analogs to two of these extrasolar planets: the third companion to υ And and the companion to 14 Her. Although the efficiency is low, we do have sensitivity to planets with masses as small as $\sim 0.01 M_J$, considerably smaller than can be detected via radial velocity methods. For comparison, we also show in Figure 3 the astrometric detection limit on M_p for a 1 $M_\odot$ primary at 10 pc, for an astrometric accuracy of $\sigma_A = 1$ mas. For an astrometric campaign of 11 years, this limit extends to ~ 5 AU. Such an astrometric campaign ($\sigma_A = 1$ mas, $P = 11$ years), would be sensitive to companions similar to those excluded in our analysis of OGLE 1998-BUL-14. The proposed Space Interferometry Mission (SIM) promises ~ 4 μas astrometric accuracy, which would permit the detection of considerably smaller mass companions.

3. Combined Limits from the 1998-1999 PLANET Seasons

Clearly one cannot make any statements about the population of the extrasolar planets as a whole based on one event. Fortunately, PLANET has monitored, over the last five years, more than 100 events, a subset of which have temporal sampling and photometric accuracy similar to that of OGLE 1998-BUL-14. Here we present a preliminary analysis of these events.

We select 23 high-quality light curves from the 1998-1999 PLANET seasons and analyze these in the same manner as OGLE 1998-BUL-14 (Albrow et al. 2000). Included in this sample are five high-magnification ($A_{\max} > 10$) events: OGLE 1998-BUL-14, MACHO 1998-BLG-35, OGLE 1999-BUL-5, OGLE 1999-BUL-35 and OGLE 1999-BUL-36.

We find that all 23 events are consistent with a single lens to within our detection threshold. Using this null result, along with the detection efficiency $\epsilon_i(d, q)$ for each event i, we find a statistical upper limit to the fraction of these lenses that have a companion of a given separation and mass ratio. If the fraction of lenses with a companion as a function of d and q is $f(d, q)$, then the probability that N events with individual efficiencies $\epsilon_i(d, q)$ would give a null

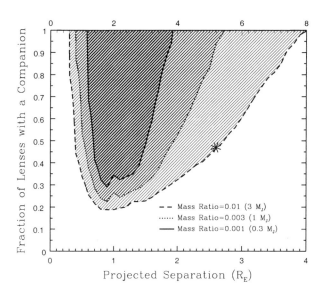

Figure 4. Upper limits to the fraction of primary lenses with a companion as a function of the projected separation in r_E for three different mass ratios, $q = 10^{-2}$ (dashed), $q = 10^{-2.5}$ (dotted) and $q = 10^{-3.0}$ (solid). The projected separation in AU assuming all primary lenses are at 6 kpc and have masses of 0.3 $M_\odot$ is shown on the top axis. Similarly, masses of the secondary are shown in parentheses. These upper limits are at the 95% confidence level and are based on 23 events.

result (no detections) is,

$$P = \Pi_{i=1}^{N}[1 - f(d,q)\epsilon_i(d,q)]. \qquad (4)$$

The 95% confidence level (c.l.) upper limit to $f(d,q)$ is found by setting $P = 5\%$.

In Figure 4 we show the 95% c.l. upper limits to $f(d,q)$ for separations $0 \leq d \leq 4$ and $q = 10^{-2}, 10^{-2.5}$, and 10^{-3}. We convert these to limits on the fraction of lenses with companions of a given mass and physical separation by assuming that all the primaries have mass $M = 1 M_\odot$ and distance $D_L = 6$ kpc. We find that < 33% of these lenses have Jupiter-mass planets with separations of 1.5-3 AU. Similarly, < 33% have planets of mass $M_p \geq 3$ M_J with separations of 1-4 AU. Although we cannot place an interesting limit on Jupiter analogs, we do find that < 50% of lenses have 3 M_J planets at the separation of Jupiter (5.2AU).

4. Conclusions

Microlensing offers a unique and complementary method of detecting extrasolar planets. Although many light curves have been monitored in the hopes of detecting the short-duration signature of planetary companions to the primary lenses, no convincing planetary detections have yet been made, despite the fact that data of sufficient quality are being acquired to detect such companions. These null results indicate that Jupiter-mass companions with separations in the "lensing zone", 1.5 − 3 AU, occur in less than 1/3 of systems.

The potential for this field is enormous. Current microlensing searches for planets will continue to monitor events alerted toward the bulge, and either push these limits down to levels probed by radial velocity surveys ($\sim 5\%$), or finally detect planets, and measure the frequency of companions at separations more relevant to our solar system. Next generation microlensing planet searches have the promise of obtaining a robust statistical estimate of the fraction of stars with planets of mass as low as that of the Earth.

Acknowledgments. We thank the MACHO, OGLE and EROS collaborations for providing real-time alerts. We are especially grateful to the observatories that support our science (Canopus, CTIO, Perth and SAAO) via the generous allocations of time that make this work possible. This work was supported by grants AST 97-27520 and AST 95-30619 from the NSF, by grant NAG5-7589 from NASA, by a grant from the Dutch ASTRON foundation through ASTRON 781.76.018, by a Marie Curie Fellowship from the European Union, by "coup de pouce 1999" award from Ministère de l'Éducation nationale, de la Recherche et de la Technologie, and by a Presidential Fellowship from the Ohio State University.

References

Albrow, M., et al. 1998, ApJ, 509, 687
Albrow, M., et al. 2000, 535, 176
Alcock, C., et al. 1996, ApJ, 463, L67
Alcock, C., et al. 1997, ApJ, 491, 436
Bennett, D. & Rhie, S. H. 1996, ApJ, 472, 660
Cumming, A., Marcy, G., & Butler, R.P. 1999, ApJ, 526, 890
Gaudi, B. S., Naber, R. M., & Sackett, P. D. 1998, ApJ, 502, L33
Gaudi, B. S., & Sackett, P. D. 2000, ApJ, 529, 56
Gould, A., & Loeb, A. 1992, ApJ, 396, 104
Griest, K., & Safizadeh, N. 1998, ApJ, 500, 37
Mao, S., & Paczyński, B. 1991, ApJ, 374, 37
Rhie, S.H., et al. 1999, ApJ, 522, 1037
Udalski, A., et al. 1994, Acta Astron., 44, 227

Discussion

Glicenstein: It seems that the light curve of OB99033 has a dip compared to the fitted point-lens point-source expectation. Is it significant?

Gaudi: No. The light curve is consistent with a PSPL event. The apparent deviation is caused by poor photometry.

Han: 1) Is your detection efficiency computation based on somewhat average photometric precision and frequency or on actual values?
2) Secondly, have you tried to compute efficiency with different observational conditions? If you did, can you provide preliminary results?

Gaudi: 1) The detection efficiency is calculated using the actual data. Thus the actual photometric precision is automatically taken into account.
2) Yes, results are forthcoming; Poisson-limited photometry should roughly triple the number of possible detections.

Hansen: 1) Do you find binary stars very often?
2) If there is a significant binary lens contribution, what fraction of your planet parameter space is potentially unstable due to undetected binary lenses?

Gaudi: 1) Yes.
2) The magnitude of this effect is unknown, but likely small due to the small fraction of binary systems in the parameter range where the effect would be important.

Gould: Re Hansen's question: the range of binary separations where the planets would be unstable is a factor of 2. Duquennoy and Mayor have shown this range to be populated by only a fraction $\lesssim 10\%$ of stars.

Planet Detection via Microlensing: Consequences of Resolving the Source

Pierre Vermaak

Department of Astronomy, University of Cape Town, Private Bag, Rondebosch, 7701, South Africa

Abstract. In this work I present the results of a simulation to determine the effects of resolving the source on planet detection via Gravitational Microlensing (GM). The probability of planet detection was calculated as a function of the binary geometry for mass ratios of $q = 10^{-3} - 10^{-5}$, taking resolved sources and unlensed light (blending) into account. Source radii up to $R_s = 0.3\ \theta_E$ were considered at which point detection probability becomes negligible. Small ($q < 10^{-4}$) mass ratio planets become undetectable at source radii of $R_s = 0.08\ \theta_E$. Blending has a slight adverse effect on planet detection and is included in the calculation but not investigated in detail here. An alternative to current GM follow-up observations was considered, where only the peaks of high amplification events are followed. Such a strategy promises to be at least twice as efficient at detecting planets as current observations but requires a large number of high amplification events.

1. Introduction

The probability of detecting the planets around a lensing star has been calculated several times (Mao & Paczyński 1991; Gould & Loeb 1992; Bolatto & Falco 1994) starting with the assumption of point sources and point mass lenses in linear relative motion. Later refinements included non-linear motion in the form of the Earth's movement and binary lens rotation (Dominik 1998) and the effects of a resolved source (Bennett & Rhie 1996; Wambsganss 1997). The aim of this work was to calculate the severity of resolved source effects on planet detection. The investigation focused on numerical simulation and examined resolved source sizes up to where planet detection becomes impossible with typical follow up observations, exceeding the parameter space of previous works on planet detection with resolved sources. Although a typical main sequence source in the Galactic Bulge has a radius of about $R_s = 0.005\ \theta_E$ and giant sources sizes are typically $R_s = 0.03\ \theta_E$, any resolved source size is possible as the Einstein radius is dependent on the event geometry.

1.1. Lensing Formalism

At Galactic distances where the thin lens approximation is valid, lensing by a point source is well described by

$$\xi = z + \frac{1}{\overline{z_1} - \overline{z}} + \sum_{i=2}^{N} \frac{q_i}{\overline{z_i} - \overline{z}} \qquad (1)$$

with z_i and q_i signifying respectively the projected lens positions and mass ratio of the primary lens to the secondary lenses i. ξ, z and z_i are in units of the angular Einstein radius of the lensing object, the radius of the ring image that would be seen were the source and lens to be precisely aligned with the observer.

During a typical lensing event, several images are formed but their separation is too small to be resolved so that in practice the combined flux of the images is observed as an amplification of the original source flux. The amplification due to each image at the position z is obtained from the Jacobian of equation 1,

$$A = \sum_i A_i = \sum_i \frac{1}{|\det J_i|} \qquad (2)$$

and summed to yield the total amplification.

Equation 2 gives rise to the existence of caustic curves: closed curves in the source plane where the amplification is infinite, wherever $\det J = 0$. The amplification is finite in practice due to the finite size of the source, but caustic curves still define regions in the sky mostly within one Einstein radius of the primary where the binary lens amplification deviates most from the comparison single lens amplification. When the point source crosses a caustic curve, the amplification rises sharply before dropping to a plateau level within the caustic region. Amplification peaks again as the source exits the caustic region. Most planet detections are therefore expected in regions close to caustics, depending on the level of disturbance the caustic causes outside the caustic region. Fig. 1 shows an example of this behaviour. Panel (a) on the left shows the relative motion of the source across the lens plane and caustics. Note the differing scales on the axes. Panel (b) on the left displays the light curves corresponding to the trajectory in (a) of a point source (solid line) and an $R_s = 0.01$ resolved source (dashed line). The difference in observed magnitude for light curves corresponding to this binary geometry and a single lens with mass equal to the primary is shown in panel (c) on the left.

If the effects of small secondary lenses are assumed to be independent, the detectability of companions by the perturbation they cause to the single lens light curve can be quantified by calculating the difference in amplification between a single lens light curve and a binary lens light curve with primary of the same mass as the single lens. Therefore it is assumed that the detection probability of multiple planets in one system should be simply related to the detection probability of a binary system. This approximation is likely to break down for high amplification events or higher ($q > 10^{-3}$) mass ratio planets (Gaudi, Naber & Sackett 1998). In most binary lens cases there are two distinct caustic areas for small mass ratio planets; a central caustic close to the primary lens and an outside caustic anywhere on the line connecting the primary and secondary lenses. For larger mass ratio planets or projected orbital separations

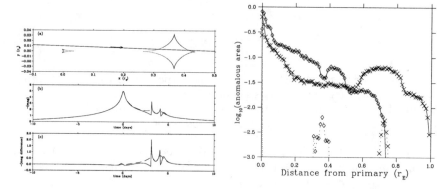

Figure 1. An example (LEFT) of binary lens caustic geometry ($a = 1.2$, $q = 10^{-3}$) and an associated light curve ($b = 0.01, \theta = 89°$) and difference curve. (RIGHT) The fraction of anomalous area to total area of a ring of width Δu centred on the primary as a function of the distance from the primary, u. The solid lines are for $q = 10^{-3}$ and dotted lines are for $q = 10^{-5}$. Crosses indicate $a = 0.7$ while diamonds are for $a = 1.2$.

close to 1 θ_E, the caustics may merge and the distinction becomes artificial. The positions of the caustic areas are discussed in more detail below.

1.2. Resolved Sources and Blended Light

A source star is considered to be resolved in the context of GM when the point source approximation fails. The actual amplification must then be determined by integrating the point source amplification over the source profile. The resolved source effect is negligible for most GM events towards the Galactic Bulge, but comes into play at very high amplification by reducing the point source amplification peak which is theoretically infinite when the source crosses a caustic, to a finite value. The importance of resolved source effects depends on the lensing geometry. Detection of planets in the presence of a resolved source was first explored by Bennett & Rhie (1996). Their calculations suggested a mostly negative influence on planet detection efficiency for planets with mass ratios q of 10^{-4} and 10^{-5} in the lensing zone (see also Wambsganss 1997).

Gravitationally magnified light from the source star may be diluted with unmagnified light from an unresolved background star in close proximity to the source star in the sky. A mathematically equivalent effect occurs if the lens itself is luminous. This is referred to as blending and theoretically affects all microlensing events (Alard 1997; Lee & Han 1997). If L_s and L_x are the apparent fluxes from the lensed and unlensed stars respectively, we define the blending parameter f as

$$f = \frac{L_s}{L_s + L_x} \qquad (3)$$

Blending has a detrimental effect on planet detection and is included in these calculations but not discussed in detail.

Planet Detection via Microlensing: Resolving the Source

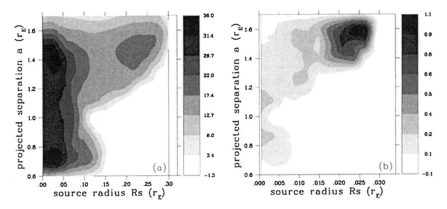

Figure 2. Contours of detection probability (P_d) as a function of source size (R_s) and projected binary separation (a) for two different mass ratios. Panel (a) is for a mass ratio $q = 10^{-3}$ and panel (b) for mass ratio $q = 10^{-5}$. Note the differing scales for the two panels. Both panels show P_d for the TF model with no blending $f = 1.0$.

Planet detection criteria for the simulation were chosen to represent two possible observational models. The Typical Follow-Up (TF) model corresponded to the sampling frequency and photometry of follow-up observational groups such as PLANET (Albrow et al. 1998), GMAN (Alcock C. et al. 1997) and MPS (Rhie S.H. et al. 1999). The second model, Peaks Only (PO), corresponded to observations made close to the peak of GM events only, where the planet detection probability is expected to be very high (Griest & Safizadeh 1997). The the resulting quantity from detection calculations, P_d, is defined as the percentage of binary events where the presence of the secondary lens would be detected during the follow up observations described by these models.

2. Results

2.1. Source Radius (R_s) and Projected Planetary Separation (a)

In Fig. 2 which shows detection probability (P_d) as a function of source size (R_s) and projected binary separation (a), the region $0.62 < a < 1.62$ yields the greatest detection probability by far. The existence of this zone has been discussed before (e.g. Gould & Loeb 1992; Bennett & Rhie 1996; Wambsganss 1997). For this range of projected orbital separation, all the binary lens caustics are within one Einstein radius of the primary lens and fall within the detection radius of the TF model. There is a drop in P_d around $a = 1$ because as $a \to 1$ the central and outside caustics merge and are only approached during high amplification events.

Resolved Source Smoothing A resolved source does not cross a caustic curve all at once so that only a small fraction of the source flux is magnified to any great extent by the caustic at any one time. This leads to flattening and broadening of short time scale light curve features by spreading the amplification peaks over a larger area of the sky while reducing the difference between the single and binary lens amplification. Caustic curves yield strong positive amplification but may be surrounded by large regions where the binary amplification is less than the single lens amplification. These regions further reduce the total source amplification when the source radius is large enough to include them. As a result, regions of fine caustic detail are cancelled out by smaller source radii than regions where the caustic curves enclose a larger area. For a small q, binaries with $a < 1$ have two outside caustics located close together whereas binaries with $a > 1$ have only one larger, continuous outside caustic. The geometry with two separate outside caustics is more susceptible to a reduction in detection probability because a resolved source that includes both caustics in its convolution area must also include the region between the caustics where binary amplification is in fact less than single lens amplification. P_d mostly declines with increasing R_s, with the obvious exception in Fig. 2(b) discussed below. For larger planets the caustic curves enclose larger continuous areas, leading to a gradual decline with R_s. In addition to the inclusion of areas of negative difference, there is also a sharper decline in P_d with R_s for $a < 1$ thanks to the smaller extent of the caustic regions.

For small q, there are two factors working against each other to determine the detection probability. For very small sources, the caustics enclose such a small area that TF observations do not detect the required anomalous data points before the source crosses over the caustic. An increase in R_s should therefore increase P_d as the anomalous area widens. On the other hand, too much widening flattens the anomaly to below the detection threshold. These two factors lead to a maximum detection probability that is reached when small caustics are widened enough to include the minimum number of data points required for detection, but not yet reduced to below the detection threshold. This effect is seen in Fig. 2(b). For $a = 1.3$, a peak in detection probability occurs at $R_s = 0.025$.

2.2. Source Radius (R_s) and Mass Ratio (q)

Fig. 3 shows the average P_d over the lensing zone (taken here as $a = 0.6$ to $a = 1.7$ in steps of 0.1) as a function of R_s and q for two different blending values, $f = 1.0$ and $f = 0.5$. These calculations confirm the precipitous drop in detection probability with decreasing mass ratio (q) that has been seen in previous works on GM planet detection. The steeper drop in P_d with R_s for smaller values of q is to be expected following the discussion in 2.1. regarding the increasingly fine caustic structure as q becomes smaller.

For comparison P_d for a moderately blended event is indicated in Fig. 3 by the dashed line.

2.3. Observing Peaks Only and the Role of the Central Caustic

Regardless of the shape or position of the outside caustics (except when $a \approx 1$ and the central and outside caustics merge) there is always a triangular central caustic structure close to the primary pointing towards the secondary for small mass ratio systems. For larger mass ratios, the primary caustic can connect with the outer caustics in some cases (Schneider & Weiß 1986).

Planet Detection via Microlensing: Resolving the Source

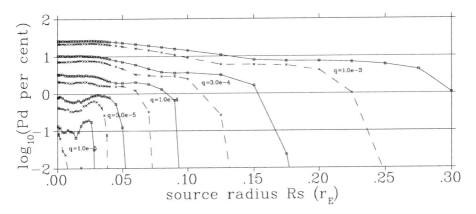

Figure 3. P_d averaged over the lensing zone as a function of R_s for various values of q and f. Solid lines indicate $f = 1.0$ while dashed lines have $f = 0.5$.

Fig. 1 illustrates the importance of the central caustic to planet detection for different mass ratios by examining a difference map for the fraction of anomalous area to total area of a ring of width Δu centred on the primary as a function of the distance from the primary, u. For larger mass ratios ($10^{-3} < q < 10^{-2}$) the central caustic plays a major part, while it plays almost no part at all in the detection of small mass ratio planets except during events with extremely high amplification. The high detection rate close to the primary for larger planets suggests the plausibility of an alternative follow-up strategy to microlensing events, where only high-amplification events are followed and only close to their peaks. Such a strategy would have a high efficiency due to the smaller number of observations needed. Unfortunately high amplification events ($A > 10$) are not common: roughly 1 in 10 of microlensing events have $A > 10$. Surveys that would detect more events of high amplification even at the expense of low amplification events are feasible making use of existing technology (Crotts 1992).

Fig 4 shows detection probability as a function of a and R_s for a $q = 10^{-3}$ system with no blending under the PO observational strategy on the left. As opposed to the TF model, the peak detection probability occurs at $a = 1$. This is due to the position of the outside caustic close to the central caustic and within $R_c = 0.1$. The detection probability exceeds 70% in this zone, more than double that of the best TF detection probability. The increase in P_d with increasing R_s in the region $0.2 < R_s < 0.4$ for the range $0.6 < a < 1.6$ is due to the effect described in 2.1. above, where fine features are widened to allow detection. As opposed to the positive effect of a resolved source on P_d for $q = 10^{-3}$, the PO model yields very low detection probability for small planets. $q = 10^{-5}$ planets are only detectable at the 1%-2% level for $R_s < 0.002$ and $0.9 < a < 1.1$ and completely undetectable outside of these regions with the PO detection criteria. On the right side of Fig. 4 P_d is shown as a function of R_s averaged over the lensing zone for various values of q. For small planets the difference between binary and single lens amplification is not enough to survive the smoothing effect that a resolved source has on the anomalous area

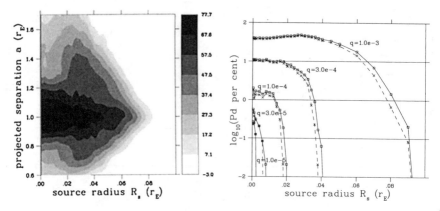

Figure 4. Detection probability (P_D) as a function of a and R_s (LEFT) for a $q = 10^{-3}$ system with no blending under the PO observational strategy. (RIGHT) P_d as a function of R_s averaged over the lensing zone for various values of q. Solid lines refer to $f = 1.0$ and dashed lines to $f = 0.5$.

and binary features are suppressed to below the detection threshold. As was the case in Fig. 3, P_d declines dramatically with decreasing mass ratio. The planet detection probability for the blending value of $f = 0.5$ is slightly lower than is the case without blending and drops to zero at smaller values of R_s.

3. Conclusions

These results impose a limit on the detection of planets via microlensing due to the resolution of the source. There is mostly a decrease in P_d with increasing R_s for typical follow up observations but a sharp peak in P_d does occur for small mass ratios thanks to the broadening of binary features in the light curve. For a mode of observation where only peaks of high amplification events are observed, a similar increase in P_d is observed as well as a substantial broadening of the lensing zone. Small planets are almost undetectable close to the peaks of GM events when the source is resolved. The PO mode of observation promises to be a highly efficient method of detecting larger mass ratio planets provided that enough high amplification events can be found.

4. Acknowledgments

This work was supported by a grant from the NRF and the SAAO. I would like to thank N. Chetty for the use of UNP facilities and J. Menzies, A. Gould and B. Erêche for helpful comments and constructive criticism.

References

Alard C. 1997, A&A, 321, 424

Albrow M. et al. 1998, ApJ, 509, 687
Alcock C. et al. 1997, ApJ, 491, 436
Bennett D. P., Rhie S. H. 1996, ApJ, 472, 660
Bolatto A. D., Falco E. E. 1994, ApJ, 436, 112
Crotts A. P. S. 1992, ApJ, 399, L43
Dominik M. 1998, A&A, 329, 361
Gaudi B. S., Naber R. M., Sackett P. D. 1998, ApJ, 502, L33
Gould A., Loeb A. 1992. ApJ, 396, 104
Griest K., Safizadeh N. 1998, ApJ, 500, 37
Lee S., Han C. 1997, preprint astro-ph/9707178
Mao S., Paczyński B. 1991. ApJ, 371, L63
Rhie S.H. et al. 1999, ApJ, 522, 1037
Schneider P., Weiß A. 1986, A&A, 164, 237
Wambsganss J. 1997, MNRAS, 284, 172

Discussion

Gaudi: I agree that the peaks should be sampled more frequently than the wings; however I don't feel that it is a good idea to monitor only the peaks, since we won't have any idea of the parameters of the primary events.

Vermaak: Absolutely. Sampling the peaks only would be an option if we have more events than we can follow. In that case we should definitely cut down on the wings to increase peak sampling.

Evans: What is the effect of multiple planets? Is the binary lens equation sufficiently accurate for calculation of detection efficiencies if – as we heard earlier from Debra Fischer – multiple systems are likely to be common?

Vermaak: The binary lens equation is fine as long as most of the detections are expected from the "outside" caustics. That is because the outside caustics behave almost independently. This will not be the case for detections at high amplification or when the outside caustics actually overlap (possible in, e.g. edge-on systems with overlapping projected planet positions).

Gaudi: I agree.

Bennett: For the solar system, the binary lens model will be a poor approximation for the central caustic when Saturn is in the lensing zone because Jupiter can cause a significant perturbation from outside the lensing zone.

Vermaak: Yes, certainly. In current Galactic lensing observations I expect most detections to be due to the outside caustics because of the relatively low amplitude of the events being followed. For the few high-A events that have been followed, it would be critical to have maximum sampling frequency at peak to possibly fit for multiple planets.

Evans: What is the geometrical interpretation of the detection probabilities? Are they related to the area or to the length of the caustic network?

Bennett: There is not a direct relationship between the size of the caustics and the detection probability. The central caustics, in particular, cause perturbations over an area much larger than their own area.

Vermaak: They are concerned with *crossing* the caustic curves more than the area they enclose, but it is not a simple relation. It depends on the caustic geometry which can vary a lot depending on the binary separation.

Microlensing 2000: A New Era of Microlensing Astrophysics
ASP Conference Series, Vol. 239, 2000
John Menzies and Penny D. Sackett, eds.

Planetary Microlensing Signatures in the High Magnification Events MACHO 98-BLG-35 and MACHO 99-LMC-2

Ian Bond (MOA Collaboration)

University of Auckland, Private Bag 92019, Auckland, New Zealand

Abstract. High magnification microlensing events have been identified as promising hunting grounds for extrasolar planet searches. In such events excursions due to planetary microlensing events occur near the times of peak amplification and at significant detection probabilities. We present intensive observations of two high magnification events obtained by the MOA collaboration. Around the times of their respective peaks, 163 observations were obtained for MACHO 98-BLG-35 and 400 observations were obtained for MACHO 99-LMC-2. The results of the analysis of the observations of 99-LMC-2 raises the new prospect of the detection of extragalactic planets.

1. Introduction

As well as survey observations, the MOA Collaboration carries out follow-up observations of selected microlensing events. A particular objective is to detect excursions from the single lens light curve due to the presence of planets in the lensing system. While excursions due to lensing by a stellar binary have been observed a number of times, the observation of fine structure due to lensing in the extreme case of a star-planet binary lens is considerably more challenging.

To obtain an observational detection of a planet by microlensing, two conditions need to be met. The source trajectory must intersect or pass very close to one of the caustics and one must be observing the event at the time of the caustic crossing. The characteristic crossing times range from 2–3 days for a Jupiter-mass planet down to a few hours for an Earth-mass planet. When compared with the typical Einstein ring crossing times of ~ 40 days, one can appreciate the difficulty in meeting these conditions. Detection probabilities for planetary microlensing have been the subject of several studies (e.g. Mao & Paczyński 1991; Gould & Loeb 1992, Wambsganss 1997).

2. High Magnification Microlensing Events

It would be of tremendous advantage observationally if it could be possible to know if and when an ongoing microlensing event will undergo a planetary caustic crossing. A study by Griest and Safizadeh (1997) pointed out that such a situation can occur in the special case of an ongoing high magnification microlensing event. The lens plane magnification map corresponding to any

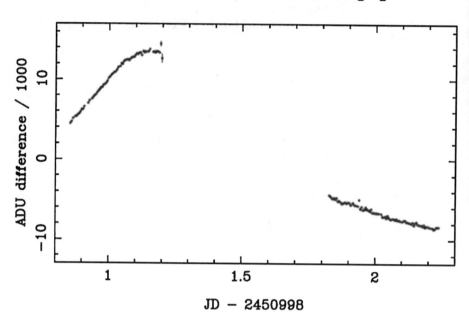

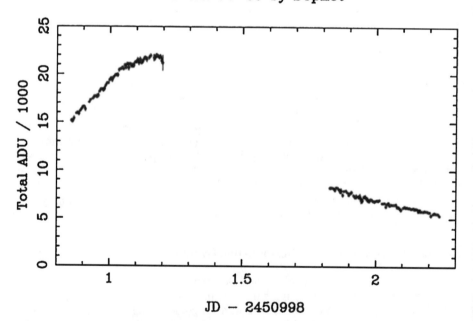

Figure 1. Light curve of MOA observations of MACHO 98-BLG-35 obtained by DoPHOT and image subtraction.

binary lens always contains a small "primary" caustic region at the centre of the map. An ongoing microlensing event with small impact parameter, i.e., high magnification, should show caustic deviations around the peak of the single lens light curve if the lensing system contained planets. Griest and Safizadeh (1997) show that if the lensing system contains a Jupiter-mass planet, the probability of detection in a high magnification event with $A_{max} \sim 100$ which is well sampled near the event peak is close to 100%.

Thanks to the online analysis capabilities of the survey/alert groups, MACHO, OGLE, and EROS, reliable estimates of the times and sizes of peak amplification of ongoing microlensing events can be determined well before the event peak. Thus if an ongoing event is known to be of high amplification, intensive observations can be planned accordingly. Such intensive observations were carried out by MOA on two high magnification events, 98-BLG-35 and 99-LMC-2, which were alerted by the MACHO Collaboration. These are described in the following sections.

3. MOA Observations of MACHO-98-BLG-35

The microlensing event MACHO-98-BLG-35 attained a peak amplification of 70.3 on 1998 July 4.6 UT. A total of 163 measurements were obtained by the MOA group over two nights around the time of peak amplification. On the night of the peak itself, 65 measurements were taken over an 8-hour period with a median sampling time of five minutes. This event was also observed intensely by the Microlensing Planet Search (MPS) Collaboration. The results of a detailed analysis of the combined MOA and MPS data were presented by Rhie et al. (2000). Large planets of around Jupiter masses were excluded within the lensing zone. Furthermore a deviation from the single lens light curve was detected and attributed to a planet in the lensing system with a mass between that of the Earth and Neptune. The detection was statistically significant at 4.5σ.

The published light curve of MACHO 98-BLG-35 was derived from the analysis of images based on profile-fitting photometry using the DoPHOT package (Schechter et al. 1993). The MOA observations of this event have also been analysed using image subtraction photometry. After geometrically aligning all images to a common coordinate system, a reference image was constructed by stacking 69 of the best seeing images. A data set of subtracted images was then generated using the techniques described elsewhere (Bond, these proceedings). An image subtraction light curve for this event was then generated by simply collecting aperture photometry measurements at the source position from all the subtracted images.

A comparison of the light curves of 98-BLG-35 obtained by the two methods is shown in Figure 1. The image subtraction light curve appears markedly smoother that the light curve derived from profile fitting photometry.

4. MOA Observations of MACHO 99-LMC-2

The microlensing event MACHO 99-LMC-2 attained a peak amplification of 43.3 on 1999 June 7.7 UT. This event was intensely observed by MOA over five

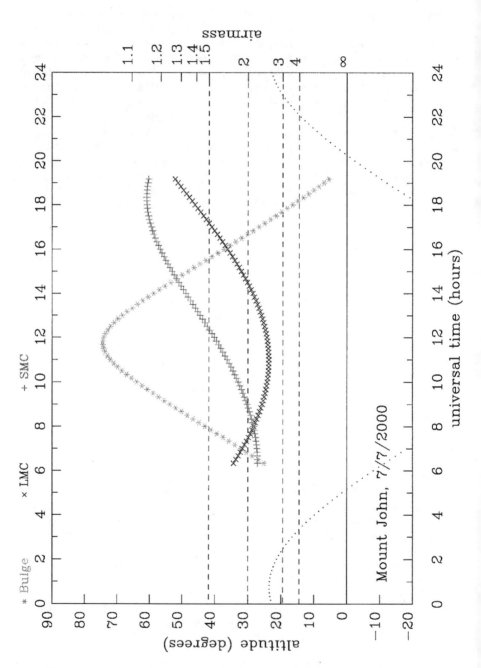

Figure 2. Altitude and airmass of selected Southern Hemisphere targets visible from the Mt John Observatory for July 7.

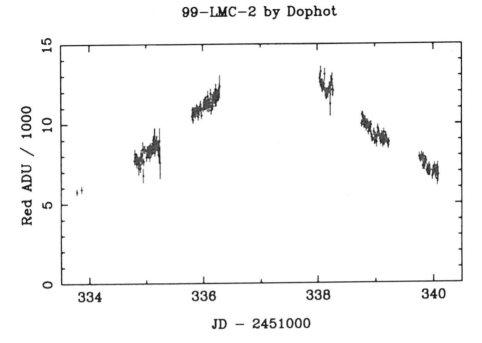

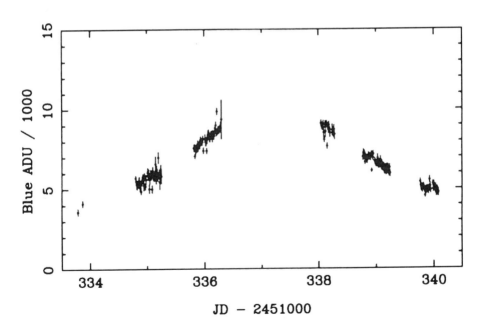

Figure 3. Light curve of MOA observations of MACHO 99-LMC-2 obtained around the peak of the event in the red and blue passbands.

nights around the time of peak amplification. A total of 200 measurements were made in each of the red and blue passbands with a median sampling time of 10 minutes per passband. Unfortunately, cloud prevented observations on the night of the peak. Otherwise, weather permitting, it is possible to observe targets in the Magellanic Clouds continuously for 12 hours during the winter nights in New Zealand (Fig. 2).

The light curve of 99-LMC-2 obtained by MOA and derived using DoPHOT is shown in Figure 3. A re-analysis of these observations using image subtraction is also underway. These observations will be used to determine how one can constrain the configuration of planetary configurations of the corresponding lensing system. A joint analysis of data on this event obtained by all the microlensing groups would be even more constraining. Because of its high magnification, this event is a good prospect for obtaining information on the distance to the lensing system through an analysis of parallax signatures. Furthermore, if the source star turns out to be a spectroscopic binary it may be possible to study the inverse effect of parallax in the light curve – the so-called "xallarap" effect (Bennett 2000).

If subsequent constraints on the distance to the lensing system show 99-LMC-2 to be an LMC self-lensing effect, this raises the prospect of constraining or studying the configurations of extragalactic planetary systems.

5. MOA Planet Hunting Strategy

There are several shortcomings associated with the use of observations of the primary caustic in the detection of planetary events. If r is the star–planet separation, the primary caustic has the same shape at the conjugate separations of r and $\frac{1}{r}$. This degeneracy introduces difficulties in modelling any observed planetary microlensing event. However, the observational advantage of knowing when to look for a planetary caustic far outweighs any shortcomings.

High amplification events appear to be the best hunting grounds for extrasolar planets through microlensing follow-up observations. The planet hunting strategy for MOA is to concentrate on agressively observing high magnification events around the times of their peak amplifications. Alerts for such events would come from the established survey groups, OGLE and EROS (MACHO has now finished). From the southern winter of 2000, MOA will endeavour to issue alerts for microlensing events based on online analysis using image subtraction (Bond, these proceedings).

References

Bennett, D. 2000, XXth Moriond Astrophysics Meeting, Les Arcs, Savoie, France, March 11–18, 2000
Gould, A. & Loeb, A. 1992, ApJ, 396, 104
Griest, K. & Safizadeh, N., ApJ, 500, 37
Mao, S. & Paczyński, B. 1991, ApJ, 374, L37
Rhie, S. et al. 2000, ApJ, 533, 340

Schechter, P. et al. 1993, PASP, 105, 1342
Wambsganss, J. 1997, MNRAS, 284, 172

Discussion

Gaudi: I do not agree that profile-fitting photometry is well-understood and I feel the prospects for understanding DIA photometry are much better.

Bond: Perhaps I should have said that the profile-fitting photometry is better understood than DIA photometry. However I am not suggesting that we should stay with profile-fitting photometry. DIA is the way of the future in my opinion, but it needs further understanding.

Bennett: We have a better understanding of the systematic errors in profile fitting than with difference imaging, but this understanding of the errors doesn't imply that we can always correct for the errors.

Gaudi: I agree that we cannot always correct for the systematic errors in profile fitting. However it is still not clear to me that our understanding of systematic errors in profile fitting is better than our understanding of systematic errors in difference imaging.

Bond: In profile fitting photometry one can use constant stars to quite effectively check for systematic errors, because PF photometry measures total fluxes. On the other hand DIA photometry measures flux differences. Systematic errors may show up in variable star light curves, which would not show up in constant stars.

Lasserre: What are the fit parameters with DoPHOT and DIA photometry (especially t_E and baseline)?
Why are the timescales so different in red and blue?
Did you look for parallax or finite source size?

Bond: The fits I showed were only there to show a quick comparison between the observed data and microlensing light curves. The fitted parameters were not supposed to be taken seriously. More detailed and careful modelling will come later.

Crotts: If you don't have an error in your code or its implementation, the most common reason why the difference image technique fails is associated with a point spread function not consistent with an elliptical Gaussian, e.g. due to image trailing or bad focus.

Bond: Difference imaging very rarely "fails". Rather one obtains DIs of varying degrees of quality. I suggested a method of assessing systematic effects in the subtracted images. In "ideal" subtractions these systematic effects are zero or well below statistical fluctuations. In less than ideal subtractions, systematic effects of typically 1-3% are present. I would be amazed if one is able to get ideal subtractions every time and I think it is essential that systematic effects in the subtraction process are assessed when extracting photometry and making light curves.

Abundance of Terrestrial Planets by Microlensing

Philip Yock

Faculty of Science, University of Auckland, Auckland, New Zealand

Abstract. Terrestrial planets may be detected using the gravitational microlensing technique. This was demonstrated in the high magnification event MACHO 98-BLG-35. Observing strategies aimed at measuring the abundance of terrestrial planets are discussed, using both existing telescopes and planned telescopes.

1. Introduction

Significant advances have been made in recent years in the study of extrasolar planets using the radial velocity technique and, more recently, the transit technique (Mayor, these proceedings; Fischer, these proceedings). Approximately thirty Jupiter-mass planets have been detected by the radial velocity technique. Initial indications from transit measurements indicate they are gas giants similar to Jupiter. However, they have been detected in eccentric orbits at > 0.2 AU or circular orbits at < 0.2 AU, quite dissimilar to Jupiter's orbit. These surprising findings will assist our understanding of planetary systems and planetary formation.

The gravitational microlensing technique complements the above studies because it is sensitive to giant and terrestrial planets at orbital radii of a few AU. In this technique, perturbations of standard microlensing caused by planets are searched for. Two versions of the technique have been utilised to date. These involve large perturbations of low-magnification events, and small perturbations of high-magnification events, respectively. Most searches have been made in the galactic bulge where the observed rate of microlensing is maximal.

In the following sections strategies for optimising the detection rate of planets by microlensing, especially low-mass terrestrial planets, are discussed. The ultimate goal is to determine their abundances. To illustrate the technique, a high magnification event is described in the following section. Other examples of both high and low magnification events are described by Gaudi in these proceedings.

2. High-magnification event MACHO 98-BLG-35

The event MACHO 98-BLG-35 has been described elsewhere in some detail (Rhie et al. 2000). The peak magnification was ~80, the highest observed to date. Here just those features of the event that demonstrate the sensitivity of the microlensing technique to the detection of low-mass planets are described.

Abundance of Terrestrial Planets by Microlensing 161

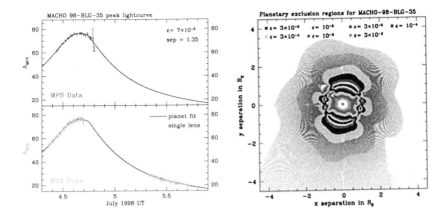

Figure 1. Light curves and "snapshot" of the planetary system in the microlens event MACHO 98-BLG-35 from Rhie et al. (2000). The pale light curves are the best fits to the data for a lens without a planet, and the heavy curves are the best fits for a lens with a single planet with mass fraction 7×10^{-5} and orbit radius 1.35 Einstein radii. The right panel shows the exclusion regions for planets with various mass fractions. The two unshaded regions at about 10 o'clock designate the possible locations of a planet at the time of the observations.

The peak of MACHO 98-BLG-35 was monitored fairly extensively from Australia and New Zealand by the MPS and MOA groups. Their light curves are reproduced in Figure 1. Some features are noteworthy. The light curves are incomplete. Nearly half of the peak was not monitored. Secondly, the photometry is accurate to $\sim 1.5\%$ only. This was achieved with the standard DoPHOT programme.

Despite the above limitations, significant information on the presence of planets orbiting the lens star for this event was obtained. This is illustrated in the right-hand panel of Figure 1. This is a two-dimensional "snap shot" of the lens's planetary system at the time of the event. It shows exclusion regions (in projection) at the 6.3σ level of confidence for planets with masses ranging from 3×10^{-6} to 3×10^{-3} times that of the lens. Assuming a most likely lens mass $\sim 0.3 M_\odot$, this range corresponds to 0.3 Earth-mass to Jupiter-mass. The exclusion region for the lowest mass is the dark, nearly circular tube at the Einstein radius ($r = 1R_E$). Here R_E is ~ 0.6-2.8 AU. The exclusion regions for heavier planets are successively thicker tubes surrounding this. The two unshaded regions just inside and outside the Einstein radius at about 10 o'clock are two possible locations (in projection) at the time of the observations of a planet with fractional mass 4×10^{-5} to 2×10^{-4}. The evidence for this planet is at the 4.5σ level of confidence.

These results indicate both the potential and the limitations of the microlensing technique. Only a 'snap-shot' is obtained. There is essentially no chance for follow-up measurements. Also, the photometry does not uniquely determine all the parameters of the system. Balanced against these limitations is the demonstrated potential of the technique to detect low-mass planets. A significant volume surrounding the lens star of MACHO 98-BLG-35 was shown to be devoid of Earth-like planets.

Improvements in the precision of the above results for MACHO 98-BLG-35 should be possible with the inclusion of more data and better photometry. Data by the PLANET group are available (J. Greenhill, private communication) but these have not been incorporated as yet. Also, a re-analysis of the data using subtraction photometry may yield higher precision (Bond, these proceedings). A "gigablaster" similar to those used for human genome sequencing (Claverie 2000; Gee 2000) is being used for this purpose.

3. Optimising the detection rate of terrestrial planets by microlensing with existing telescopes

The microlensing technique may be understood physically as follows. In high-magnification events, good alignment of the lens and source stars occurs. This causes a circular, or near-circular, image to form around the lens. If a planet is present in such an event near the (projected) Einstein radius it will perturb the image. The perturbation will be small, however, because most of the image will lie far from the planet and remain unperturbed. This is the case illustrated in the left panel of Figure 1. In low-magnification events the image consists of two, short, rotating arcs either side of the lens star. If either of these approaches a planet (in projection) during an event, a relatively large perturbation of the light curve occurs. However, in most cases, neither image will approach a planet. Moreover, the relatively low light level of low-magnification events results in photometry of poorer accuracy. In both high and low magnification events, the Einstein radius is typically $\sim$ a few AU. This enables planets at these orbital radii to be detected.

To date, the most fruitful microlensing events for planetary searches have been the high magnification ones. This is primarily because of the high probability of detectable planetary perturbations occurring in the peaks of these events, if planets are present. This was first predicted by Griest and Safizadeh (1998). Also, the short durations of high-magnification peaks permit them to be sampled densely without undue difficulty. This enhances the probability of detecting low-mass planets for which the planetary perturbation is only brief. Additionally, the high light-levels attained in high magnification events enable good photometry to be performed with small telescopes.

If no drastic changes occur in the observing strategies of the current microlensing groups in the next few years, it may be anticipated that the current results will be extended somewhat, but not greatly. A tighter constraint on the abundance of giant planets at a few AU than that reported by Gaudi in these proceedings can be expected, and possibly some detections. Also, further constraints and/or results on the presence of terrestrial planets can be expected from events like MACHO 98-BLG-35. In the same period, the abundance of giant planets at a few AU is likely to be determined by the radial velocity technique.

Further information on terrestrial planets would be obtained if more telescopes were devoted to the microlensing technique. Currently there are some

Abundance of Terrestrial Planets by Microlensing   163

ten 1-m class telescopes in Chile, South Africa and Australasia that are being used for microlensing. If all these were dedicated to the relentless observation of the peaks of high magnification events, thus minimising gaps in light curves caused by inclement weather, tighter constraints and/or results on the presence of terrestrial planets would be obtained. If, furthermore, rapid on-line analysis was employed by more survey groups to alert on high magnification events in progress, the detection rate would be additionally enhanced. A first rough indication of whether or not terrestrial planets are common could then be possible.

4. Measuring the abundance of terrestrial planets by microlensing with new telescopes

High-quality planetary abundance measurements could be made with a single 2-m class wide-field telescope monitoring a few square degrees of the galactic bulge at a sampling rate of a few observations per hour from the Antarctic or from space (Sahu 1998; Yock 2000; Bennett & Rhie 2000). The rate of detection of microlensing events with such an instrument would significantly exceed the rate currently being detected, especially if a near infrared passband was used. This would enable the galactic centre to be monitored where the microlensing rate is expected to be highest (Gould 1995). Additionally, the high sampling rate would result in both high and low magnification events being densely sampled, increasing the detection rate of low-mass planets. A simple scaling of the results obtained to date suggests that useful planetary abundance measurements could be made in a few years from the Antarctic. Specific predictions require further site testing to be carried out at the Antarctic. This is currently in progress at the high-altitude site known as Dome-C (Burton 1998). Probably the ideal is the proposal that was made recently by Bennett & Rhie (2000) for a space-based ~ 1.5-m diffraction-limited wide-field telescope known as GEST. This would have the capability to detect ~ 100 Earth-like planets in 2.5 years if such planets are common.

5. References

Bennett, D.P. & Rhie, S.H. 2000, in Disks, Planetesimals and Planets, ASP Conference Series, ed. F. Garzsn, C. Eiroa, D. de Winter & T.J. Mahoney (San Francisco: ASP)
Burton, M.G., 1998, in Astrophysics from Antarctica, ASP Conf. Series 141, ed. G. Novak & R. Landsberg (San Francisco: ASP), p.3
Claverie, J-M. 2000, Nature 403, 12
Gee, H. 2000, Nature 404, 214
Griest, K. & Safizadeh, N. 1998, ApJ, 500, 37
Gould, A. 1995, ApJ, 446, L71
Rhie, S.H. et al. 2000, ApJ, 533,
Sahu, K., 1998, in Astrophysics from Antarctica, ASP Conf. Series 141, ed. G. Novak & R. Landsberg (San Francisco: ASP), p.179
Yock, P. 2000, PASA, 17, 35

Discussion Session I:
Mass and Orbital Characteristics of Binaries and Planets

edited by John Menzies[1] and Penny D. Sackett[2]

[1] SAAO, PO Box 9, Observatory 7935, South Africa

[2] Kapteyn Institute, 9700 AV Groningen, The Netherlands

1. Introduction

The first of two full-meeting discussion sections was held the afternoon of 22 February 2000, with Neill Reid acting as moderator. Participants had been given a list of thought-provoking questions before lunch, and were encouraged to use these as a starting point for a free-ranging discussion on the merits and drawbacks of microlensing as a technique for the study of stellar binaries and extrasolar planets, its current successes, and possible methods for improving its future prospects. The moderator was given the privilege of opening the session. Editorial remarks are indicated by a slanting typeface. (Manual transcription and interpretation of an audio tape may have produced imperfect results in reflecting precisely the remarks of all speakers.)

2. Discussion

Reid: I think we now know that extrasolar planetary systems exist. In most studies there are two phases (as with brown dwarfs): first of all we must find out if they exist, and then, if they do, what are their properties? Microlensing has the capability to detect planets, but I think that is not enough. The main questions for which we want to get answers are the frequency of planets, the frequency of binary systems among stars, and how they form. How does microlensing fit into that picture? From personal experience, I know this is something not often addressed by the microlensing community. I offer this as a first gambit, to get the discussion going. These are things that need to be thought about – for funding purposes apart from anything else. I invite other comments from the audience.

Bennett: I think it is pretty clear what microlensing can provide that other techniques cannot. One is its sensitivity to planets around 2 to 5 AU.

(*interjection:* Why is that interesting?)

The latest book currently being reviewed in the popular press argues that we are alone in the Universe because there are no Jupiter-mass planets found around other stars at the right radii to save their planetary systems from bombardment by asteroids, etc. In fact radial velocity search techniques are not sensitive to

Jupiter-mass planets and a little below, while microlensing can find such objects. Microlensing can push much lower and reach a little below Neptune mass from the ground, while more ambitious satellite projects are expecting to find Earth masses. But microlensing is the only techniques that is sensitive to objects in extremely large orbits or ejected from planetary systems. There is a fair amount of parameter space that only microlensing can probe, and we are trying to focus on that in our work.

Bradbury: If ejected from their parent systems Earth-mass planets are not going to be harbouring life anyway. You need to find Jupiters orbiting stars that would protect 'Earths' from comets, etc. These discoveries are as likely to come from astrometric or photometric searches as from microlensing. It is really a question of who gets the satellite first – the astrometric community or the microlensing one.

Bennett: I disagree. It is wrong to concentrate on arguments that maintain that only systems like our own are interesting. It is silly to argue that life would be preferred in such systems. We really know nothing about life except experimentally; it is impossible to predict theoretically where life will occur. We should start by trying to understand the physics of formation of planets. We should concentrate on the parts of parameter space that microlensing can probe. Some space missions like TTF can probe parts of parameter space that microlensing cannot, but they are extremely expensive. Microlensing is much cheaper.

Gaudi: For once I agree entirely with Dave! *(Laughter.)* I think the implications of Neill Reid's comments are rather short-sighted. The microlensing community do know what information they can provide that other techniques cannot. You have to walk before you can run. Microlensing as a technique is beginning to mature as a way of detecting planets. What has been done so far is quite amazing, but we are just coming to terms with the capabilities of the technique. We *can* constrain the frequency of binaries, at least in the ranges to which we are sensitive, and constrain the distribution of mass ratios, and so on. This has not yet been done but it will be. It will just be a matter of time.

Reid: But how is it to be done? That is part of the focus of the question we are addressing.

Gaudi: Well, take the binary statistics, for example. You need a well-defined sample, which already exists. It needs to be analysed self-consistently – determination of light curves, detection efficiencies, mass ratios – and these need to be folded together. It is not conceptually difficult, it just requires a lot of hard work. A start has been made in a recent paper by MACHO dealing with binaries they discovered. It is crude, by their own admission, and needs to be done better, but it is a start. I expect this will be done in the next four years or so.

Hansen: To address Neill Reid's challenge at the level implied by the strength of his comments it will be necessary to characterise the lensing star population. If some of the correlations produced by the radial velocity surveys are to be believed, they may not be sampling the same host star population as the microlensing surveys. It will be necessary to get more input, from other than microlensing, in this area.

Gould: I don't think it is clear that we will be able to characterise the binary population with the kind of data collection we have at present. We have lists of objects that we have detected, that we know are binaries, but going from there to characterising the population, while conceptually simple is computationally complex. It might require us to acquire data more comprehensively and consistently than we do now, so the entire binary light curve is covered. We have lots of binary light curves partially covered and some solutions, but without routinely covering the first caustic it will be difficult to make progress. With planets the situation is likely to be a lot clearer. It is possible to characterise a planet based on a relatively short time span of data. The alert for looking for a planet is that a microlensing event is in progress. For a binary it is often that the event has already started but by then it is too late to get the first caustic crossing.

Mayor: Regarding the field of the duplicity of binary stars, one domain where it will be very difficult to get a systematic determination of the mass distribution is that of brown dwarfs. At present we are just picking a few objects, bright young stars to search for brown dwarfs. A systematic survey for these objects by microlensing would be an area where no other technique could compete.

Sackett: Binary brown dwarfs over a certain range of separations are easy for us to detect. The difficulty is: can we recognise such an object as a brown dwarf? We don't independently know the mass of the binary without other information. The challenge is to learn more about the lens population, as has been said before. One way, in the era of 8–10 m telescopes, like SALT, is to detect light from the lens itself spectroscopically. This has been discussed in a paper by Shude Mao. The lens and source star would have different radial velocities, different lines or line ratios, etc. If this could be done it would push microlensing into new fields, and should be given more attention.

Zinnecker: As I said in my talk, another aspect that might be worth considering for microlensing is infrared observations. With a large field of view infrared telescope we could probe a large sample of stars by looking towards the Galactic Centre. In the optical, with 13 magnitudes of extinction, we can see nothing, but there the star density is very high and the extinction could be penetrated in the infrared. This would give the best chance for planet detection.

Mayor: Can somebody compare the detection efficiency of the astrometric and microlensing techniques for planet detection? Comparisons of efficiency are always done with the radial velocity technique. Astrometry is sensitive to objects from about 1 to a few AU. With SIM and FAME you would be looking at a few Earth masses.

Gaudi: I showed a plot of this in my talk. SIM will be able to detect very low-mass planets, and should be better than microlensing for an ensemble of

Discussion Session I: Binaries and Planets 167

microlensing events. But it will only be a 5-year mission and will never be able to detect planets at large separations. Microlensing can do this and does not need a long time for the job.

Han: I once investigated whether astrometric planet searches might be worthwhile. After two weeks of serious computational work I realised that there is no use in trying to use astrometry for planet searches. You can increase the interval over which there is a deviation from a single-lens behaviour – maybe you can get a doubling of the interval – but it does not seem worth using very expensive instruments to do it.

Gould: We are not talking about astrometric microlensing, but about using astrometry to detect planetary companions to stars.

Han: Sorry, I misunderstood.

Zinnecker: We might consider the prospects of interferometry in resolving planets. Some new large telescopes are expected to come on stream between 2005 and 2010. Several 8-m telescopes, combined interferometrically and working at $10\mu m$ may resolve some systems. We might be able to follow up microlensing-detected planets by means of interferometry if it is at the right distance. There is a problem in the community with accepting the one-off nature of events detected by microlensing, though I think they are believable given that many groups are now doing it. But the uniqueness has been used to criticise the results. There is a small hope that interferometry might be able to follow up some events, even as far away as 1 kpc, in the not too distant future.

Gaudi: On a related note: one of the main criticisms of microlensing is that we don't know what we are looking at in terms of the mass and distance of the primary lens. In the future, SIM or SIM-like satellites will allow a complete solution of microlensing systems and we will know the masses and distances of the lenses. SIM won't have a long enough lifetime to do it, but some future one may. In ten years, microlensing will be a different field.

Gyuk: Microlensing is really complementary to other techniques. Microlensing will eventually turn up planets, and as time goes by more and more objects will be found, thus improving the statistics. Radial velocity and astrometric observations give better and more detailed information with repeated observations of local events.

Gould: There has been some discussion in the meeting of the prospects for finding extragalactic planets. They may or may not be detectable by microlensing, but certainly not by any other means – apart perhaps by signals sent by residents there!

Feast: Neill Reid is trying to divorce planet detection from anything else. Microlensing observations are made for a variety of reasons; one is to find planets. You cannot tell what is going to be found until you have made the observations of the particular event you are working on.

Gould: There is a major problem for microlensing with getting the orbital characteristics of a binary. You can get the projected separation reasonably easily, and in favourable cases the projected velocity of a planet, but it will never be possible to get the eccentricity, and only statistical data on the semi-major axis. How valuable is this type of information when combined with other kinds of information? We are sensitive further out and to lower mass planets. Given that we had a large number of detections at say 8 – 12 AU where others cannot get information, how would that help with understanding planetary systems as a whole? Would it help with understanding the physics of planetary formation?

Reid: Bulge lenses are typically going to be M dwarfs at a few kpc. Microlensing is sampling a different mass range than all the radial velocity surveys. Only a few hundred M dwarfs have been observed so far by the local surveys. We know there is a difference in binary star frequency when you integrate over the whole range of separations in M dwarfs and G dwarfs. So there is an area where you can address questions that are not going to be addressed, just for lack of photons, by the Doppler community for a while.

Carr: The radial velocity searches produced a major surprise, which is why they seemed to be so interesting. What kinds of surprises will come from microlensing?

Gaudi: From the preliminary results I showed today, it is clear that planets are not common – not 100% of systems have Jupiters at these separations. Is that surprising? Previous to radial velocity searches it may have been thought that all stars had planetary systems. The radial velocity results showed we did not understand the problem.

Sackett: I have a question for those studying binarity. I am more optimistic than Andy Gould about whether microlensing can say something interesting about a specific class of binaries. Microlensing can detect systems with mass ratios in the range from 1 to about 0.01, with separations of the order of a few AU. Getting the mass ratio is the easiest thing for us to do. It is difficult to get the total mass. How interesting is that?

Han: Astrometric observations can say something useful for wide binary systems, say with separations of up to 100 AU.

Vermaak: With regard to lensing in M31, it is not clear whether it will be practicable or not to detect such events, but if it is, it will probe a different range of orbital parameters than Galactic microlensing. The central caustic will be hit for a much larger range of projected orbital characteristics. It will probe a larger projected orbital separation range than in the Galaxy.

Kerins: I have not studied planetary detection in M31 in any detail, but there seem to be some problems in that case. Current sampling frequencies are too low

Discussion Session I: Binaries and Planets

to allow planet detection. Even if the frequency could be raised, the practicalities of image reduction will make it hard to alert others for follow-up observations. Unless you are prepared to monitor fields continuously without worrying until some later time about what is going on or whether any events are occurring it will be difficult to make progress in this area. The other point is that you will usually only see the peak of an event, and it will be difficult to get any useful information out of that.

Gaudi: I have worked this out, and can say that if you only observe the peak you can determine only the mass ratio divided by the square of the impact parameter, and the separation multiplied by the impact parameter, so the solution is completely degenerate. I would still like to hear the answer to Penny's question.

Fischer: It would be interesting to see a mass ratio distribution. As an outsider to the microlensing group, it still seems to me that you are looking at planets you can't see orbiting stars you can't see. Why not join up with the transit people?

Sackett: We have thought about that, and it is both encouraging and discouraging. The recent detection of a transit is very encouraging. On the other hand, our fields are in the direction of the Galactic Centre, where there is lots of extinction, and we would need better data than we usually get, at least 1% photometry. To get higher precision we need lots of photons, which means we need to observe giants (since the fields are at 8 kpc). But a transit of a Jupiter-sized planet across a solar-type star produces only a 1% effect and it would be much less with a giant. The Galactic Centre might not be the right place to look, and it may be that there are different fields where transit-type experiments could be done. It is worth thinking more about.

Zinnecker: Observing in the infrared might help. But as to the question: there are indeed merits in having the distribution of mass ratios in the range down to 0.1 or even to 0.01. Around a solar-type star, objects with a mass ratio in the range from 0.1 down to 0.01 or so can form in the disc around the star. There are interesting issues relating to discs around young stars, which have already been found, where objects at a few AU in the range from 0.01 to 0.1 could form through gravitational instabilities. Objects in the range 10 to 100 M_J have not been probed in the same way as for lower masses, where, as Mayor told us, there is a 'desert'. It is important to see if in another range of separations there is also a 'desert'.

Reid: It is worth pointing out that if a planet is found around a halo MACHO, the chances are that the latter would not be a white dwarf. The white dwarf would have incinerated the planet.

Bradbury: There have been a number of proposals from different people here for astronomical telescopes in space to look for planets. The need to have 24-hour coverage from the ground has also been noted. Are the space telescopes going to give 24 hours a day coverage? Would it be better to spend the large sums of money needed for satellites on large numbers of 2- or 3-m telescopes on the ground to do the same kind of intensive monitoring?

Bennett: I have a poster at this meeting describing one of these space missions intended to search for microlensing. It is actually more cost-effective to go into space if you are searching for planets of less than Earth mass. Part of the reason concerns the requirement for continuous coverage. On Earth you have to observe at several longitudes or go to the South Pole, and not all locations are good observing sites, so you don't get 24-hour coverage anyway. Then, to search for low-mass planets you need to observe main sequence stars, which are faint, with good photometric precision, which is very difficult to do.

Mayor: A photometric accuracy of 10^{-4} mag is talked about in connection with proposed space missions like COROT, Kepler, etc, compared with only about 10^{-2} mag from the ground. Even with this high precision, are there still problems with interpreting the light curves?

Gould: I suspect not, but I don't really know.

Mayor: Given that this factor of 10^2 improvement in photometric precision is available, why do you not apply for target of opportunity time for events known to be in progress based on ground-based observations?

Bennett: I obtained HST target of opportunity time for the longest timescale event that has ever been observed. I tried to get this with a turnaround time of a few weeks, but that turned out not to be feasible. The kind of turnaround time needed for target of opportunity planet follow-up is much shorter than that and is just not practical. Besides, the false alarm rate is unlikely to go below about 90%. The HST is just not set up work in the required kind of target of opportunity mode.

Popowski: On the subject of precise photometry from spacecraft: all the missions planned to do this kind of work will have rather bright limiting magnitudes, which will be completely unsuitable for microlensing work.

Gould: Where binaries are concerned we would go a long way towards eliminating degeneracies by having continuous coverage of the light curves. The rest could be solved by astrometry. For the MACHO binaries, there are uncertainties that are often due to poor sampling or data quality. In the PLANET data there are a number of objects that are probably binaries about which we won't be able to say much, again because of the coverage. In one event, MACHO 98-SMC-1, when the data from several groups were combined it proved more or less possible to resolve all the degeneracies, though not completely. With the PLANET data alone, we found a huge sea of degenerate solutions. I think the problem is soluble, but I think we will have to reorganise the way we do microlensing surveys. To generate a good statistical database we need dedicated telescopes on the ground that look at the same spot repeatedly searching for planets. We

Discussion Session I: Binaries and Planets   171

would still need space observations to get masses rather than mass ratios.

Reid: Does the microlensing community see planet and binary detection as a major priority, now that dark matter detection seems to fading into the background?

Gaudi: Microlensing has divided itself into two areas, one of which is the continuing search for dark matter and understanding what is going on towards the LMC. The other is the general astrophysical application of microlensing, not only searching for planets, but also stellar atmospheres, detection of black holes, and even using it as a giant magnifying glass allowing us to take spectra of objects not otherwise accessible. In some quarters microlensing is seen as some kind of 'cult', but we see it as a tool that can be used for a variety of projects.

Cook: The community, if it does exist, has to get out and raise a lot of money because all of the projects and hopes revolve around new 2-m telescopes or continuing surveys. We have heard that EROS is ending in 2002, MACHO has already ended, and although OGLE would like to carry on, the future is uncertain. We need to make our case in a better way, and more clearly, and preferably to speak with one voice.

Reid: I agree. The case, and I believe there is a good one, needs to be made more strongly – and preferably with one voice.

Zinnecker: It is worth considering using the NGST, a 6- to 8-m space telescope projected for about 2010. It will have a resolution of 50 mas at 2μm, allowing it to look right into the Galactic Centre where there are many potential source stars, and crowding is not a problem. It is expected to operate in a kind of staring mode, where it will observe a particular field for a week or so. It will have an 8k x 8k detector, which may not be large enough, but at least it is worth considering for monitoring a field for some time to detect and follow events.

Gould: This infrared idea keeps coming up. I proposed in 1995 that it would be a good thing to do. I still think it is worth considering. The idea was to look at the Galactic Centre, not for the planet detection, but to take advantage of the very short timescale events resulting from Bulge-Bulge lensing to give information about Galactic structure rather than planets. The problem is really one of cost. Optical arrays are much cheaper than infrared ones, so new projects are mostly going the optical route where it is relatively cheap to have billions of pixels. It really depends on the state of infrared detector technology and probably needs some interest from the military to precipitate more research work.

Bradbury: There is a fair amount of progress in micro-bolometer research by a group in Texas. This is a mid-infrared detector array that relies on changes of resistance due to heating by infrared radiation. It may not be possible to get millions of pixels.

Evans: To return to an earlier theme, it is important to give surprises to the rest of the astronomical community. The main surprises have come from Galactic microlensing where the results towards the Galactic Centre and the LMC have been difficult to interpret. When we go towards M31 it is likely there will be more surprises. It is by no means certain that the baryon fraction will not change as a function of Hubble type – everything else does. It is possible that the results for M31 will differ from what we have found in the Galaxy.

Part 3: Stellar Astrophysics

Stellar Atmospheres

Peter H. Hauschildt
Dept. of Physics and Astronomy & Center for Simulational Physics, University of Georgia, Athens, GA 30602-2451

France Allard
C.R.A.L (UML 5574) Ecole Normale Superieure, 69364 Lyon Cedex 7, France

Jason Aufdenberg
Dept. of Physics and Astronomy, Arizona State University, Tempe, AZ 852870-2504

Travis Barman & Andreas Schweitzer
Dept. of Physics and Astronomy, University of Georgia, Athens, GA 30602-2451

E. Baron
Dept. of Physics and Astronomy, University of Oklahoma 440 W. Brooks, Rm 131, Norman, OK 73019-0225

Abstract. We give an overview about the state-of-the-art in stellar (and substellar) atmosphere simulations. Recent developments in numerical methods and parallel supercomputers, as well as advances in the quality of input data such as atomic and molecular line lists have led to substantial improvements in the quality of synthetic spectra when compared to multi-wavelength observations. A wide range of objects from giants to main-sequence stars down to substellar objects is considered. We discuss effects such as limb darkening, atomic and molecular NLTE (and) line blanketing, winds, external irradiation, formation and opacities of dust particles and clouds; each of these affects the structure of the atmospheres and their spectra. Current models can simultaneously fit many of the observed features of a given star with a single model atmosphere. However, a number of problems remain unsolved and will have to be addressed in the future, in particular for very low mass stars and substellar objects.

1. Introduction

Stellar atmosphere modeling has experienced a renaissance in the past decade with the advent of better algorithms and faster computers. This has allowed research groups to remove or relax many of the "standard" assumptions that were made in the 70s through 80s and that had become accepted wisdom over the years. Surprisingly (or not) the new calculations show that many of these assumptions are actually quite bad and can lead to spurious results or incorrect interpretations of observed spectra. The intricate connection between geometry (plane parallel or spherical), line blanketing (atomic and/or molecular) and non-LTE effects (using small to extremely large model atoms and molecules) began to emerge slowly as crucial ingredients for physically correct and meaningful interpretations and analyses of stellar spectra. Unfortunately, easy and simple solutions do not really work for stellar atmospheres (although everybody likes the easy way out and some of them are useful for teaching purposes) and have actually hindered progress and reduced the reliability of results.

Table 1. Selected molecules considered in the EOS

NH	C_2	CN	CO	MgH	CaH	SiH	TiO	H_2O	H_2
N_2	NO	CO_2	O_2	ZrO	VO	MgS	SiO	AlH	HCl
HF	HS	TiH	AlO	BO	CrO	LaO	MgO	ScO	YO
SiF	NaCl	CaOH	HCN	C_2H_2	CH_4	CH_2	C_2H	HCO	NH_2
LiOH	C_2O	AlOF	NaOH	MgOH	AlO_2	Al_2O	AlOH	SiH_2	SiO_2
H_2S	OCS	KOH	TiO_2	TiOCl	VO_2	FeF_2	YO_2	ZrO_2	BaOH
LaO_2	C_2H_4	C_3	SiC_2	CH_3	C_3H	NH_3	C_2N_2	C_2N	CaF_2
AlOCl	Si_2C	CS_2	$CaCl_2$	AlF	CaF	Si_2	SiS	CS	AlCl
KCl	CaCl	TiS	TiCl	SiN	AlS	AL_2	FeO	SiC	TiF_2
FeH	LiCl	NS	NaH	SO	S_2	$AlBO_2$	AlClF	$AlCl_2$	AlF_2
$AlOF_2$	AlO_2H	Al_2O_2	$BeBO_2$	OBF	HBO	HBO_2	HBS	BH_2	BO_2H
BH_3	H_3BO_3	KBO_2	$LiBO_2$	$NaBO_2$	BO_2	$BaCl_2$	BaF_2	BaO_2H_2	BaClF
$BeCl_2$	BeF_2	BeOH	BeH_2	BeH_2O_2	Be_2O	Be_3O_3	ClCN	ChCl	CHF
CHP	CH_3Cl	KCN	NaCN	BeC_2	C_2HCl	C_2HF	$(NaCN)_2$	C_4	C_5
CaO_2H_2	MgClF	SiH_3Cl	$FeCl_2$	K_2Cl_2	$MgCl_2$	Na_2Cl_2	$TiOCl_2$	$SrCl_2$	$TiCl_2$
$ZrCl_2$	$TiCl_3$	$ZrCl_3$	$ZrCl_4$	CrO_2	SrOH	SiH_3F	OTiF	SiH_2F_2	MgF_2
ZrF_2	TiF_3	ZrF_4	FeO_2H_2	$(LiOH)_2$	$(KOH)_2$	$(NaOH)_2$	MgO_2H_2	$(NaOH)_2$	SrO_2H_2
PH_2	PH_3	SiH_4	Si_2N	PO_2	SO_2	P_4	Si_3	NO_2	NO_3
C_3N	C_2H_3	C_4H	HC_3N	C_4H_2	CH_3CN	HC_5N	C_6H	C_4H_2	C_6H_2
HC_7N	C_4H_4S	C_4H_4O	C_4H_6	C_6H_4	HC_9N	C_5H_5N	C_6H_5O	C_6H_6	C_6H_6O
$HC_{11}N$	OH^-	CH^-	C_2^-	OH	CH	CN	SiH^-	H_2^-	HS^-
CS^-	FeO^-	BO^-	$AlCl_2^-$	AlF_2^-	$AlOF_2^-$	$AlOH^-$	CO_2^-	NO^+	H_2^+
TiO^+	ZrO^+	$AlOH^+$	$BaOH^+$	HCO^+	$CaOH^+$	$SrOH^+$	H_3O^+	H_3^+	

New observational techniques opened and continue to open up new areas of stellar atmosphere research. Most importantly of these are the observations of very low mass stars and, after decades of searching, brown dwarfs and extrasolar giant planets. Modeling these objects requires sophisticated stellar atmosphere type modeling with complex equations of state and hundreds of millions of molecular spectral lines in order to even approximately reproduce the observed spectra. These new observations have prompted further evolution of stellar atmosphere modeling and helped rejuvenate the field in general.

In the following we will briefly introduce the numerical methods that modern stellar atmosphere research employs (there are also plenty of legacy applications and codes that are still widely used) and then discuss some results that are of interest in the context of this meeting.

2. Methods and Models

For our model calculations, we use our multi-purpose stellar atmosphere code PHOENIX (version 10.9 Hauschildt, Baron, & Allard, 1997; Baron & Hauschildt, 1998; Hauschildt, Allard, & Baron, 1999; Hauschildt et al., 1999; Hauschildt & Baron, 1999). Details of the numerical methods are given in the above references, so we do not repeat the description here.

Table 2. Selected liquid/dust species considered in the EOS

Al/l	B/l	Ba/l	Be/l	Ca/l	Cr/l	Cu/l	Fe/l	K/l
Li/l	Mg/l	Mn/l	Na/l	Nb/l	Ni/l	P/l	S/l	Si/l
Sr/l	Ti/l	V/l	Zn/l	Zr/l	BeO/l	ClK/l	NbO/l	OSr/l
ClNa/l	VO/l	B_2Ti/l	$BaCl_2$/l	$CaCl_2$/l	Cl_2Fe/l	Cl_2Sr/l	O_2Si/l	Li_2O/l
Mg_2Si/l	Cu_2O/l	Cl_3Fe/l	Cr_2O_3/l	NiS_2/l	$BLiO_2$/l	Cl_2S_2/l	Ni_3S_2/l	Al_2O_3/l
O_3V_2/l	Cl_5Nb/l	Nb_2O_5/l	$B_4K_2O_7$	$B_4Na_2O_7$	Li_2O_3Si	$B_4Li_2O_7$	$Mg_3O_8P_2$	$Al_3F_{14}N$
B_5H_9/l	$H_{10}O_8S$	$B_8K_2O_{13}$	$B_{10}H_{14}$	Al	B	Ba	Be	C
Ca	Co	Cr	Cu	Fe	Li	Mg	Mn	Na
Nb	Ni	P	S	Si	Sr	Ti	V	Zn
Zr	MgO	FeS	CaO	CaS	MgS	TiN	AlN	NiS
MnS	TiO	VO	CuO	FeO	TiC	SiC	ZrC	H_2O
TiO_2	ZrO_2	SiO_2	FeS_2	NiS_2	Mg_3N_2	Ni_3S_2	Ti_2O_3	Ti_3O_5
Ti_4O_7	V_2O_3	Al_2O_3	Al_2O_3	Al_2O_3	Al_2O_3	Al_2S_3	Cr_2O_3	$CaTiO_3$
$MgTiO_3$	$MgSiO_3$	$CaSiO_3$	$MnSiO_3$	Na_2SiO_3	K_2SiO_5	Fe_2SiO_4	Ca_2SiO_4	Mg_2SiO_4
$ZrSiO_4$	Fe_2O_3	Fe_3O_4	$MgAl_2O_4$	$MgTi_2O_5$	Al_2SiO_5	$CaMgSi_2$	Ca_2MgSi	Ca_2Al_2S
$CaAl_2Si$	$KAlSi_3O$	$NaAlSi_3$	Al_6Si_2O	MgC_2	Cr_3C_2	Mg_2C_3	Al_4C_3	Cr_7C_3
$Cr_{23}C_6$								

Table 3. Complete listing of PHOENIX (version 10) atomic NLTE species. The table entries are of the form N/L, where N is the number of NLTE levels and L the number of primary NLTE lines for each model atom.

	I	II	III	IV	V	VI	VII
H	80/3160						
He	19/37	10/45					
Li	57/333	55/124					
C	228/1387	85/336	79/365	35/171			
N	252/2313	152/1110	87/266	80/388	39/206	15/23	
O	36/66	171/1304	137/765	134/415	97/452	39/196	
Ne	26/37						
Na	53/142	35/171	69/353	46/110	64/187	102/375	
Mg	273/835	72/340	91/656	54/169	53/133	78/180	
Al	111/250	188/1674	58/297	31/142	49/77	40/93	
Si	329/1871	93/436	155/1027	52/292	35/125	36/49	
P	229/903	89/760	51/145	50/174	40/204	10/9	
S	146/439	84/444	41/170	28/50	19/41	31/144	
K	73/210	22/66	38/178	24/57	29/75		
Ca	194/1029	87/455	150/1661	67/122	39/91	23/37	26/59
Ti	395/5279	204/2399					
Fe	494/6903	617/13675	566/9721	243/2592	132/961	87/551	
Co	316/4428	255/2725	213/2248				
Ni	153/1690	429/7445	259/3672	189/1845	245/2638	246/2868	
Rb	29/76						
Sr	52/74	32/90					
Cs	38/75						
Ba	76/114	51/121					
Total	10991/103196						

One of the most important recent improvements of cool stellar atmosphere models is that new molecular line data have become available. This includes the addition of the HITRAN92 (Rothman et al., 1992) data base, first incorporated in our "Extended" model grid (Allard & Hauschildt, unpublished). For the molecules with lines available from other sources (e.g. water vapor and CO) it is usually better to use other sources in model calculations because the HITRAN92 lists are fairly incomplete (they have been prepared for the conditions of the Earth's atmosphere) although the line data are of very high quality. The OH molecule is an exception because the vibrational bands present in the HITRAN92 database are *not* present in the Kurucz CD 15 (Kurucz, 1993) dataset listing electronic transitions of OH. Therefore, we use both HITRAN92 and CD 15 OH line data simultaneously in the model calculations.

The most recent CO line list was calculated by Goorvich (Goorvitch & Chackerian, 1994a,b). This list is more accurate than older CO line data. How-

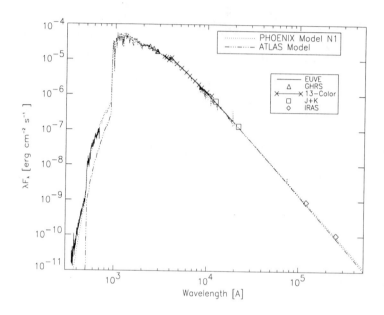

Figure 1. The spectral energy distribution of ϵ CMa from 350 Å to 25 μm compared with the 21750 K non-LTE model (N1) and the ATLAS9 21000 K model. The $EUVE$ data are corrected for the interstellar neutral hydrogen column assuming a neutral hydrogen column density of $N(H\,\textsc{i}) = 5 \times 10^{17}$ cm^{-2}.

ever, we have kept the electronic CO transitions available on Kurucz's CD 15. Similarly, we have replaced the Kurucz CN line list (CD 15) with a more recent calculation by Jørgensen (Jørgensen & Larsson, 1990). CN lines are comparatively weak in most of the models presented here. In addition, we include line lists for YO (Littleton, 1987) and ZrO (Littleton & Davis, 1985) as well as H_3^+ partition functions and lines (Neale & Tennyson, 1995; Neale, Miller, & Tennyson, 1996).

A long-standing problem with M dwarf models was that TiO line lists were incomplete at high temperatures. The use of "straight means" (AH95 models) helped to block flux which otherwise escapes between lines in the incomplete list. But these models also blocked too much flux in most cases, and were only appropriate in late-type M dwarfs when TiO bands are very strong. A more complete line list was needed to model stars from the onset of TiO formation to its gradual disappearance from the gas phase in brown dwarfs. The AMES-TiO list (Schwenke, 1998) largely resolved this issue. We found that the list provides more opacity in most bands and adequately suppresses flux between

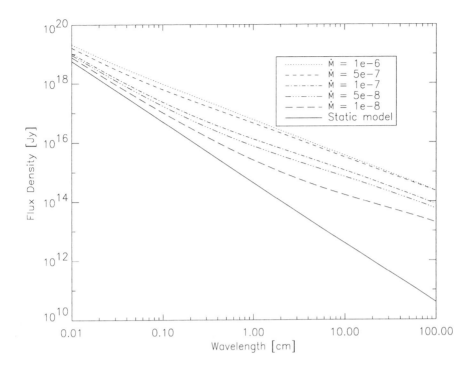

Figure 2. A comparison of synthetic radio spectra from A-type supergiant, expanding model atmospheres which have different mass-loss rates but otherwise fixed parameters (T_{eff} = 9500 K, $\log(g)$ = 1.5, v_∞ = 260 km s^{-1}, β-law velocity field with β = 3.0). Mass-loss rates have units of $M_\odot$ yr^{-1}.

bands (Allard, Hauschildt, & Schwenke, 2000). New, smaller oscillator strength values also play an important role by systematically assigning cooler models (at least for early type M dwarfs) to a given star, and in this way contribute to broader bands and lower inter-band flux as well. The combination of these effects helps to resolve most of the previously observed discrepancy between models and observations in the optical spectral energy distribution (SED) and photometry of M stars. Leinert et al. (1999) note, however, that flux excess remains substantial in the visual spectrum, suggesting some further incompleteness or f_{el} inaccuracies of the new TiO in the a-f system.

The absorption coefficient of water has been a problem for modelers of cool dwarf atmospheres for decades. In our Base model grid we used the straight means of the Ludwig (1971) water opacity tables. These tables are known to overestimate the water opacity significantly at higher (and more importantly for M dwarfs) temperatures (Schryber, Miller, & Tennyson, 1995), but they are accurate at lower temperatures. The currently available list of water lines of Miller & Tennyson (MT-H$_2$O, Allard et al., 1994) is incomplete and therefore

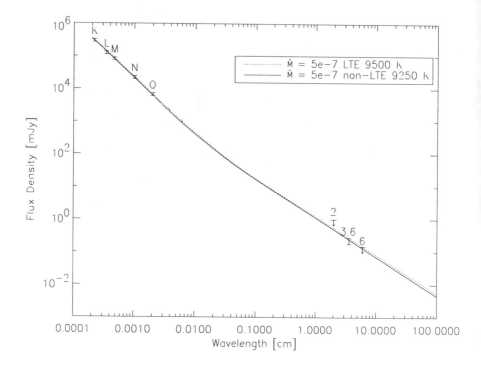

Figure 3. Synthetic continua compared with IR spectophotometric data (Abbott, Telesco, & Wolff, 1984), the 3.6 cm detection (Howarth, 1999), and radio upper limits at 2 cm and 6 cm (Drake & Linsky, 1989) for Deneb. A higher degree of ionization in the cooler model (T_{eff} = 9250 K) with non-LTE metal line-blanketing produces a similar synthetic radio spectrum to the hotter model (T_{eff} = 9500 K) with LTE metal-line blanketing.

some water opacity will be missing in models with $T_{\text{eff}} < 4000$ K. Comparison with high-resolution observations in spectral regions where the MT-H$_2$O list should be complete showis that its accuracy is high (Jones et al., 1995). For most cool star models presented here, we have replaced the UCL water vapor line-list (Miller et al., 1994; Schryber, Miller, & Tennyson, 1995) used in Hauschildt, Allard, & Baron (1999) with the AMES water line-list (Partridge & Schwenke, 1997, hereafter: AMES-H$_2$O). This list includes about 307 million lines of water vapor. The introduction of the AMES-H$_2$O opacities brings solid improvements of the near-infrared SED of late-type dwarfs, but fails, as the AH95 models did, to reproduce adequately the $J - K$ colors of hotter stars. Water vapor is a more important factor for the structure of the atmosphere than TiO because its overall opacity is larger and its lines are closer to the peak of the SED than the TiO bands, so the flux blocking effect of water vapor is more important for the temperature structure than that of TiO opacities for these low temperatures.

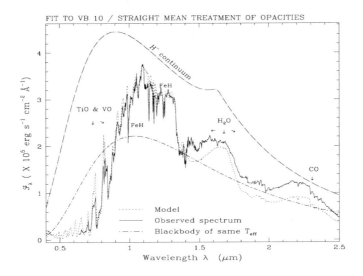

Figure 4. Best fit of Allard & Hauschildt (1995b) to the spectrum of the dM8e star VB 10 (Allard & Hauschildt, 1995a). The corresponding H^- continuum obtained by neglecting molecular opacities only in the radiative transfer (long dot-dashed) reveals the magnitude of these opacities in a typical late-type M dwarf. The Planck distribution of the same $T_{\rm eff}$ is also shown for comparison. From Allard et al. (1997).

Schwenke and collaborators at NASA AMES are preparing a new dipole moment function for H_2O, which may change the high temperature high overtone water bands, and help resolve this discrepancy in the near future. Until a revised version of the AMES-H_2O line list becomes available, we will maintain two sets of model atmospheres for cool stars which allow the investigation of these issues: the AMES grid based on the new TiO and H_2O lists, and the AMES-MT grid which rely on the AMES-TiO and Schryber, Miller, & Tennyson (1995) line lists.

Our combined molecular line list includes about 550 million molecular lines. The lines are selected for every model from the master line list at the beginning of each model iteration to account for changes in the model structure (see below). Both atomic and molecular lines are treated with a direct opacity sampling method (dOS). We do *not* use pre-computed opacity sampling tables, but instead dynamically select the relevant LTE background lines from master line lists at the beginning of each iteration for every model and sum the contribution of every line within a search window to compute the total line opacity at *arbitrary*

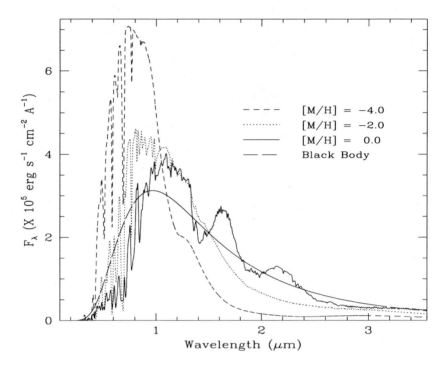

Figure 5. Spectral distributions of emerging fluxes at the stellar surface for 3,000 K models with metallicities corresponding roughly to the solar neighborhood ([M/H] = 0.0), halo ([M/H] = −2.0), and Population III ([M/H] = −4.0) stars. A black-body of the same effective temperature (smooth curve) is shown for comparison. From Allard et al. (1997).

wavelength points. The latter feature is crucial in NLTE calculations in which the wavelength grid is both irregular and variable from iteration to iteration due to changes in the physical conditions. This approach also allows detailed and depth-dependent line profiles to be used during the iterations. Although the direct line treatment seems at first glance computationally prohibitive, it can lead to more accurate models. This is due to the fact that the line forming regions in cool stars and planets span a huge range in pressure and temperature so that the line wings form in very different layers than the line cores. Therefore, the physics of the line formation is best modeled by an approach that treats the variation of the line profile and the level excitation as accurately as possible. To make this method computationally more efficient, we employ modern numerical techniques, e.g. vectorized and parallelized block algorithms with high data locality (Hauschildt, Baron, & Allard, 1997), and we use high-end workstations or parallel supercomputers for the model calculations.

Stellar Atmospheres 183

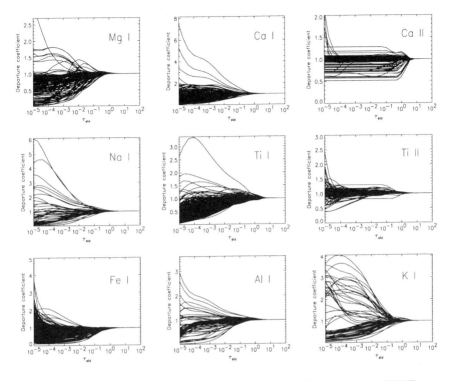

Figure 6. Overview over selected departure coefficients for a NLTE model with $T_{\mathrm{eff}} = 4000\,\mathrm{K}$, $\log(g) = 0.0$, and solar abundances.

In the calculations presented in this contribution, we have included a constant statistical velocity field, $\xi = 2\,\mathrm{km\,s^{-1}}$, which is treated like a microturbulence. The choice of lines is dictated by whether they are stronger than a threshold $\Gamma \equiv \chi_l/\kappa_c = 10^{-4}$, where χ_l is the extinction coefficient of the line at the line center and κ_c is the local b-f absorption coefficient (see Hauschildt, Allard, & Baron, 1999, for details of the line selection process). This typically leads to about $10 - 250 \times 10^6$ lines which are selected from master line lists. The profiles of these lines are assumed to be depth-dependent Voigt or Doppler profiles (for very weak lines). Details of the computation of the damping constants and the line profiles are given in Schweitzer et al. (1996). We have verified in test calculations that the details of the line profiles and the threshold Γ do not have a significant effect on either the model structure or the synthetic spectra. In addition, we include about 2000 photo-ionization cross sections for atoms and ions (Mathisen, 1984; Verner & Yakovlev, 1995).

The equation of state (EOS) is an enlarged and enhanced version of the EOS used in AH95. We include about 500 species (atoms, ions and molecules) in the EOS, cf Table 1. This set of EOS species was determined in test calculations. The EOS calculations themselves follow the method discussed in AH95. For effective temperatures, $T_{\mathrm{eff}} < 2500\,\mathrm{K}$, the formation of dust particles has

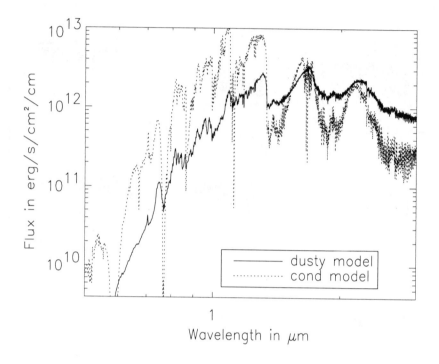

Figure 7. This plot shows the difference between synthetic spectra calculated for a model atmospheres assuming complete settling of all formed dust particles below the layers where spectrum forms ("cond" model) and for a model that assumes that the dust particles remain close to the layer in which they formed ("dusty" model) for $T_{\rm eff} = 1700\,{\rm K}$, $\log(g) = 4.5$ and solar abundances.

to be considered in the EOS. In our models we allow for the formation (and dissolution) of a variety of grain species, see Table 2 for a selection.

The NLTE treatment of large model atoms or molecules such as H_2O and TiO which have several million transitions is a formidable problem which requires an efficient method for the numerical solution of the multi-level NLTE radiative transfer problem. Classical techniques, such as the complete linearization or the Equivalent Two Level Atom method, are computationally prohibitive for large model atoms and molecules. Currently, the operator splitting or approximate Λ-operator iteration (ALI) method (e.g., Cannon, 1973; Rybicki, 1972, 1984; Scharmer, 1984) seems to be the most effective way of treating complex NLTE radiative transfer and rate equation problems. Variants of the ALI method have been developed to handle complex model atoms, e.g. Anderson's multi-group scheme (Anderson, 1987, 1989) or extensions of the opacity distribution function method (Hubeny & Lanz, 1995). However, these methods have problems if line overlaps are complex or if the line opacity changes rapidly with optical depth,

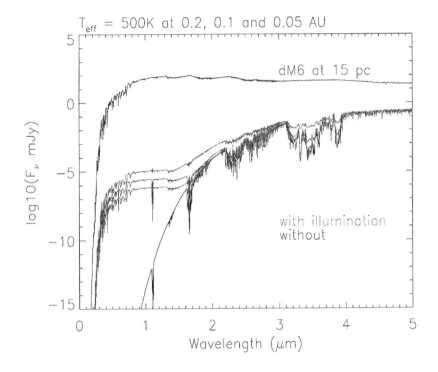

Figure 8. Effects of irradition of a dusty $T_{\text{eff}} = 500\,\text{K}$, $\log(g) = 3.5$, $R = 0.1 R_\odot$ giant planet by a dM6 star ($T_{\text{eff}} = 3000\,\text{K}$, $\log(g) = 5.0$, $R = 0.1 R_\odot$) for varying distances between the star and the planet (as indicated). The abundances are assumed to be solar and the spectra show the irradiated hemisphere of the planet.

a situation which occurs in cool stellar atmospheres. The ALI rate operator formalism, on the other hand, has been used successfully to treat very large model atoms such as Fe directly and efficiently (cf. Hauschildt & Baron, 1995; Hauschildt et al., 1996; Baron et al., 1996). It allows us to currently treat the species listed in Table 3 in direct NLTE.

3. Results

In the following sections we will give a few representative results that highlight important new developments in stellar atmospheres.

3.1. Hot Stars

Historically, the assumption of plane parallel geometry was thought to be a good approximation for main sequence stars and many models for giants and supergiants did not use spherical geometry. However, a closer look on the effects of spherical geometry coupled to line blanketing shows that the situation is far

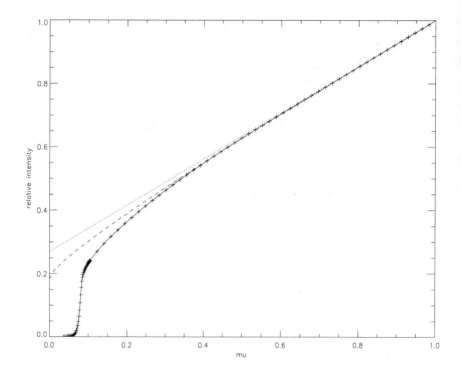

Figure 9. Example of limb darkening for a subgiant model (calculated using spherical geometry and spherically symmetric radiative transfer) with $T_{\text{eff}} = 5000\,\text{K}$, $\log(g) = 2.5$ and solar abundances. The computed (normalized to the center of the stellar disk) monochromatic intensities at $\approx 5000\text{Å}$ are shown (+ symbols) and compared to linear (dotted line) and square root (dashed line) limb darkening laws.

more complex. As an example, we consider the EUV spectrum of the B2 II star ϵ CMa. Spectroscopy of ϵ CMa (HD 52089, HR 2618) (Cassinelli et al. (1995)) below the Lyman edge by the *Extreme Ultraviolet Explorer (EUVE)* satellite is possible due to the extremely low neutral hydrogen column density toward this star. At a *HIPPARCOS* distance of 132^{+11}_{-9} pc along an exceptionally rarefied interstellar tunnel extending out from the Local Bubble (Welsh, 1989), ϵ CMa is attenuated by a neutral hydrogen column density of less than 5×10^{17} cm^{-2} (Gry, York, & Vidal-Madjar, 1985). Consequently, ϵ CMa is an extremely important star for its contribution to the local interstellar hydrogen ionization, producing more hydrogen ionizing flux than all nearby stars combined (Vallerga & Welsh, 1996). The large EUV flux from ϵ CMa affects the ionization state of the Local Cloud, the region of neutral hydrogen concentrated within a few parsecs of the Sun, in which the solar system is embedded (Bruhweiler (1996)). However, the EUV spectrum of ϵ CMa could not be fit using standard plane parallel model atmospheres; the observed spectrum displays an apparent EUV excess compared to such models (Fig. 1).

We found significant differences in the strength of the predicted EUV flux between our line blanketed *spherical* and line blanketed plane-parallel models. A more realistic model treatment of early B giants with a spherical geometry and NLTE metal line blanketing results in the prediction of significantly larger EUV fluxes compared with plane-parallel models. This result appears to explain a large part of the reported discrepancy between the observed EUV flux of early B giants such as ϵ CMa and the EUV flux predicted by plane-parallel LTE and NLTE line blanketed model atmospheres. It is important to note that the relative extension (e.g. the ratio of the outer to the inner radius of the model atmosphere) is small. This effect on the spectrum is the result of a change in the temperature structure of the spherical models compared to the plane parallel models and only occurs when line blanketing (and thus backwarming) is included (see Aufdenberg et al., 1998, 1999, for details).

The mass-loss rates of OB supergiants are often approximated using a fully ionized, uniform, spherically symmetric mass flow model where the terminal velocity and radio flux density are measured quantities (Scuderi et al., 1998; Drake & Linsky, 1989). This model predicts a frequency dependence for the flux density of $S_\nu \propto \nu^{0.67}$. The winds of A-supergiants are, however, only partially ionized. Thermal radio emission from partially ionized circumstellar environments and winds has been studied by Simon et al. (1983). They show that the power-law radio spectrum varies from the fully ionized, optically thin case ($S_\nu \propto \nu^{0.6}$) to the optically thick, Rayleigh-Jeans case ($S_\nu \propto \nu^2$). The steepness of the radio spectrum depends on the degree of ionization and radial extent of the ionized envelope. As the extended envelope becomes less ionized, the spectral slope steepens and the radio flux drops. Our extended, expanding model atmospheres assume neither a constant temperature distribution nor constant ionization.

Figure 2 shows model IR-radio spectra of A-type supergiants with a range of mass-loss rates from zero to 10^{-6} $M_\odot$ yr^{-1}. The frequency dependence of the synthetic radio spectra differ significantly from $S_\nu \propto \nu^{0.67}$. The static model shows the expected optically thick Rayleigh-Jeans spectrum ($S_\nu \propto \nu^2$) while the wind model spectra beyond ~ 0.1 cm follow $S_\nu \propto \nu^{1.1}$. The flux density longward of 0.01 cm increases significantly with $\dot{M}$ because the total density, including the ion density, is increased in the wind.

We also find that departures from LTE are quite important. Predicted flux densities in LTE are about an order of magnitude lower than corresponding non-LTE predictions with otherwise identical model parameters. In the non-LTE models hydrogen is ionized out to a radius ~ 4 times larger than in the LTE models. The temperature structure at the formation depths of the radio continuum is quite insensitive to T_{eff}. As a result, the radio flux densities predicted by the LTE models, where the ionization structures are determined purely by the local electron temperature, are also quite insensitive to T_{eff}. In contrast, temperature structure in the hydrostatic zones immediately below the wind *is* sensitive to T_{eff} and in non-LTE, it is the Balmer continuum radiation field from these hydrostatic layers which controls the degree of hydrogen ionization in the outer wind.

For the A-type supergiant Deneb, IR spectrophotometry, the radio measurement at 3.6 cm, and upper limits at 2 cm and 6 cm are compared with two synthetic radio spectra in Figure 3. The non-LTE model, which is our best fitting model to the UV-optical-IR SED, predicts radio fluxes consistent with the upper limits and the 3.6 cm detection. While the LTE model has a very similar radio spectrum and provides a good fit to Deneb's UV-optical continuum, comparisons with the observed high-dispersion UV spectrum reveal that the non-LTE model is the overall better representation of Deneb's SED.

3.2. Cool Stars

Line blanketing The number of molecular lines that are important in M-dwarf (and later) atmospheres are quite large. About 215 million molecular lines are selected (see above) for a typical giant model with $T_{\rm eff} \approx 3000\,{\rm K}$ whereas about 130 million molecular lines have to be considered for a dwarf model with the same effective temperature. The large "density" (in wavelength space) of molecular lines causes large line blanketing effects as illustrated in Figure 4 for a very simple case. The nearly complete coverage of the optical spectrum by TiO lines and of the near IR spectrum by H_2O lines effectively locks the peak of the spectral energy distribution in place at around $1.1\mu m$ even for substantially different $T_{\rm eff}$, in stark contrast to the behavior expected for blackbodies. Line blanketing also produces a strong metallicity effect on the spectra as illustrated in Figure 5. Lowering the metal abundances reduces both the TiO and H_2O opacities by roughly the same amount. However, the H_2O opacity in the near IR is replaced by increasingly (with lower metallicity) stronger collision-induced opacities (due to the larger pressures in the spectrum forming regions). Therefore, the spectrum gets *bluer* with lower metallicities, even for comparatively low effective temperatures.

NLTE effects Due to their very low electron temperatures, the electron density is extremely low in M stars; the absolute electron densities are even lower than found in low density atmospheres, such as those of novae and SNe. Collisional rates due to collisions with electrons, which tend to restore LTE, are thus very small in cool stars. This in turn could significantly increase the importance of NLTE effects in M stars when compared to, e.g., solar type stars with much higher electron densities and temperatures. Collisions with molecular hydrogen and helium will at least in part compensate for the diminished electron collisions, but cross-sections for these processes are not very well known. Therefore, the assumption of LTE for atoms and molecules in cool stars is by no means certain and needs to be verified for each species individually. Therefore, we have performed test calculations to place an upper limit on the importance of atomic NLTE effects in cool stars by only considering electron collisions.

We have calculated a small number of NLTE models in order to investigate the importance of NLTE effects on the structure of the model atmospheres. The results for cooler models were discussed in Hauschildt et al. (1997) and are not repeated here. Figure 6 shows an overview of selected NLTE species for models with $T_{\rm eff} = 4000\,{\rm K}$, $\log(g) = 0.0$ and solar abundances. The total number of NLTE levels in each model is 4532 with a total of 47993 primary NLTE lines (see Hauschildt & Baron, 1999; Hauschildt et al., 1999, and references therein

Stellar Atmospheres   189

for details). The following species (and number of levels) were treated in NLTE:
H I (30), Mg I (273), Mg II (72), Ca I (194), Ca II (87), Fe I (494), Fe II (617),
O I (36), O II (171), Ti I (395), Ti II (204), C I (228), C II (85), N I (252),
N II (152), Si I (329), Si II (93), S I (146), S II (84), Al I (111), Al II (188), K I
(73), K II (22), Na I (53), and Na II (35). For most of the species, the departure
coefficients are always close to unity, in particular for species with resonance
lines and photoionization edges in the UV part of the spectrum. The species
shown in Figure 6 are species with the most pronounced departures from LTE.
The departures are generally too small to significantly affect the structure of the
atmospheres. Results for NLTE calculations for the CO molecule show that the
high cross-sections of H_2 and He collisions restore LTE very successfully in the
case of dwarfs stars (Schweitzer, Hauschildt, & Allard, 2000).

Dust and cloud formation   The effects of dust formation on the atmosphere
are mainly (a) the removal of important opacity sources (e.g., TiO, VO) from
the gas phase and a corresponding weakening of their spectral lines; and (b) the
presence of additional opacities produced by the grains themselves. The latter
depends on the behavior of the macroscopic dust particles:

a) They might remain in the layers where the dust originally formed and thus
   cause strong optical and IR opacities ("Dusty" models).

b) They could rain out and settle below the line and continuum forming
   regions, in this case no grain opacities would be detectable in the spectrum
   ("Cond" models).

c) They can form clouds in the atmosphere so that dust opacities would only
   be present in the cloud layers but not in all the layers where the dust
   originally formed ("Cloudy" models).

Observational evidence suggests that case (a) is realized for $T_{\text{eff}} > 1700\,\text{K}$
(late M dwarfs to early L dwarfs) whereas for both giant planets and extreme
T dwarfs case (b) appears to be more appropriate. In the intermediate regime,
clouds appear to form and case (c) is the best approximation. Figure 7 illustrates the general trend for an example model atmosphere. At the present time,
this sequence is still very tentative and the physical models of dust and cloud
formation have to be refined to obtain a truly physical picture of brown dwarfs
and extrasolar giant planets.

Irradiation   One feature of the extrasolar giant planets that have been discovered so far is that they are close to a more massive and hotter companion. Under
such conditions, the structure of the atmosphere of the cooler companion (secondary) will be strongly affected by the presence of the strong radiation field
of the hotter (primary) component. This will give rise to strong NLTE effects
in the secondary's atmosphere. If the separation of the two components is relatively large (but still small enough that the primary's radiation field has an
effect on the secondary), the calculation of an irradiated atmosphere is straightforward. If the separation is small, then the irradiated part of the secondary's
surface can be approximated by a patchwork of plane parallel surface elements.
In the case of cool stars with strong convection in the lower atmosphere, the situation is somewhat more complex and entropy-matching procedures need to be

performed (Brett & Smith, 1993). Ultimately, the radiation transfer problem in the case of an irradiated atmosphere has to be solved using a full 3D approach. This is presently not feasible with the same complex input physics that can be included in 1D models, but simplified 3D models are possible and over the next decade or so computers should become powerful enough to handle the full 3D problem.

An important question concerns the *input* radiation field used in the irradiation modeling. A simple approximation would be to use blackbody radiation fields with a given color temperature. However, stellar radiation fields are *very* different from Planckian fields, in particular in the optical and UV spectral region where the differences can be orders of magnitude because of the very complex wavelength dependence of the stellar opacity. Blackbody radiation fields will only be useful for the simplest tests. Therefore, we calculate the input radiation fields with PHOENIX and thus have full control over the wavelength and angular grid on which the input field will be available. As an example, we show in Figure 8 the effects of irradiation on the emitted spectrum of a $T_{\rm eff} = 500\,\rm K$ dusty giant planet with a dM6 primary. Although this is a simple example of an irradiated atmosphere, it clearly shows the combined effects of reflection and heating of the outer atmosphere of the planet by the external radiation field.

3.3. Microlensing and Stars

One of the purposes for this meeting is to bring together different areas that contribute to our understanding of microlensing and what we can learn from it. From the stellar atmosphere point of view there are (at least!) two important areas that can be addressed by microlensing observations: limb darkening and star spots. Limb darkening is an important tool for the investigation of the structure of stellar atmospheres, in particular for subgiant and giants. In Figure 9 we show the limb darking in the optical (~ 5000Å) for a cool giant with $T_{\rm eff} = 5000\,\rm K$ and $\log(g) = 2.5$ (solar abundances). For comparison, linear and square-root limb darkening laws are also shown. The form of the limb darkening and the deviations of the computed intensities from the simple limb darkening laws changes substantially with wavelength and model parameters (see also Orosz & Hauschildt, 2000). This is caused by the effects of spherical radiation transport, which tends to concentrate the emitted intensities closer to the center of the stellar disk than the plane parallel approximation. This is due to the fact that curvature close to the rim of the stellar disk actually reduces the total emissivity and optical depth and thus produces less emitted radiation than plane parallel models in which the pathlength and optical depth close to the rim actually approach infinity. This illustrates that in order to obtain reliable and useful stellar atmosphere information from microlensing observations we need to use rather sophisticated modeling; simple approximation will in most cases not work. Using detailed limb darkening models as shown here has helped Orosz & Hauschildt (2000) to address a number of problems regarding lightcurves of eclipsing binaries. The effects of star spots on microlensing observations is discussed in detail in the paper by P. Sackett (this volume).

4. Summary and Conclusions

In this paper we have discussed a few new results of stellar atmosphere modeling that have helped to resolve some outstanding problems with understanding and interpreting observed stellar spectra. During the last decade, progress was made by breakthroughs in both methodology and computer technology, which has led to substantially improved models and synthetic spectra. In many cases, even our current "best effort" models cannot reproduce observed spectra statisfactorly; this is in particular the case for L and T dwarfs. However, this is due to physical effects that we "know" but we cannot currently describe well enough (e.g. incomplete line lists for key molecules or dust and cloud formation processes). Another area that requires much more work is our detailed understanding of winds from both hot and cool stars. There is currently a lot of effort being put into the solution of these key problems, although it is clear that once they are solved, others will pop up in unexpected places.

Acknowledgments. This work was supported in part by the CNRS, INSU and by NSF grant AST-9720704, NASA ATP grant NAG 5-8425 and LTSA grant NAG 5-3619, as well as NASA/JPL grant 961582 to the University of Georgia, NASA LTSA grant NAG5-3435 and NASA EPSCoR grant NCCS-168 to Wichita State University. This work was supported in part by the Pôle Scientifique de Modélisation Numérique at ENS-Lyon. Some of the calculations presented in this paper were performed on the CNUSC IBM SP2, the IBM SP2 and SGI Origin 2000 of the UGA UCNS, on the IBM SP "Blue Horizon" of the San Diego Supercomputer Center (SDSC), with support from the National Science Foundation, and on the IBM SP and the Cray T3E of the NERSC with support from the DoE. We thank all these institutions for a generous allocation of computer time.

REFERENCES

Abbott, D. C., Telesco, C. M., & Wolff, S. C. 1984, ApJ, 279, 225.
Allard, F., & Hauschildt, P. H. 1995a, In *The Bottom of the Main Sequence - and Beyond*, C. Tinney, editor, ESO Astrophysics Symposia, pages 32–44, Springer-Verlag Berlin Heidelberg.
Allard, F., & Hauschildt, P. H. 1995b, ApJ, 445, 433.
Allard, F., Hauschildt, P. H., Miller, S., & Tennyson, J. 1994, ApJL, 426, 39.
Allard, F., Hauschildt, P. H., Alexander, D. R., & Starrfield, S. 1997, ARAA, 35, 137.
Allard, F., Hauschildt, P. H., & Schwenke, D. 2000, ApJ, in press.
Anderson, L. S. 1987, In *Numerical Radiative Transfer*, W. Kalkofen, editor, page 163, Cambridge University Press.
Anderson, L. S. 1989, ApJ, 339, 559.
Aufdenberg, J. P., Hauschildt, P. H., Shore, S. N., & Baron, E. 1998, ApJ, 498, 837.
Aufdenberg, J. P., Hauschildt, P. H., Sankrit, R., & Baron, E. 1999, MNRAS, 302, 599.
Baron, E., & Hauschildt, P. H. 1998, ApJ, 495, 370.
Baron, E., Hauschildt, P. H., Nugent, P., & Branch, D. 1996, MNRAS, 283, 297.

Brett, J. M., & Smith, R. C. 1993, MNRAS, 264, 641.
Bruhweiler, F. C. 1996, In *Astrophysics in the Extreme Ultraviolet*, S. Bowyer and R. F. Malina, editors, pages 261–268, Dordrecht: Kluwer.
Cannon, C. J. 1973, JQSRT, 13, 627.
Cassinelli, J., Cohen, D., MacFarlane, J.J. Drew, J., Lynas-Gray, A., Hoare, M., Vallerga, J., Welsh, B., Vedder, P., Hubeny, I., & Lanz, T. 1995, ApJ, 438, 932.
Drake, S. A., & Linsky, J. L. 1989, AJ, 98, 1831.
Goorvitch, D., & Chackerian, C. 1994a, ApJS, 91, 483.
Goorvitch, D., & Chackerian, C. 1994b, ApJS, 92, 311.
Gry, C., York, D. G., & Vidal-Madjar, A. 1985, ApJ, 296, 593.
Hauschildt, P., Baron, E., Starrfield, S., & Allard, F. 1996, ApJ, 462, 386.
Hauschildt, P., Allard, F., Ferguson, J., Baron, E., & Alexander, D. R. 1999, ApJ, 525, 871.
Hauschildt, P. H., & Baron, E. 1995, JQSRT, 54, 987.
Hauschildt, P. H., & Baron, E. 1999, Journal of Computational and Applied Mathematics, 102, 41–63.
Hauschildt, P. H., Allard, F., Alexander, D. R., & Baron, E. 1997, ApJ, 488, 428.
Hauschildt, P. H., Baron, E., & Allard, F. 1997, ApJ, 483, 390.
Hauschildt, P. H., Allard, F., & Baron, E. 1999, ApJ, 512, 377.
Howarth, I. D. 1999, private communication.
Hubeny, I., & Lanz, T. 1995, ApJ, 439, 875.
Jones, H. R. A., Longmore, A. J., Allard, F., Hauschildt, P. H., Miller, S., & Tennyson, J. 1995, MNRAS, 277, 767.
Jørgensen, U. G., & Larsson, M. 1990, A&A, 238, 424.
Kurucz, R. L. 1993, Molecular data for opacity calculations, Kurucz CD-ROM No. 15.
Leinert, C., Allard, F., Richichi, A., & Hauschildt, P. H. 1999, A&A, submitted.
Littleton, J. E. 1987, private communication.
Littleton, J. E., & Davis, S. P. 1985, ApJ, 296, 152.
Ludwig, C. B. 1971, Applied Optics, Vol 10, 5, 1057.
Mathisen, R. 1984, Photo cross-sections for stellar atmosphere calculations — compilation of references and data, Inst. of Theoret. Astrophys. Univ. of Oslo, Publ. Series No. 1.
Miller, S., Tennyson, J., Jones, H. R. A., & Longmore, A. J. 1994, In *Molecules in the Stellar Environment*, U. G. Jørgensen, editor, Lecture Notes in Physics, pages 296–309, Springer-Verlag Berlin Heidelberg.
Neale, L., & Tennyson, J. 1995, Astrophys. J. Letters, 454, L169.
Neale, L., Miller, S., & Tennyson, J. 1996, Astrophys. J., 464, 516.
Orosz, J. A., & Hauschildt, P. H. 2000, A&A, submitted.
Partridge, H., & Schwenke, D. W. 1997, J. Chem. Phys., 106, 4618.
Rothman, L. S., Gamache, R. R., Tipping, R. H., Rinsland, C. P., Smith, M. A. H., Chris Benner, D., Malathy Devi, V., Flaub, J.-M., Camy-Peyret, C., Perrin, A., Goldman, A., Massie, S. T., Brown, L., & Toth, R. A. 1992, JQSRT, 48, 469.
Rybicki, G. B. 1972, In *Line Formation in the Presence of Magnetic Fields*, G. Athay, L. L. House, and G. Newkirk Jr., editors, page 145, High Altitude Observatory, Boulder.
Rybicki, G. B. 1984, In *Methods in Radiative Transfer*, W. Kalkofen, editor, page 21, Cambridge Univ. Press.

Scharmer, G. B. 1984, In *Methods in Radiative Transfer*, W. Kalkofen, editor, page 173, Cambridge Univ. Press.
Schryber, H., Miller, S., & Tennyson, J. 1995, JQSRT, 53, 373.
Schweitzer, A., Hauschildt, P. H., Allard, F., & Basri, G. 1996, MNRAS, 283, 821.
Schweitzer, S., Hauschildt, P. H., & Allard, F. 2000, ApJ, submitted.
Schwenke, D. W. 1998, Chemistry and Physics of Molecules and Grains in Space. Faraday Discussion, 109, 321.
Scuderi, S., Panagia, N., Stanghellini, C., Trigilio, C., & Umana, G. 1998, A&A, 332, 251.
Simon, M., Felli, M., Cassar, L., Fischer, J., & Massi, M. 1983, ApJ, 266, 623.
Vallerga, J. V., & Welsh, B. Y. 1996, In *Astrophysics in the Extreme Ultraviolet*, S. Bowyer and R. F. Malina, editors, pages 277–282, Dordrecht: Kluwer.
Verner, D. A., & Yakovlev, D. G. 1995, A&AS, 109, 125.
Welsh, B. Y. 1989, ApJ, 373, 556.

Discussion

Gray: The red giant limb darkening law you displayed showed a very sharp drop towards the limb. How does this vary with different parameters? Is it sensitive? Does it keep the same sharp edge?

Hauschildt: The edge is very dependent on log g until reaching the dwarf regime (when the standard limb-darkening law is recovered). The dependence on T_{eff} is less pronounced and much weaker. The location of the edge is wavelength dependent. The shape is slowly changing towards the dwarf regime.

Zinnecker: Can you summarise again how much the dust opacity affects the spectral appearance of M, L, T dwarfs and ultimately Jupiter-like planets?

Hauschildt: A very approximate and tentative picture is like this: Early M dwarfs: too hot for dust.
Late M dwarfs ($T_{eff} < 2800K$): dust begins to form, small effect on spectra.
"early" L dwarfs: dust opacity becomes strong; shallow H_2O bands by back-warming.
"Mid-late" L dwarfs ($T_{eff} < 2200K$): strong effect of dust opacities; "full dust".
late L dwarfs ($T_{eff} = 1500K?$): Less deep opacity, possibly due to retraction of convection zone deeper into atmosphere; cloud formation.
T dwarfs, "Jovian"planets: even less dust opacity, but still important; formation of clouds; the atmospheres appear "clearing" with low T_{eff}.

Gould: Is spectral resolution of the atmosphere of any interest to you? Would it help resolve theoretical questions?

Hauschildt: Yes. Different lines, better spectral features, originate in different segments of the atmospheres. Extracting this information from microlensing events could help to constrain the structure of the atmosphere predicted by models.

Sackett: If you could choose, what wavelength range should be probed at what resolution for giants?

Hauschildt: There is no easy answer for this that I can think of. This can be determined by looking at the models (and the possible observations). I think that the NIR range around band heads and close to resonance lines is most likely to be useful. The actual "best range" will also be different for dwarfs and giants.

Kerins: In your spectral energy distribution fit to Kelu-1 most of the discrepancies with the real spectrum involve overestimates of the flux. However, just longward of 1μm the fit significantly underestimates the flux. Is there any significance to this?

Hauschildt: Yes, this is (mostly) due to known deficiencies of the molecular line data for H_2O (IR) and TiO. The latter have been addressed in new TiO line lists by Schwenke(1998) which effectively "kills" the excess peaks in the visible. New H_2O line lists are in preparation.

Gaudi: How well do your models agree with models of other groups?

Hauschildt: In general comparisons are hard due to the fact that groups use different input physics. For similar setups, results agree within reasonable limits, i.e. quite well.

Microlensing Extended Stellar Sources

H. M. Bryce and M. A. Hendry

Department of Physics and Astronomy, University of Glasgow, Glasgow, G12 8QQ, UK

D. Valls-Gabaud

Laboratoire d'Astrophysique, UMR CNRS 5572, Observatoire Midi-Pyrenees, 31400 Toulouse, France

Abstract. We have developed a code to compute multi-colour microlensing light curves for extended sources, including the effects of limb darkening and photospheric star spots as a function of spot temperature, position, size and lens trajectory. Our model also includes the effect of structure within the spot and rotation of the stellar source.

Our results indicate that star spots generally give a clear signature only for transit events. Moreover, this signature is strongly suppressed by limb darkening for spots close to the limb – although such spots can be detected by favourable lens trajectories.

1. Introduction

The amplification for a point source microlensing event is dependent only on the projected separation between lens and source. However, if the source size is comparable to the Einstein radius and the projected separation is small, it is necessary to treat the source as finite, with the amplification being calculated as an integral over the source star. Thus, the microlensing light curve also contains information about the source surface brightness profile.

Finite source effects were considered by Gould (1994) for the case in which a lens *transits* the star, allowing measurement of the lens proper motion and thus distinguishing between Magellanic Cloud (MC) and Galactic MACHOs. Witt & Mao (1994) showed that the magnification profile of an extended source with a constant surface brightness profile was significantly different from the point source treatment for a 100 $R_\odot$ source microlensed by a 0.1 $M_\odot$ MACHO. The ability to recover the lens mass, transverse velocity and the source size are considerable assets in extended source microlensing, however rare the events may be.

The opportunity to use such events to provide information about the source star has also been investigated. Simmons et al. (1995) modelled the polarisation signature of an extended source microlensing event, assuming a pure electron scattering grey model atmosphere. They showed that including polarisation information significantly improves the accuracy of estimates of the source radius, as well as better constraining the radial surface brightness profile of the source.

Gould (1997) suggested the use of extended source events to determine the rotation speed of red giants. The use of extended sources to estimate source limb darkening parameters has also been discussed by several authors (cf. Hendry et al. 1998; Valls-Gabaud 1998). Such an event produces a chromatic signature as the lens effectively *sees* a star of different radius in different photometric colour bands. This provides an unambiguous signature of microlensing, as opposed to that from a variable star (Valls-Gabaud 1998).

2. Microlensing spots

Recent literature has turned attention to the microlensing of extended sources with non-radially symmetric surface brightness profiles (cf. Hendry, Bryce & Valls-Gabaud 2000; Heyrovský & Sasselov 2000 (hereafter HS00); Kim et al. 2000). Microlensing provides an almost unique probe of such features: Doppler imaging is inapplicable to late type stars, due principally to their long rotation periods, and direct optical imaging of stellar photospheres remains beyond the scope of current technology for all but a handful of stars. There is, nonetheless, already direct evidence for the presence of a large 'hot spot' on α Orionis, which shows strong limb darkening (Uitenbroek, Dupree & Gilliland 1998). Hot spots due to convection cells are likely to be present on the photospheres of red supergiants, and may be associated with non-radial oscillations. *Gravitational imaging* of such features on late type giants and supergiants would provide valuable constraints on stellar atmosphere models, placing limits on e.g. the form of the limb darkening law and the role of rotation.

While one can, of course, also consider the decremental effect of a *cool* spot, it seems pragmatic – in terms of spot detectability – to consider the impact of a hot spot on a cool, massive star. HS00 have demonstrated that hot central spots with a radius of $0.2R_*$ can produce a change in the microlensed flux $\geq 2\%$. We have improved their model in several important respects.

- We have incorporated the geometric foreshortening of circular spots on the photosphere close to the stellar limb, instead of placing circular disks on the star. Circular spots close to the limb appear increasingly elliptical.

- We have computed spot detectability both in terms of the percentage deviation from the unspotted flux and the goodness of fit of realistic, sparsely-sampled observations – with magnitudes and errors appropriate to current MC and Bulge data – to the model of an unspotted star.

- We have considered linear, logarithmic and square root limb darkening (LD) laws, with band-dependent LD coefficients from LTE stellar atmosphere models (Claret & Gimenez 1990) for U to K bands.

- Finally, we have considered the observational signatures of multiple spots, spots with temperature structure and spots on rotating stars – features which make our model somewhat more realistic.

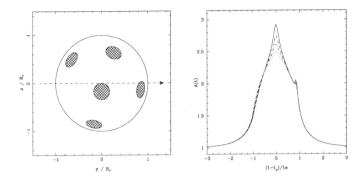

Figure 1. The microlensing light curve produced by 5 hot (+800K) circular spots on a 4000K $\log g = 2$ star with stellar radius equal to the angular Einstein radius (AER) of the lens, being lensed at zero impact parameter. Note that the spots close to the limb appear foreshortened and do not contribute significantly to the microlensed flux. The unbroken, dashed, dot-dashed and dotted lines represent B, V, R and I bands respectively.

3. Results

We have investigated the microlensing signatures of a range of spot parameters: radius 5°, 10°, 20°; spot temperature relative to stellar effective temperature ranging from $+200$ K $\to$ $+1000$ K and -200 K $\to$ -1000 K; spot geometry – i.e. varying the spot position on photosphere.

We summarise first our qualitative conclusions. As in HS00 our results show that central hot spots are detectable. Furthermore spots close to the limb can also be detected if the lens passes close to the feature, especially at minimum impact parameter (see Figs. 3 and 4 below). As the impact parameter increases, the central area where spots are detectable contracts, and only very large spots will produce detectable signals if the lens does not transit the source. A spot close to the limb produces only a small signal, as it is suppressed by the geometric foreshortening and the effects of the limb darkening. The stellar radius also affects the detectability of spots; however even on a small star ($R_* \sim 10^{-5}$ AER) a plausible hot spot can be resolved at small impact parameter. The effect of a spot is clearly *localised* and of short duration, however. For example, a central 10° spot will produce a significant deviation for about 15% of the Einstein radius crossing time, with no effect in the light curve wings. Thus, excellent temporal sampling ($\sim$ hours) is required to detect spots – even for surveys with high photometric precision.

3.1. Spot temperature and structure

The percentage deviation from the unspotted flux increases with the temperature of the spot (see Fig. 2), although the dependence on temperature is less strong than on spot position. Cool spots produce depressions in the light curve, but are less detectable than a hot spot with the same position, radius and (absolute) temperature difference. Not only is the light curve deviation smaller for a cool

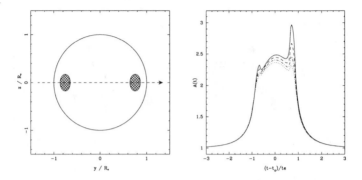

Figure 2. The same star as above, with the same colour bands represented but with a different maculation. The spot on the left is at 4500K and produces a 3σ signal for 10% of the lensing event, for typical magnitude errors. The spot on the right is at 5000K and produces a 3σ signal for 30% of the event, as well as an higher amplification.

spot but it is also present over a shorter timescale. Including the effect of umbrae within spots produces additional small-scale structure in the light curve, but differing from the signature of a uniform temperature spot only over very short timescales.

3.2. Spot 'imaging'

Since the microlensing light curve is essentially a convolution integral over the extended source, it is not surprising to find that the signatures of spots are non-unique, in the sense that spots of different shape and temperature structure can give rise to light curves that are identical within observational error. Thus, the use of microlensing as a tool for resolving detailed maps of the stellar photosphere is highly limited. Nevertheless, since the effect of each spot is fairly localised, one can at least use the light curve to place constraints on spot position and "filling factor". Similarly, a *failure* to detect spot signatures from high time resolution observations of a transit event – particularly from e.g. a fold caustic crossing – can provide useful limits on surface activity for stellar atmosphere models.

3.3. Estimating stellar radii

The presence of starspots can also affect the estimation of the source parameters; specifically, neglecting the effects of maculae in the light curve can produce a biased determination of the stellar radius, and thus the lens Einstein radius and proper motion. For example, a star with a central hot spot will give a best fit to an unspotted star of smaller stellar radius; moreover, the impact of sparse sampling may "mask" the presence of the spot feature and thus yield a perfectly acceptable χ^2 to the smaller radius. The best fitting stellar radius estimated from V band data is typically 10% smaller than the true radius, for a range of plausible spot parameters. However, at longer wavelengths (e.g. I

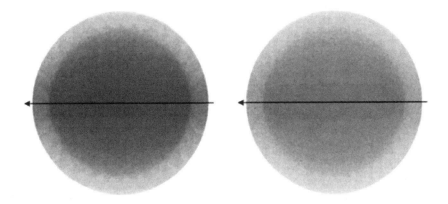

Figure 3. Grey scale map showing the detectability of a 10° hot spot of temperature contrast +1000K, on a 0.1 AER star. A linear limb darkening law is assumed. The arrow denotes the lens trajectory. Shown are the results for V band (left panel) and R band (right panel) observations.

and K bands) the bias in the estimated radius is less severe, consistent with the reduced temperature sensitivity of the Planck function at these wavelengths. Moreover, cool spots do not produce a strong bias in the estimated radius.

3.4. Stellar rotation

The area and temperature contrast of a spot essentially determine the shape (i.e. amplitude and width) of the spot profile on the microlensing light curve. Interestingly, for the case of a rotating source, one can change the ratio of amplitude to width for the spot signature, since this effectively changes the transverse velocity of the lens with respect to the *spot*, without changing the lens tranvserse velocity with respect to the star as a whole. Whether the spot feature is narrowed or broadened depends on the orientation of the lens trajectory with respect to the rotation axis of the source. Even for the long rotation periods expected for late-type giants, the effect of rotation can have an impact on the observed light curve – e.g. for a source rotation period $\tau_S = 0.05 t_e$ and a spot of radius of 15°, the rotation of the source may bias the estimation of the spot radius by up to 10%. Of course a further complicating factor could be significant evolution in the surface distribution of spots during the lensing event – a problem more likely to be relevant for faster rotators.

3.5. Maps of spot detectability

As discussed above, the position of a spot on the stellar photosphere is the crucial factor that determines its detectability. As an illustration, Figures 3 and 4 show grey scale plots indicating the detectability of a 10° spot, of temperature 5000K, at various positions on the photosphere of a star with $T_{eff} = 4000K$, $\log g = 2$. Regions where the spot provides a larger signature are darker; in this example the darkest level on the plots indicates an average deviation of $\geq 10\%$ compared with the unspotted light curve, while the lightest level indicates no

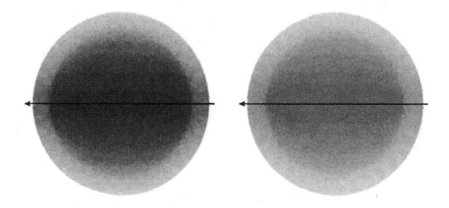

Figure 4. Same as for Fig. 3, but for a 0.01 AER star.

significant deviation. Note that in the outer annulus of the stellar disk the spot is barely detectable, due to the effects of limb darkening and geometrical foreshortening, although this effect is more pronounced at high stellar latitudes, far from the lens trajectory. Note also that spots are more detectable in the V band – consistent with the greater bias in the determination of the stellar radius at shorter wavelengths.

4. Small scale radial oscillations

We have developed a model of the microlensing of stars displaying radial oscillations. Although intrinsic variability is adopted as an exclusion criterion by current microlensing surveys, this does not of course preclude the possiblity that variable stars may themselves be microlensed (c.f. EROS-2, Ansari et al. 1995) – and such an observation could provide important information on issues such as stellar pulsation and asteroseismology. Complex light curves are produced by the radial oscillations of even a simplified model – see Figure 5. Here we modelled a sinuosoidal oscillation only in the stellar radius, neglecting chromatic variations other than those due to limb darkening. The period of oscillation was 5 days, which would typically correspond to about four complete cycles over the timescale of an event with small Einstein radius. Higher order non-radial oscillations can be modelled as a series of spots; however, detecting such features would be limited by the same criteria as discussed above – with the additional possible limitation of evolution during the event. Note that, due to the effect of limb darkening, for a fixed stellar radius the oscillation is more detectable at shorter wavelengths. Unlike spots, a small level of oscillation can be detected without the need for a transit or near-transit event. For example, if the impact parameter ≤ 0.2 AER, a 2% change in the radius of a 0.001 AER radius star produces a highly significant deviation from the light curve modelled for a static star in the Bulge, for more than 25% of the event duration. For a well-sampled light curve, however, the deviations do not significantly bias the estimation of e.g. the stellar radius, although a poor or *aliased* sampling rate

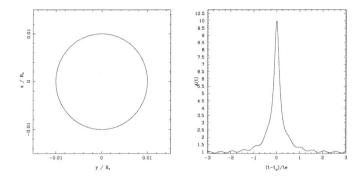

Figure 5. The microlensing light curve produced by a 6000K, $\log g = 3$ star, varying in radius by 3% from a mean radius of 0.01 AER being lensed with a minimum impact parameter 0.1. Most of the light curve gives a clear residual compared to the same star without the oscillation. The colour bands are superimposed on each other in this case as there is no strong chromatic effect for such a small stellar radius.

could introduce such a bias. Thus, detection of small scale oscillations will only occur if aggressive temporal sampling strategies are employed.

5. Spectroscopic signatures

Spectroscopic studies of transit microlensing events can provide strong constraints on stellar atmosphere models, (cf Heyrovský, Sasselov & Loeb 2000). We expect to see equivalent widths of spectral lines vary as the lens *gravitationally images* the source. Strong resonant lines such as Lyα, CaII and MgII appear brighter at the limb. Thus, in transit microlensing events spectroscopic targeting of such lines would yield good estimates of the stellar radius and hence the Einstein radius – in a manner analogous to the polarimetric signature modelled by Simmons et al. (1995). Like the chromatic signature of extended source microlensing, the variation of line profile shape produces a unique discriminant between microlensing and intrinsic variability. To image spots, we need to probe lines sensitive to temperature structure: for example Hα will be sensitive to active regions and molecular lines in the atmospheres of late-type giants will be sensitive to the presence of hot spots. However the atmospheres of red giants are poorly constrained by models: their limb darkening laws are extrapolated from limited data and are complicated by the presence of many molecules and by variations in metallicity. Moreover, many late-type giants possess extended circumstellar envelopes, for which microlensing would also present a powerful diagnostic tool. For example, the spectroscopic signature of a thin spherical shell of microlensed circumstellar material can provide an important probe of the density and velocity structure of the envelope (c.f. Ignace & Hendry 1999).

6. Conclusions

With the advent of automated "alert" status for candidate events, the prospects for using microlensing as a tool for gravitational imaging and investigating stellar atmospheres are dramatically improved. Of crucial importance in such studies are coordinated global observations to achieve the high level of sampling required to constrain models effectively. Observations of this quality are *also* precisely what is required to extract useful information from the microlensing of extended sources – making detailed modelling of such events a timely issue. Moreover, the intensive observation of second caustic crossings in binary events – currently being pursued with the principal goal of detecting planetary companions – would also provide powerful constraints on the range of spot-related phenomena discussed in this paper. We are currently, therefore, extending our analyses to consider the photometric, polarimetric and spectroscopic signatures of star spots lensed during line caustic crossing events.

References

Ansari, R. et al, 1995, A&A, 299, 21
Alcock, C. et al, 1997, ApJ, 491, 436
Claret and Gimenez, 1990, Ap&SS, 169, 223
Gould, A., 1994, ApJ, 421, 71
Gould, A., 1997, ApJ, 483, 98
Hendry, M.A., et al, 1998, New Astron. Rev., 42, 125
Hendry, M.A., Bryce, H.M. & Valls-Gabaud, D., 2000, in prep.
Heyrovský, D., Sasselov, D., and Loeb, A., 2000, ApJ, submitted (astro-ph/9902273)
Heyrovský, D. and Sasselov, D., 2000, ApJ, 529, 69
Ignace, R. and Hendry, M. A., 1999, A&A, 341, 201
Kim, H. et al, 2000, MNRAS, submitted
Simmons, J. F. L. et al, 1995, MNRAS, 276, 182
Uitenbroek, H., Dupree, A.K. & Gilliland, R.L., 1998, ApJ, 116, 2501
Valls-Gabaud, D., 1998., MNRAS, 294, 747
Witt, H.J. and Mao, S., 1994, ApJ, 430, 50

Discussion

Fischer: 1) Is there degeneracy between models for spots and planet crossings?
2) M dwarfs often have flaring. What would that signature look like and are you suggesting this phenomenon occurs in late-type giants?
Bryce: 1) Yes, possibly, but it would be necessary to microlens an extended source by a lens with a planet. They do produce similar narrow peaks despite being entirely different.
2) A flare or "clumpy bit" of wind would produce a similar signature.

Vermaak: You have examined only large sources (compared to the Einstein radius) here. Could you measure spots on any transit provided you could get the needed sampling frequency?

Bryce: A large source compared to the Einstein radius does produce a spot signature – but the amplification is below 1.34 ... Although such a large source provides a larger sample of lens.

Source Reconstruction as an Inverse Problem

Norman Gray

Department of Physics and Astronomy, University of Glasgow, Glasgow, G12 8QQ, UK

Iain J Coleman

British Antarctic Survey, Madingley Road, Cambridge, CB3 0ET, UK

Abstract. Inverse Problem techniques offer powerful tools which deal naturally with marginal data and asymmetric or strongly smoothing kernels, in cases where parameter-fitting methods may be used only with some caution. Although they are typically subject to some bias, they can invert data without requiring one to assume a particular model for the source. The Backus-Gilbert method in particular concentrates on the tradeoff between resolution and stability, and allows one to select an optimal compromise between them. We use these tools to analyse the problem of reconstructing features of the source star in a microlensing event, show that it should be possible to obtain useful information about the star with reasonably obtainable data, and note that the quality of the reconstruction is more sensitive to the number of data points than to the quality of individual ones.

1. Introduction

Where once all the interest in microlensing was in the details of the lensing population, there is now an increasing interest in lensing events as probes of the stellar sources. From this point of view, once the lens's geometrical details have been worked out, the event can be used as a "super-telescope", providing otherwise completely unattainable resolution of the surfaces of distant stellar disks.

Initially, analyses assumed that the microlensing source star could be taken to be a point source, and the first discussion of "finite source effects" was in the context of a problem – Witt & Mao (1994) asked at what point the point-source approximation would break down; Nemiroff & Wickramasinghe (1994) and Gould (1994) discuss the issue more positively, pointing out that finite-source effects would cause distortions to the light curve which would break the degeneracy in lensing parameters, and thus allow a more accurate estimate of the lens transverse velocity and mass.

It is not merely information about the lens which can be obtained from events. As well as suggesting the recovery of limb-darkening parameters from intensity data, Simmons, Newsam, & Willis (1995) noted that a microlensing event would break any rotational symmetry on a source star's disk, and so produce a polarization signature, which would in turn allow one to recover source limb-polarization; as well, one could obtain the (projected) direction of the lens velocity from the variation in the direction of polarization. Newsam et al. (1998) show how even relatively poor polarization data can substantially improve fits of source parameters.

Although the basic microlensing effect is achromatic, the fact that stars have different limb-darkening profiles in different colours means that a lens differentially amplifying the disk will produce a chromatic effect. Several workers (Simmons et al. 1995; Valls-Gabaud 1995; Sasselov 1996; Valls-Gabaud 1998) have discussed how one might obtain such chromaticity information. It is even possible to discuss how one might observe the signatures of stellar spots (Heyrovský & Sasselov 2000; Bryce & Hendry 2000).

The usual way in which source structure is detected is by applying a model-fitting (equivalently, parameter-fitting) algorithm to the observed data, to obtain the best-fit parameters of a suitable limb-darkening model; this is the approach used, for example, by the MACHO collaboration (Alcock et al. 1997) and the PLANET collaboration (Albrow et al. 1999) to make the first detections of limb-darkening in microlensing events. It is also the approach which underlies the insightful error analysis by Gaudi & Gould (1999).

A parameter-fitting algorithm essentially consists of a mechanism for systematically moving through parameter space, repeatedly solving the "forward problem" – calculating the data to be predicted from a given limb-darkening profile – until the predicted data is optimally close to the data actually observed. Here we want to suggest that, because of the fact that the underlying source function is convolved through a broad and asymmetric amplification kernel, a model-fitting approach is potentially problematic, and that this recovery problem is more naturally addressed using the well-established technology of inverse problems.

We plan to discuss the merits of inverse problem techniques in general, and the Backus-Gilbert method in particular, and exemplify the possibilities by inverting simulated microlensing data to recover limb-darkening and limb-polarization effects. We will see that we are able to discuss explicitly and robustly the tradeoff between resolution and stability which is implicit in any such inversion, including recoveries obtained by model-fitting.

2. Background

The geometry we consider is as shown in Figure 1. The amplification function is the familiar one:

$$A(\xi) = \frac{1}{2}\left(\xi + \frac{1}{\xi}\right), \quad \xi = \left(1 + \frac{4}{\zeta^2}\right)^{1/2}, \quad \zeta^2 = r^2 + s^2 - 2rs\cos(\chi - \phi). \quad (1)$$

Denote the intensity on the stellar surface $I(r)$, and the Stokes parameter by $Q(r,\chi) = -P(r)\cos 2\chi$, where $P(r)$ is the polarization of the stellar surface and we are assuming that the surface is rotationally symmetric. In the case of a microlensing event, we cannot directly resolve details of the lensed source,

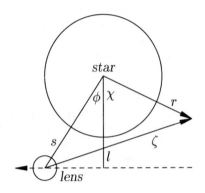

Figure 1. Geometry of a lensing event. The projected path of the lens has impact parameter l, and the path is parameterised by polar coordinates $s(t)$ and $\phi(t)$, relative to the centre of the source, projected into the lens plane. Any point in that plane can be given in polar coordinates r and χ, and this point is a distance ζ from the centre of the lens. All dimensions are normalised to the Einstein radius in the source plane. The angles ϕ and χ are taken with respect to the line joining the source to the lens's point of closest approach.

and must therefore measure integrals over the source surface. We immediately obtain

$$I(s(t),\phi(t)) = \int_0^\infty I(r)\tilde{A}_I(r;s,\phi)\,\mathrm{d}r \qquad (2)$$

$$Q(s(t),\phi(t)) = \int_0^\infty P(r)\tilde{A}_Q(r;s,\phi)\,\mathrm{d}r \qquad (3)$$

where the amplification kernels are

$$\tilde{A}_I = r\int_0^{2\pi} A(r,\chi;s,\phi)\,\mathrm{d}\chi \qquad (4)$$

$$\tilde{A}_Q = -r\int_0^{2\pi} \cos 2\chi A(r,\chi;s,\phi)\,\mathrm{d}\chi. \qquad (5)$$

Note that the kernel $\tilde{A}_I$ is a factor $2\pi r$ times the angular average of the amplification function, and the functions $I(s,\phi)$ and $Q(s,\phi)$ have the dimensions of flux rather than intensity.

Analytic expressions for these angle-averaged amplification functions have been obtained by numerous people. Schneider & Wagoner (1987) and Gaudi & Gould (1999) perform the average for approximate forms of the amplification function, Heyrovský & Loeb (1997) deal with an elliptical source for a slightly restricted class of source functions. Gray (2000) produced integrals for the exact amplification function, for axisymmetric sources, obtaining

$$\tilde{A}_I(r;s,\phi) = r(2\pi + I_1 - I_2), \qquad \tilde{A}_Q(r;s,\phi) = -r\cos 2\phi(Q_1 - Q_2)$$

where I_1, I_2, Q_1 and Q_2 are elliptic integrals whose arguments have an algebraic dependence on r and s. Although it is helpful, it is not in fact necessary to have analytic forms for the angle-averaged amplification functions, and the following analysis would be just as successful if these could be obtained only numerically.

Equations (2) and (3) are one-dimensional integral equations – the classic form of an inverse problem. The possibility of treating the source reconstruction problem as an inverse problem was first suggested by Simmons (1995). In this context, the function $I(r)$ is termed the underlying or source function, and the function $\tilde{A}_I(r;s)$ the kernel. The data is the set of values $I(s(t_i), \phi(t_i))$ for some set of times t_i.

The use of inverse problem techniques is relatively rare in this branch of astrophysics. Mineshige & Yonehara (1999) used the technique to map the Einstein Cross accretion disk using a hypothetical caustic crossing, and Wambsganss (2000) points out a number of other uses in the field of cosmological microlensing. Coleman, Gray, & Simmons (1998) use a similar technique with eclipsing binaries. The method described in this paper is discussed at greater length in Gray & Coleman (2000).

3. Inverse problem techniques vs. parameter fitting

Parameter fitting is the most appropriate technique in the case where (a) there is no doubt about the most appropriate model to use, so that the aim is simply to recover model parameters, and (b) when the problem kernel is not *ill-conditioned*. When the model itself is open to dispute, or the observational situation means that the kernel is ill-conditioned, then any parameter fit must be done extremely cautiously if it is not to be deceptive.

Ill-conditioning can be characterised in several ways. Fundamentally, an ill-conditioned kernel maps a large volume of parameter space to a small volume of data space; a strongly smoothing (nearly flat) kernel would be an example of this. This implies that the inversion is highly unstable, so that a tiny, noise-induced, change in the data ($I(t)$ in eqn. (2)) could, after a naïve inversion, be taken to indicate a radically different recovered function $I(r)$.

Inverse problem techniques – also known as "non-parametric" or "model-free" techniques – dispense with a parametrised model, and instead approach the problem from the question "how much information does this kernel permit to be recovered from this data?" (see, for example, Craig & Brown (1986) for a general introduction). They are most natural in the case of marginal data, or an asymmetric or broad kernel. These techniques are typically associated with Bayesian approaches to data analysis, and deal with the instability of the inversion by adding prior information, such as the supposition that the underlying function be smooth (in the case of inversion by regularisation) or otherwise featureless (in the case of maximum-entropy inversion). This explicit addition of prior information inevitably makes inverse problem techniques suffer from bias; in a parameter-fitting algorithm, the model acts as implicit prior information, so that route is not free of bias either.

3.1. Backus-Gilbert

The particular technique we have used is the Backus-Gilbert method. This works by allowing us to explicitly trade off recovery resolution against stability. We very briefly outline the method here; there is a fuller account in, for example, Parker (1977), and an astrophysical example in Loredo & Epstein (1989).

Given a kernel $K(r; s_i)$, underlying function $u(r)$, data $F(s_i)$, and noise n_i, the general 1-d inverse problem can be written as

$$F(s_i) = \int u(r)K(r, s_i)\,\mathrm{d}r + n_i.$$

(compare eqn. (2) and (3)). We suppose that we can find "response kernels" $q_i(r)$ which permit us to form an estimate $\hat{u}$ of the underlying function as a weighted average of the data:

$$\hat{u}(r) = \sum_i q_i(r)F(s_i). \qquad (6)$$

This is a random variable, but we can relate its mean to the underlying function through an "averaging kernel" Δ:

$$\langle \hat{u}(r) \rangle = \int \Delta(r, r')u(r')\,\mathrm{d}r'. \qquad (7)$$

Ideally, this kernel would be the Dirac delta function, and the estimator $\hat{u}$ would perfectly track the underlying function. Since the underlying function is (of course) unknown, we cannot use equations (7) directly, but this definition of $\Delta(r, r')$ allows us to define Width[Δ] and Var[$\hat{u}(r)$], which depend only on the kernel $K(r; s_i)$ and the noise n_i. This means that we can explicitly trade off improved recovery resolution (narrower Width[Δ]) against improved stability (smaller Var[$\hat{u}(r)$]), with the relative weighting parametrised by smoothing parameter λ. For each value of λ we can analytically obtain a set of coefficients $q_i(r)$ for equations (6).

Note that this technique is an analysis of the *kernel*, rather than a particular data set, which means (a) the understanding we gain of how much information is available from a data set is portable both to other inverse problem techniques, and also to parameter-fitting approaches, and (b) the analysis can be done, and an optimal λ selected, *prior to any data being collected*, given a set of demands on the required resolution and stability. A feature of the Backus-Gilbert method is that the response kernels $q_i(r)$ have a dependence on the radial parameter r, so that the analysis needs to be redone for each $\hat{u}(r)$ we wish to recover (at a cost of potentially large matrix inversions each time). This means that the Backus-Gilbert method (at least in its simplest form) is not an obvious choice for a data reduction pipeline, but it has the compensation that the optimal tradeoff can be chosen differently for different values of r.

4. Recovery of limb-darkening and limb-polarization profiles

In Figure 2, we show the tradeoff curves for the polarization kernel, equations (3), for a variety of numbers of data points, and choices of smoothing parameter λ

Source Reconstruction as an Inverse Problem 209

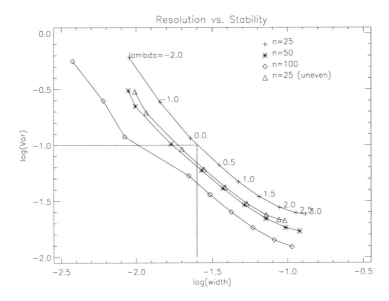

Figure 2. Resolution versus stability for various numbers of data points, and choices of smoothing parameter λ. The box in the lower left hand corner shows the region of 'acceptable' recoveries.

(there is a similar curve for limb-darkening, eqn. (2)). Recovery quality can be improved either by increasing the number of data points, or by adjusting the times at which data is taken. In the figure, the line marked '$n = 25$ (uneven)' shows the tradeoff curve for data taken unevenly, with observations clustered at points the analysis shows to be particularly sensitive; it is clear that the quality of the recovery is sensitive to both the number of data points and the manner in which they are obtained.

The bottom left of the plot corresponds to high resolution and high stability, the top left to recoveries with high resolution and low stability (in the limit, picking a single data point and scaling it), and the bottom right to low resolution and high stability (in the limit, simply averaging all the data).

The question of what counts as "adequate" resolution and stability can only be decided in the context of a particular stellar model (though this does not make the Backus-Gilbert analysis, or the trade-off diagram in Fig. 2, model-dependent). For a pure electron atmosphere, Chandrasekhar calculated a limb-polarization of 11%, suggesting that the variance of the limb-polarization recovery may be of order 10% at most. Estimating the width of a typical limb-polarization profile suggests a similar upper bound for the resolution in Figure 2. The figure therefore indicates what value of the tradeoff parameter λ we should pick, and therefore what is the best resolution and variance we can expect to achieve with a given number of data points.

In Figure 3, we show two recoveries of a limb-darkening profile from synthetic data with different noise standard-deviation. Note (a) that the optimal

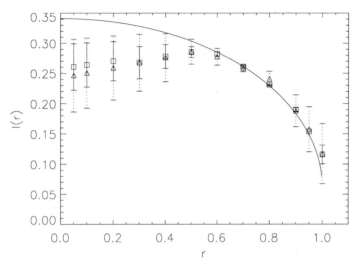

Figure 3. Recoveries of a limb-darkening profile (solid line) with 100 data points, and impact parameter, Einstein radius and stellar radius all equal. The points plotted with boxes and solid error bars are recovered from simulated data with $\sigma = 0.002$, and the points with triangles and dotted error bars from data with $\sigma = 0.1$. After Coleman (1998).

recovered variance varies across the disk; (b) that the recovery is biased – towards a flat profile in this case – with the bias being greatest at the centre of the disk and at the limb; (c) that both the bias and the variance are least around 70% of the projected radius, which is particularly fortunate since it is in this region that limb-darkening profiles tend to be most sensitive to parameters such as g, and where the profiles of giants differ most from those of normal stars (Hauschildt 2000); and (d) that the quality of the recovery is surprisingly insensitive to the quality of the data.

5. Discussion

One of the aims of this paper is to emphasise the seriousness of the ill-conditioning of the source reconstruction problem, and hence the desirability of using an analytical technique which starts by examining that ill-conditioning, goes on to discuss what information is nonetheless recoverable, and only then produces the numerical information which is the point of the exercise. Of course, the same questions can be asked using a parameter-fit approach, but less naturally, since such approaches assume, in a sense, that the information is recoverable, with the result that problems can only be uncovered *post hoc*, by an intelligent examination of goodness-of-fit measures, or by tracing the propagation of errors through a calculation.

The second aim is to use this inverse problem approach to analyse the kernel which turns the underlying limb-darkening and limb-polarization functions into microlensing intensity and polarization data. It turns out that the information is indeed recoverable with adequate uncertainties but, depending on the quality of the data available, one many have to make significant compromises over the resolution one is prepared to accept. It is also possible to use this approach to analyse the effect of modifications in the way the data is collected, and discover that this can indeed significantly improve the data's effective quality.

Acknowledgments. The authors would like to thank J F L Simmons for helpful comments on the manuscript.

References

Albrow, M. D., Beaulieu, J.-P., Caldwell, J. A. R., Dominik, M., Greenhill, J., Hill, K., Kane, S., Martin, R., et al., 1999, ApJ, 522, 1011

Alcock, C., Allsman, R. A., Alves, D., Axelrod, T. S., Becker, A. C., Bennett, D. P., Cook, K. H., Freeman, K. C., et al., 1997, ApJ, 491, L11

Bryce, H. & Hendry, M. A. 2000, in these proceedings

Coleman, I. J. 1998, PhD thesis, University of Glasgow

Coleman, I. J., Gray, N., & Simmons, J. F. L. 1998, A&AS, 131, 187

Craig, I. J. D. & Brown, J. C. 1986, Inverse Problems in Astronomy (Adam Hilger)

Gaudi, B. S. & Gould, A. 1999, ApJ, 513, 619

Gould, A. 1994, ApJ, 421, L71

Gray, N. 2000, (astro-ph/0001359)

Gray, N. & Coleman, I. J. 2000, Gravitational Microlensing Source Limb Darkening and Limb Polarization, II: The Inverse Problem, in preparation

Hauschildt, P. H. 2000, in these proceedings

Heyrovský, D. & Loeb, A. 1997, ApJ, 490, 38

Heyrovský, D. & Sasselov, D. 2000, ApJ, 529, 69

Loredo, T. J. & Epstein, R. I. 1989, ApJ, 336, 896

Mineshige, S. & Yonehara, A. 1999, PASJ, 51, 497

Nemiroff, R. J. & Wickramasinghe, W. A. D. T. 1994, ApJ, 424, L21

Newsam, A. M., Simmons, J. F. L., Hendry, M. A., & Coleman, I. J. 1998, New Astron. Reviews, 42, 121

Parker, R. L. 1977, Ann. Rev. Earth Planet. Sci., 5, 35

Sasselov, D. D. 1996, in 12th IAP Astrophysics Colloquium, July 1996, Paris, eds. R. Ferlet, J-P. Maillard & B. Raban (Gif-sur-Yvette: Editions Frontieres), p. 141

Schneider, P. & Wagoner, R. V. 1987, ApJ, 314, 154

Simmons, J. F. L. 1995, personal communication

Simmons, J. F. L., Newsam, A. M., & Willis, J. P. 1995, MNRAS, 276, 182

Valls-Gabaud, D. 1995, in Large scale structure in the universe, Proceedings of an international workshop, Potsdam, Germany, edited by J.P. Muecket, S. Gottloeber and V. Mueller (World Scientific Co.), p. 326

Valls-Gabaud, D. 1998, MNRAS, 294, 747
Wambsganss, J. 2000, in these proceedings
Witt, H. J. & Mao, S. 1994, ApJ, 430, 505

Discussion

Bennett: I just wanted to point out that most of the microlensing data relevant to limb darkening measurements come from binary caustic and cusp crossing events, where there is no analytic formula for the lens magnification.

Gray: Agreed. But though an analytic form makes things neater, it isn't actually required, so the technique would pay dividends in that context also.

Gould: Gaudi & Gould found that $\sigma \sim N^{3/2}$ where N is the number of radial bins. Can this power law be explained from within the Backus-Gilbert framework?

Gaudi: I believe our conclusion basically implies that the slope of the variance vs width relation in Gray's plot should be 3/2.

Gray: The $N^{3/2}$ behaviour should certainly be linked somehow with the degradation in the best available resolution as you go towards the source center. It might be that the link is via an increase in the size of bins toward the center. The plot I showed has width against variance rather than source radius, so I wouldn't expect the $N^{3/2}$ behaviour to be so directly reflected.

Microlensing and the Physics of Stellar Atmospheres

Penny D. Sackett

Kapteyn Institute, 9700 AV Groningen, The Netherlands
Anglo-Australian Observatory, P.O. Box 296, Epping NSW 1710, Australia

Abstract. The simple physics of microlensing provides a well understood tool with which to probe the atmospheres of distant stars in the Galaxy and Local Group with high magnification and resolution. Recent results in measuring stellar surface structure through broad band photometry and spectroscopy of high amplification microlensing events are reviewed, with emphasis on the dramatic expectations for future contributions of microlensing to the field of stellar atmospheres.

1. Introduction

The physics of microlensing is simple. For most current applications, the principles of geometric optics combined with one relation (for the deflection angle) from General Relativity is all that is required. For observed Galactic microlensing events, the distances between source, lens and observer are large compared to intralens distances, so that small angle approximations are valid. Although it is possible that most lenses may be multiple, ~90% of observed Galactic microlensing light curves can be modeled as being due to a single point lens. Usually, though not always (cf Albrow et al. 2000), binary lenses can be considered static throughout the duration of the event.

The magnification gradient near caustics is large, producing a sharply peaked lensing "beam" that sweeps across the source due to the relative motion between the lens and the sight line to the source (Fig. 1). Furthermore, the combined magnification of the multiple microimages (which are too close to be resolved with current techniques) is a known function of source position that is always greater than unity, so that more flux is received from the source during the lensing event. The net result is a well-understood astrophysical tool that can simultaneously deliver high resolution and high magnification of tiny background sources. In Galactic microlensing, these sources are stars at distances of a few to a few tens of kiloparsecs.

The great potential of microlensing for the study of stellar polarization (Simmons, Willis & Newsam 1995; Simmons, Newsam & Willis 1995; Newsam et al. 1998; Gray 2000), stellar spots (Heyrovský & Sasselov 2000; Bryce & Hendry 2000), and motion in circumstellar envelopes (Ignace & Hendry 1999) will not be treated here. Instead, the focus will be on how the composition of spherically-symmetric stellar atmospheres can be probed by microlensing.

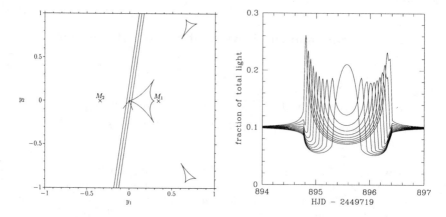

Figure 1. **Left:** The trajectory of the background source star passed over a caustic curve cusp of the binary lens MACHO 97-BLG-28. Units are θ_E. **Right:** The fractional (magnified) flux in 10 concentric rings of equal area over the stellar disk are shown as a function of time during the two-day crossing (Albrow et al. 1999a).

2. Caustic Transits

The angular radius θ_E of a typical Einstein ring is about two orders of magnitude larger than the size θ_* of a typical Galactic source star (few μas), but the gradients in magnification that generate source resolution effects are appreciable only in regions near caustics. For a single point lens, the caustic is a single point coincident with the position of the lens on the sky that must directly transit the background source in order to create a sizable finite source effect. The probability of such a point transit is of order $\rho \equiv \theta_*/\theta_E \approx 2\%$. The amount of resolving power will depend on the dimensionless impact parameter β, the distance of the source center from the point caustic in units of θ_E. The first clear point caustic transit was observed in event MACHO 95-BLG-30 (Alcock et al. 1997).

Lensing stellar binaries with mass ratios $0.1 \lesssim q \equiv m_2/m_1 \lesssim 1$ and separations $0.6 \lesssim d \equiv \theta_{\rm sep}/\theta_E \lesssim 1.6$ generate extended caustic structures that cover a sizable fraction of the Einstein ring (see, eg., Gould 2000). Since events generally are not alerted unless the source lies inside the Einstein ring, any alerted binary event with q and d in these ranges is highly likely to result in a caustic crossing. If the source crosses the caustic at a position at which the derivative of the caustic curve is discontinuous, it is said to have been transited by a cusp. For a given lensing binary, the probability of a cusp transit is of order $\rho\, N_{\rm cusps} \approx 10\%$. Since $\sim 10\%$ of all events are observed to be lensing stellar binaries, the total cusp-transit probability is $\sim 1\%$. To date, two cusp-crossing events have been observed, MACHO 97-BLG-28 (Albrow et al. 1999a) and MACHO 97-BLG-41 (Albrow et al. 2000). The remaining caustic crossings are transits of simple fold (line) caustics, which are observed in $\lesssim 10\%$ of all events. Caustics thus present a non-negligible cross section to background stellar sources, with fold caustic transits being most likely by a factor of ~ 5.

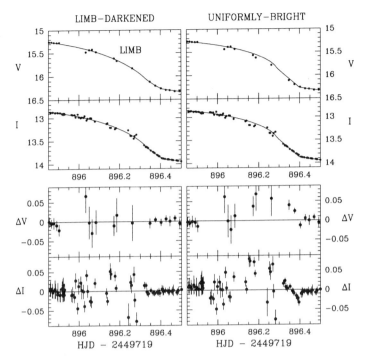

Figure 2. **Top:** The V- and I-band light curves of MACHO 97-BLG-28 plummet as the trailing limb of the source exits the caustic. **Bottom:** Residuals (in magnitudes) from the best models using a uniformly-bright (right) and limb-darkened (left) stellar disk. The same models are superposed on the light curves (Albrow et al. 1999a). The varying scatter reflects different conditions at the three telescopes.

The largest effect of a caustic crossing over an extended source is a broadening and reduction of the light curve peak at transit that depends on the finite size ($\rho \neq 0$) of the source. If the angular size θ_* of the source star can be estimated independently (e.g. from color-surface brightness relations), then the time required for the source to travel its own radius, and thus its proper motion μ relative to the lens, can be determined from the light curve shape. Conversely, unless an independent method is available (cf Han 2000) to measure μ or θ_E, photometric microlensing cannot translate knowledge of the dimensionless parameter ρ into a measurement of source radius. What photometric or spectroscopic data alone *can* yield is a characterization of how the source profile differs from that of a uniform disk (Fig. 2). Microlensing has already yielded such information for stars as distant as the Galactic Bulge and the SMC.

3. Recent Contributions of Microlensing to Stellar Physics

The potential to recover profiles of stellar atmospheres from microlensing has been recognized for several years (Bogdanov & Cherepashchuk 1995; Loeb &

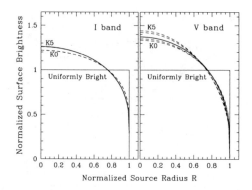

Figure 3. Stellar profiles deduced from microlensing light curve data (bold lines) and from atmospheric models (dashed lines) for the K-giant source star of MACHO 97-BLG-28 (Albrow et al. 1999a).

Sasselov 1995; Valls-Gabaud 1995), but made possible only recently, due to the improved photometry and especially temporal sampling now obtained for a large number of events by worldwide monitoring networks. For only a few stars, most of which are supergiants or very nearby, has limb darkening been observationally determined by any technique. Microlensing has the advantage that: (1) many types of stars can be studied, including those quite distant; (2) the probe is decoupled from the source; (3) the signal is amplified (not eclipsed); and (4) intensive observations need only occur over one night.

3.1. Limb Darkening

The first cusp crossing was observed in MACHO 97-BLG-28, and led to the first limb-darkening measurement of a Galactic Bulge star (Albrow et al 1999a). As the source crossed the caustic cusp, a characteristic anomalous bump was generated in the otherwise smooth light curve. First the leading limb, then the center, and finally the trailing limb of the stellar disk were differentially magnified (Fig. 1). Analysis of the light curve shape during the limb crossing allowed departures from a uniformly-bright stellar disk to be quantified (Fig. 2) and translated into a surface brightness profile in the V and I passbands. A two-parameter limb-darkened model provided a marginally better fit than a linear model. Spectra provided an independent typing of the source as a KIII giant. The stellar profile reconstructed from the microlensing light curve alone is in good agreement with those from stellar atmosphere models (van Hamme 1993; Claret, Diaz-Cordoves, & Gimenez 1995; Diaz-Cordoves, Claret, & Gimenez) for K giants fitted to the same two-parameter (square-root) law (Fig. 3).

This first microlensing measurement of limb darkening was encouraging, but constructing realistic error bars for the results proved awkward. In traditional parametrizations for limb darkening the coefficients c_λ and d_λ defined by

$$I_\lambda(\theta) = I_\lambda(0) \left[1 - c_\lambda(1 - \cos\theta) - d_\lambda(1 - \cos^n \theta)\right] \quad \text{where } n = 0, 1/2, 2 \quad (1)$$

are correlated not only with each another, but also with other parameters in the microlensing fit because they carry information about the total flux F of

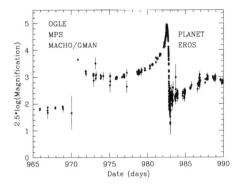
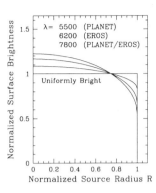

Figure 4. **Left:** The binary lens event MACHO 98-SMC-1 was exceptionally well-sampled due to the efforts of five microlensing teams. **Right:** Data over the second fold caustic crossing allowed one-parameter profiles to be deduced at several wavelengths for this A-dwarf source star in the SMC. (Based on Afonso et al. 2000.)

the source. (Here θ is the angle between the normal to the stellar surface and the line of sight.) A different parametrization was therefore constructed for the analysis of fold caustic crossings (Albrow et al. 1999b),

$$I_\lambda(\theta) = \langle I_\lambda \rangle \left[1 - \Gamma_\lambda (1 - \frac{3}{2}\cos\theta) \right] \quad \text{where} \quad \langle I_\lambda \rangle \equiv \frac{F}{\pi \theta_*^2} \quad , \quad (2)$$

which decouples the limb-darkening parameter Γ_λ from the source flux. To first order, $c_\lambda = 3\Gamma_\lambda/(\Gamma_\lambda + 2)$. This form was implemented in the analysis of the multiband data collected by five teams (Fig. 4) for the fold caustic crossing event MACHO 98-SMC-1 (Afonso et al. 2000). The source star was typed from spectra to be an A dwarf in the Small Magellanic Cloud (and thus a radius $\theta_* = 80$ nanoarcsec!). As expected, limb darkening decreases with increasing wavelength and at given wavelength is smaller for a hot dwarf than a cool giant (Figs. 3 & 4). Unfortunately, no models of metal-poor A-dwarf stars were available for direct comparison with the 98-SMC-1 limb-darkening measurements.

A third microlensing limb-darkening measurement was made for the cool (4750 K) giant source star in the Galactic Bulge event MACHO 97-BLG-41. This was a cusp-crossing event in which rotation of a binary lens was measured for the first time (Albrow et al. 2000, see also Menzies et al. 2000). The linear parameter $\Gamma_I = 0.42 \pm 0.09$, corresponding to $c_I = 0.52 \pm 0.10$, determined from the light curve agreed well with that of $c_I \approx 0.56$ from atmospheric models (Claret et al. 1995) of stars of the appropriate temperature and gravity.

The photometric precision required to recover linear limb-darkening coefficients with 10% accuracy has been estimated recently by Rhie & Bennett (2000), and found to be about 1% (relative photometry). As inspection of the residuals (Fig. 2) clearly reveals, a worldwide network of 1m-class telescopes is quite capable of this precision. At the current rate of observed caustic crossing events, the community can thus expect that microlensing will provide 2 or 3 limb-darkening measurements per year; indeed more results are now in preparation.

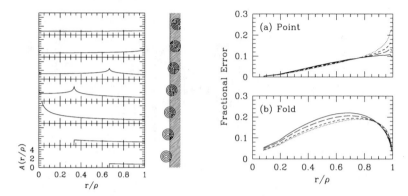

Figure 5. **Left:** The azimuthally-averaged magnification as a function of fractional position r/ρ on stellar disk as a source (with $\rho = 1$) moves (from top to bottom) out of a caustic over a fold singularity. **Right:** Fractional error in the recovered intensity profile ($\delta I/I$) as a function of r/ρ for a direct point transit (top) and a fold crossing (bottom). Typical results are displayed for a 2-m telescope, including limb-darkening effects for VIK passbands (Gaudi & Gould 1999).

3.2. Spectroscopy

The magnification boost provided by microlensing can yield higher S/N spectra of faint sources than would otherwise be possible. Lennon et al. (1996) used microlensing to measure the effective temperature, gravity and metallicity of a G dwarf in the Bulge; at the time of the caustic boost, the 3.5-m NTT had the collecting power of a 17.5-m aperture. In another case, lithium was detected in a Bulge turn-off star using Keck and microlensing (Minitti et al. 1998).

Attempts have been made to perform time-resolved spectroscopy during caustic crossings in order to detect the varying spectral signatures expected (Loeb & Sasselov 1995; Valls-Gabaud 1998) as light from different positions across the stellar disk (and thus different optical and physical depths) is differentially magnified. The caustic alert provided by the MACHO team (Alcock et al. 2000) allowed Lennon and colleagues (1996, 1997) to take spectra over the peak of the fold crossing in MACHO 96-BLG-3, though these did not extend far enough down the decline to detect spectral differences. Temporal coverage was also insufficient to detect strong spectral changes during the point caustic transit in MACHO 95-BLG-30, although slight equivalent width variations in TiO (Alcock et al. 1997) and Hα (Sasselov 1998) lines may have been seen.

4. Stellar Tomography: A New Era of Stellar Atmosphere Physics

The ability of current microlensing collaborations to predict (fold) caustic crossings ~1-3 days in advance opens a new era for stellar atmosphere physics. Follow-up teams, separately or in collaboration with existing microlensing networks, could obtain spectrophotometric data on auxiliary telescopes in order to take advantage of the magnification and resolution afforded by microlens caustics.

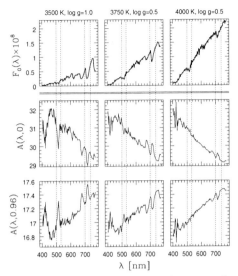

Figure 6. Unlensed (top) and microlensed spectra for three different model atmospheres of a cool giant. Microlensed spectra are for a point caustic at the center (middle) and limb (bottom) of the source. Vertical dotted lines mark TiO bands (Heyrovský, Sasselov & Loeb 2000).

Theoretical expectations of how microlensing may contribute to our understanding of stellar atmospheres over the next decade are encouraging. Gaudi & Gould (1999) have simulated a 2m-class telescope with a 1Å-resolution spectrograph continuously observing a typical $V = 17$ ($\rho = 0.02$) Bulge source undergoing a fold caustic crossing of 7 hours duration. They find that the intensity profile can be recovered to a precision of 10-20% using a spatial resolution of 10% across the star for most wavelengths. Most of the spatial resolution generated by fold caustics comes from the period in which the trailing limb is exiting the caustic curve. Direct transits of point caustics could provide even more reliable intensity profiles (Fig. 5), but due to geometric factors are much more unlikely to occur. These results appear to agree with the estimates derived from different approaches (Hendry et al. 1998; Gray & Coleman 2000).

Loeb & Sasselov (1995) noted that the "light curves" of atmospheric emission lines that are most prominent in the cool outer layers of giants will experience sharp peaks when the limb (rather than the center) of the source crosses a caustic. Recently, this work has been extended (Heyrovský, Sasselov & Loeb 2000) to lines across the whole optical spectrum to show how time-resolved spectroscopy during a caustic crossing can discriminate between different atmospheric models. For a given impact parameter, a particular line may appear in emission or absorption depending on the temperature structure of the star (Fig. 6). Since the duration of the caustic crossing is limited, 4-m and 8-m telescopes will be required to obtain the highest spectral resolution.

Experience and modeling thus indicate that any aperture can perform microlensing tomography of stellar atmospheres in the next decade, *provided that the community is willing to reschedule telescope access on a few days' notice.*

Acknowledgements

The author gratefully acknowledges support from the Nederlandse Organisatie voor Wetenschappelijk Onderzoek (GBE 614-21-009) and the meeting LOC.

References

Afonso, C., et al. (EROS, MACHO, MPS, OGLE & PLANET Collaborations) 2000, ApJ, 532, 340
Albrow, M.D., et al. (PLANET Collaboration) 1999a, ApJ, 522, 1011
Albrow, M.D., et al. (PLANET Collaboration) 1999b, ApJ, 522, 1022
Albrow, M.D., et al. (PLANET Collaboration) 2000, ApJ, 534, 000
Alcock, C., et al. (MACHO Collaboration) 1997, ApJ, 491, 436
Alcock, C., et al. (MACHO Collaboration) 2000, ApJ, submitted (astro-ph/9907369)
Bogdanov, M.B. & Cherepashchuk, A.M. 1995, Astronomy Letters, 21, 505
Bryce, H.M., & Hendry, M.A. 2000, these proceedings
Claret, A., Diaz-Cordoves J., & Gimenez, A., 1995, A&AS, 114, 247
Diaz-Cordoves J., Claret, A., & Gimenez, A., 1995, A&AS, 110, 329
Gaudi, B.S., & Gould, A. 1999, ApJ, 513, 619
Gould, A. 2000, these proceedings (astro-ph/0004042)
Gray, N. 2000, MNRAS, submitted (astro-ph/0001359)
Gray, N. 2000 & Coleman, I.J. 2000, these proceedings (astro-ph/0004200)
Han, C. 2000, these proceedings (astro-ph/0003369)
Hendry, M.A., Coleman, I.J., Gray, N., Newsam, A.M., & Simmons, J.F.L. 1998, New Astron. Reviews, 42, 125
Heyrovský, D., & Sasselov, D. 2000, ApJ, 529, 69
Heyrovský, D, Sasselov, D., & Loeb, A. 2000, ApJ, submitted (astro-ph/9902273)
Ignace, R., & Hendry, M.A. 1999, A&A, 341, 201
Lennon, D.J., et al. 1996, ApJ, 471, L23
Lennon, D.J., et al. 1997, ESO Messenger, 90, 30 (astro-ph/9711147)
Loeb, A. & Sasselov, D. 1995, ApJ, 491, L33
Menzies, J.W., et al. (PLANET Collaboration) 2000, these proceedings
Minitti, D. et al. 1998, ApJ, 499, L175
Newsam, A.M., Simmons, J.F.L., Hendry, M.A., & Coleman, I.J. 1998, New Astron. Reviews, 42, 121
Rhie, S. & Bennett, D. 2000, ApJ, submitted (astro-ph/9912050)
Sasselov, D. 1998, ASP Conf. Ser. 154, Cool Stars, Stellar Systems and the Sun, eds. R. A. Donahue and J.A. Bookbinder, 383
Simmons, J.F.L, Willis, J.P., & Newsam, A.M. 1995, A&A, 293, L46
Simmons, J.F.L, Newsam, A.M., & Willis, J.P. 1995, MNRAS, 276, 182
Valls-Gabaud 1995, Large Scale Structure in the Universe, eds. J.P. Muecket, S. Gottloeber and V. Mueller, 326

Valls-Gabaud, D. 1998, MNRAS, 294, 747
van Hamme, W. 1993, AJ, 106, 2096

Discussion

Feast: Can you say what is likely to be the contribution to limb darkening problems from microlensing compared with that which is in principle possible from eclipsing binaries?

Sackett: Microlensing has the advantages that the probe is detached from the source, that many different types of source stars can be studied in this way, and that intensive observations are generally required only over one night. I can't claim to have done a complete literature search, but my impression is that limb-darkening is generally assumed, not derived, from eclipsing binaries.

Hauschildt: 1. Contribution functions are good visualisation tools but in reality RT effects complicate matters; but this can be addressed by modelling.
2. What telescope size would you require for limb-darkening observations by microlensing?

Sackett: 1. I agree. The contribution function plots were only meant to be illustrative.
2. Limb-darkening observations are currently being done with 1-m telescopes. This will be even easier when the PLANET dual-beam cameras are in full operation. In order to get time-resolved spectral information, larger apertures will be required. Since the time available (several hours) during the caustic crossing is fixed, it will always be a question of desired resolution. For some problems, 4- and 8-m telescopes may be required.

Bennett: There can be some uncertainty in the duration of the caustic crossing. For MACHO-98-SMC-1, the predictions ranged from 30 minutes to 12 hours.

Sackett: That's absolutely correct, since it depends on the proper motion of the lensing system. MACHO-98-SMC-1 is a bit of a special case since the ambiguity there was whether the lens was a fast moving halo object or not. Most Bulge caustic crossings will last several hours.

Binney: When considering the merits of data obtained during microlensing events versus traditional approaches to determining fundamental data for stellar calibration (interferometry, eclipsing binaries, etc), is it a problem that the objects one gets microlensing data for are remote and faint?

Sackett: Microlensing boosts the signal from distant sources precisely at the time that the limb darkening is being measured. In the age of 8- to 10-m telescopes, follow-up spectroscopy and photometry won't really be a problem; the current limb-darkening measurements are coming from 1-m telescopes. One does need to ensure that the blend fraction is well-determined.

Gould: Ultimately we can figure out what type of star we have from spectra taken during the event, but right now we are limited in our interpretation of the event by the error in the $(V - I)$ color.

Cook: I would like to point out that the remoteness of the sources being lensed may be an advantage. For example, microlensing of Bulge main sequence stars may be used to provide more photons for high resolution spectroscopy, and I am involved in a project to study the chemical evolution of the Bulge by determining detailed abundances with Keck, which is only possible when the source is significantly magnified.

A Free-Floating Planet Population in the Galaxy?

Hans Zinnecker

Astrophysikalisches Institut Potsdam, An der Sternwarte 16, D-14482 Potsdam, Germany

Abstract. Most young low-mass stars are born as binary systems, and circumstellar disks have recently been observed around the individual components of proto-binary systems (e.g. L1551-IRS5). Thus planets and planetary systems are likely to form around the individual stellar components in sufficiently wide binary systems. However, a good fraction of planets born in binary systems will, in the long run, be subject to ejection due to gravitational perturbations. Therefore, we expect that there should exist a free-floating population of Jupiter-like or even Earth-like planets in interstellar space. There is hope to detect the free-floating Jupiters through gravitational microlensing observations towards the Galactic Bulge, especially with large-format detectors in the near-infrared (e.g. with VISTA or NGST), on timescales of a few days.

1. Speculation of Existence

Observations over the past decade have shown that most young low-mass stars are born as binary systems (e.g. Mathieu 1994, Mayor et al. 2000). Furthermore, circumstellar disks have recently been observed around the individual stellar components of protobinary systems (e.g. L1551-IRS5, Rodriguez et al. 1998; HK Tau, Menard & Stapelfeldt 2000). Thus, planets and planetary systems are likely to form around the individual components in sufficiently wide binary systems. A case in point is 16 Cygni, where a Jupiter-like companion has been detected around component B by radial velocity measurements (Cochran & Hatzes 1996).

Here we speculate that a good fraction of planets born in binary systems will be subject to dynamical ejection due to gravitational perturbations resulting from periastron passages of the stars revolving around each other, typically in rather eccentric orbits. [1] Therefore we expect a population of ejected and thus free-floating planets to exist in the Galactic Disk – and maybe in the Galactic Halo as well, as the frequency of wide visual binaries (separations > 30 AU) seems to be at least as high for halo stars as for disk stars (Köhler et al. 2000).

---

[1] even though the case of TMR-1C (Terebey et al. 2000) proved to be a background object

2. Caveat

To be fair, there are also arguments against our speculation. For example, planet formation may be inhibited by the mutual spiral shocks induced in the respective circumstellar disks which will heat the disk gas and dust (Nelson 2000). This might cause the dust grains to evaporate and be destroyed, certainly the ice mantles if not the silicate cores, too, depending on the shock temperature (bigger/smaller than 1500K).Thus grain growth and the collisional build-up of planetesimals will be prevented. Similarly, direct gravitational instability of the gas disk (another possibility to form a giant Jupiter-like planet) is impeded, if the disk is too hot (the Toomre Q parameter exceeds unity for too large a sound speed or gas temperature, even if the disk surface density is high). Finally, the Goldreich-Ward (1973) instability of a cold thin dust disk, even if it operated in a circumstellar disk around a single star, may not do so in the respective disks around binary star components, as the dust may never settle into the disk's midplane in a binary star+disk system.

While these qualitative arguments against planet formation in binary systems must be further investigated and while a lot of uncertainty remains, it is nonetheless worthwhile to proceed on the assumption that some planet formation occurs even in binary systems, especially in those which are wider than a critical separation of the stellar pair, which we take to be the separation where the mode of the semi-major axis distribution occurs – 30 AU according to Duquennoy & Mayor (1991). That is, we assume that 50% of all low-mass binaries can indeed form planets.

3. Prospects for Detection

Next we evaluate the prospects for detecting a population of Jupiter-mass free-floating planets towards the Galactic Bulge (Center). We also bracket our estimates by considering objects a factor of 10 more massive (i.e. minimum mass brown dwarfs) and a factor 10 less massive (maximum mass of Earth-type planets). We consider planets associated with stars below 1 $M_\odot$ (with main sequence lifetimes exceeding the age of the Galaxy). We assume 50% binary frequency for these low-mass stars, half of which we expect to form planets. Each planet forming binary system is assumed to form 4 planets, two around each component, on average. Finally we assume that one of the two planets per stellar component is eventually ejected.

The number density of low-mass stellar systems in the solar neighborhood is $n_* = 0.1$ per pc^3. Thus, under the above assumption, the number density of free-floating planets (Jupiters) will be $n_p = 0.05$ per pc^3. Then the surface density of free-floating planets towards the Galactic Center will be of order $N_p = 10^3$ pc^{-2}, assuming a distance to the Galactic Center of 10 kpc and an average stellar density in the inner Galaxy 2 times higher than in the local Solar Neighborhood.

Now, the probability P (also sometimes called the "optical depth") for a microlensing occurrence is given by the following area coverage factor

$$P = \pi \times R_{E,p}^2 \times N_p$$

where $R_{E,p}$ is the Einstein radius of the planet ($R_{E,p} = \sqrt{4GM_pD_sx(1-x)}/c^2$, with D_s being the distance to the source population, here of order 10 kpc; and x=D_L/D_s, the ratio of the distance of the lens population to the source population, typically x=0.5).
When normalized to $M_{Jup} = 10^{-3}$ $M_\odot$ and for x=0.5, numerically this turns out to be

$$R_{E,p} = 10^{12.5} \times \sqrt{M/M_{Jup}} \text{ cm}.$$

Hence

$$P = 3 \times 10^{-9} \times (M/M_{Jup})$$

is a reasonable estimate for the "optical depth" due to free-floating planets of mass M towards the inner Galaxy.
The timescale for Einstein ring crossing, i.e. the half-width of the lightcurve of the source magnification, is given by
$$t = R_{E,p}/v$$
where v $\sim$ 200 km/s is the relative speed of the observer with respect to the lens.
One finds
$$t = 10^{12.5}/10^{7.3} \text{ sec} = 1.5 \times 10^7 \text{ sec} \sim 2 \text{ days}$$

4. Detection Requirements

The above numbers translate into the following detection requirements: to detect a microlensing event due to free-floating Jupiters in the inner Galaxy we need to observe some 3×10^8 stellar objects in the Galactic Bulge for a few days, with a time resolution of hours. This is difficult but not impossible. For example, the VST 2.5m telescope to be installed on Paranal/Chile in 2002 with its 1 degree field-of-view in very good seeing (0.4 arcsec) corresponds to $\sim 10^8$ pixels, and may thus be able to realistically resolve some 10^7 point sources down to the confusion limit in the far-red bands (10 pixels per object). Therefore some 30 fields of 1 degree must be monitored each night. The VISTA telescope, a 4m mirror on Paranal with a 1 degree field-of-view for infrared (JHK) observations, will be even better (first light in 2005) because in the near-infrared we can penetrate the dust towards the Galactic Center, providing a higher and fainter source surface density. Thus the confusion limit will be reached more quickly and 30 fields can actually be monitored once or twice per night (good sampling). Finally, it is conceivable that the 8m Next Generation Space Telescope (NGST), to be launched in 2009, will be able to do the job, as it will mostly operate in staring mode and always at the diffraction limit (60 mas). Its field-of-view is around 4 arcmin (8k x 8k detectors with 30 mas pixels), corresponding to $10^{7.5}$ pixels. Due to the much higher spatial resolution compared to VISTA, the NGST will reach a fainter confusion limit than VISTA within a short exposure time (a few seconds, or tens of seconds, depending on the precise galactic longitude and latitude) and will likely resolve $10^{6.5}$ point sources. It then 'just' needs 100 such short exposures of the same field, sampled at intervals of a few hours, in order to monitor three hundred million objects for a microlensing variability of a few days, thus finding out if the predicted microlensing by free-floating Jupiters occurs. If so, this would herald the detection of a new class of celestial objects.

5. Acknowledgement

My first wee attempt to estimate the number of free-floating planets ejected from binary systems and their detection by gravitational microlensing goes back to a young binary star workshop in Stony Brook 1996. Since then, I benefitted enormously from the expertise of my colleague Joachim Wambsganss to whom I owe both my increasing interest and increasing knowledge of gravitational microlensing. I am also grateful to Rosanne DiStefano for insightful discussions.

References

Cochran, W.M., Hatzes, A.P. Butler, R. P., Marcy, G. 1997, ApJ, 483, 457.
DiStefano & Scalzo 1999, ApJ. 512, 564
Duquennoy & Mayor 1991, A & A 248, 485
Goldreich, P., Ward, M. 1973, Ap.J., 183, 1051
Köhler, Zinnecker, & Jahreiss 2000, in Birth and Evolution of Binary Stars (Poster Proc. IAU-Symp. 200), eds. Reipurth & Zinnecker, p. 148
Mathieu 1994, ARAA 32, 465
Mayor et al. 2000, in The Formation of Binary Stars, IAU Symp. 200, eds. Zinnecker & Mathieu, ASP Conf. Series, in press
Menard & Stapelfeldt 2000, in The Formation of Binary Stars, IAU Symp. 200, eds. Zinnecker & Mathieu, ASP Conf. Series, in press
Nelson 2000, ApJ. Lett., in press
Rodriguez et al. 1998, Nature 395, 355
Terebey et al. 2000, A.J. 119, 2341

Discussion

Carr: Both MACHO and EROS have a candidate event with the short duration anticipated. Is this consistent with your expected event rate?

Zinnecker: Maybe. I must check their shortest events. I suspect, however, that the event rate for free Jupiter lenses is still smaller but I am not sure.

Bennett: 1) I think that planet formation in single star systems is expected to eject at least some low mass planets. 2) The Galactic Exoplanet Survey Telescope (GEST), a proposed 1.5-m space telescope with a $\sim$1 sq degree field of view, should be able to see $\sim$50 isolated Earth mass planets if there is one such planet for every star in the Galaxy. 3) MACHO-95-BLG-3 appears to be an example of lensing by an isolated Jupiter mass planet. It was a 2-day event with a clump giant source and it doesn't appear to be consistent with the tail of the timescale distribution of normal events.

Graff: As J-F Glicenstein will discuss EROS has a 6-day event towards the spiral arms which could be a planet. It could also be a high velocity star. In general it will be difficult to separate these two possibilities.

Gaudi: Your idea is related to one originally suggested by Rosanne DiStefano & Scalzo (1999) regarding detections of planets in wide orbits around stars These planets would give similar signatures to "ejected" planets. So one can never be sure about "free-floating" planets.

Zinnecker: Yes, free-floating Jupiters will contribute the same signal as bound Jupiters if their orbital radii exceed 3 $R_{E,p}$ or about 1–2 AU. I must admit that I had not realized this initially. However, discrimination between bound and free-floating Jupiters is possible if most Jupiters were close in, like 51 Peg B.

Part 4: Galactic Structure and Constituents

Microlensing and Galactic Structure

James Binney

Oxford University, Theoretical Physics, Keble Road, Oxford, OX1 3NP, U.K.

Abstract. Because we know little about the Galactic force-field away from the plane, the Galactic mass distribution is very ill-determined. I show that a microlensing survey of galaxies closer than 50 Mpc would enable us to map in three dimensions the Galactic density of stellar mass, which should be strictly less than the total mass density. A lower limit can be placed on the stellar mass needed at $R < R_0$ to generate the measured optical depth towards sources in the bulge. If the Galaxy is barred, this limit is lower by a factor of up to two than in the axisymmetric case. Even our limited knowledge of the Galactic force field suffices to rule out the presence of the amount of mass an axisymmetric Galaxy needs to generate the measured optical depth. Several lines of argument imply that the Galaxy is strongly barred only at $R < 4$ kpc, and if this is the case, even barred Galaxy models cannot generate the measured optical depth without violating some constraint on the Galactic force-field. Galactic mass models that are based on the assumption that light traces mass, for which there is significant support in the inner Galaxy, yield microlensing optical depths that are smaller than the measured value by a factor of more than 2.5.

1. Introduction

Notwithstanding the difficulties to which heavy obscuration by dust gives rise, we now have a reasonable idea of how luminosity is distributed in the Galaxy. We know we live in a galaxy that has a bar, 3 – 4 kpc long, whose nearer end lies at positive longitudes. Within the bar there is a moderately flattened component in which the luminosity density rises roughly as $r^{-1.8}$ with decreasing galactocentric radius r. Around the bar there is a roughly exponential disk with scale length ~ 3 kpc, and a moderately flattened metal-poor halo in which the luminosity density varies as $\sim r^{-3.5}$.

The situation regarding the Galactic distribution of mass is very much less satisfactory. Observations of external galaxies and cosmological theory have convinced us that light is not a good tracer of mass. Hence we cannot straight-forwardly translate our models of the luminosity distribution into models of the mass distribution; in principle, we should start afresh and derive the mass distribution by tracing out the Galaxy's gravitational field $\mathbf{F}(\mathbf{r})$ and then applying the divergence operator: $\nabla \cdot \mathbf{F} = -4\pi G\rho$.

From observations of gas that flows in the Galactic plane we have a fair idea of what $\mathbf{F(r)}$ looks like at points $\mathbf{r}$ within the plane, but we have very little secure knowledge of $\mathbf{F}$ out of the plane. This deficiency is serious, because until we know $\mathbf{F}$ away from the plane, we cannot apply ∇ *anywhere*, and can say *nothing* with security about ρ *anywhere*!

What we currently know about $\mathbf{F}$ out of the plane relates to points near the Sun. Crézé et al. (1998) and Holmberg & Flynn (2000) have used proper motions of Hipparcos stars to estimate the gradient of F_z near the plane and hence infer that the local mass density is $0.076 \pm 0.015 \, \mathrm{M}_\odot \, \mathrm{pc}^{-3}$ and $0.10 \pm 0.01 \, \mathrm{M}_\odot \, \mathrm{pc}^{-3}$, respectively, while Kuijken & Gilmore (1991) have used the radial velocities of stars at the SGP to estimate F_z at $z = -1.1 \, \mathrm{kpc}$ and thus infer that the surface density of the Galaxy within 1.1 kpc of the plane is $71 \pm 6 \, \mathrm{M}_\odot \, \mathrm{pc}^{-3}$. To constrain $\mathbf{F}$ at points that lie far from the Sun, we have to study objects that move to such points. High-velocity stars are the obvious tracers to use, because they are so numerous. They strongly constrain models of $\mathbf{F}$ because stars on essentially the same orbit can be studied both locally and in situ (Binney 1994; Dehnen & Binney 1996). Unfortunately, the potential of this approach has yet to be systematically exploited. Recently, there has been considerable interest in using tidal streamers associated with disrupted satellites as tracers of $\mathbf{F}$ (Johnston et al. 1999, Helmi et al. 1999). My own view is that high-velocity stars have greater potential because (a) they are vastly more numerous and (b) they do not require approximations of the level of the (manifestly false) assumption that all elements of a streamer are on the same orbit. Moreover, extracting useful information from streamers requires space-based astrometry, the requisite observations high-velocity stars are available now, and all that's lacking is machinery for modelling them.

Gravitational microlensing directly probes the Galaxy's mass distribution, but in a very different way from classical studies of gas and stars. In fact, microlensing does not measure the smooth Galactic force-field $\mathbf{F}$ but graininess in the mass distribution. Hence, it is insensitive to the contribution to the latter from elementary particles and gas, and is therefore complementary to the traditional approach to the determination of ρ from $\mathbf{F}$, rather than competitive with it.

2. From optical depth to stellar density

The optical depth to microlensing of a stellar object at distance s_0 is

$$\tau = \frac{4\pi G}{c^2} \int_0^{s_0} \mathrm{d}s \, \rho_* \hat{s}, \tag{1}$$

where

$$\hat{s} = \left(\frac{1}{s_0 - s} + \frac{1}{s}\right)^{-1} \tag{2}$$

is the harmonic mean of the source-lens and lens-observer distances. If the source is extragalactic, $\hat{s} \simeq s$, and τ becomes proportional to $\int \mathrm{d}s \, \rho_* s$ in the direction of the source. Suppose we measure τ along many lines of sight all over the sky. Can we reconstruct ρ_* from the data?

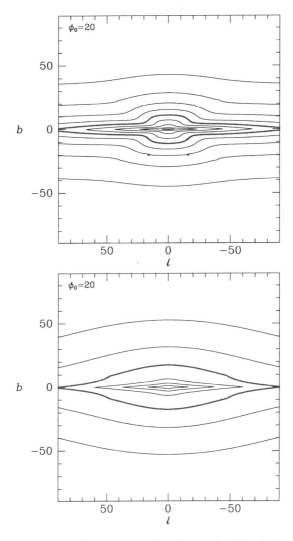

Figure 1. Contours of equal microlensing optical depth for a distant source for two Galactic models. There are two contours per decade and the heavy contour is for $\tau = 10^{-6}$. The upper panel is for the model one obtains from the luminosity model of Kent, Dame & Fazio (1991) with an assumed mass-to-light ratio $\Upsilon_K = 1$. The lower panel is for Model 1 of Dehnen & Binney (1998), and all components, including the dark halo, have been assumed to contribute fully to τ. Consequently, the optical depth at large $|b|$ is larger in the lower than in the upper panel.

With some simplifying assumptions, we can. Binney & Gerhard (1996) show that if the Galactic luminosity density $j(\mathbf{r})$ is symmetric about the Galactic plane and two other, orthogonal planes (as it would be if it were triaxially ellipsoidal), then a Richardson–Lucy algorithm can be used to recover $j(\mathbf{r})$ from its line-of-sight projection, $I(\Omega) = \int ds\, j$, at each point Ω on the celestial sphere. It is straightforward to modify the derivation of Binney & Gerhard to show that, with the same assumptions regarding symmetry, a Richardson–Lucy algorithm for the recovery of ρ_* from $\tau(\Omega)$ is

$$\rho_*^{(k+1)}(\mathbf{r}) = \rho_*^{(k)}(\mathbf{r}) \sum_{i=1}^{8} \frac{\tau(\Omega_i)}{\tau^{(k)}(\Omega_i)} \frac{1}{s(\mathbf{r}_i)} \Big/ \sum_{i=1}^{8} \frac{1}{s(\mathbf{r}_i)}. \tag{3}$$

Here the sum over i is over the eight points $\mathbf{r}_i$ that are connected to $\mathbf{r}$ by the assumed symmetry of ρ_*; the lines of sight to these points are in the directions Ω_i, and their distances from the Sun are $s(\mathbf{r}_i)$.

While it is in principle possible to recover ρ_* from τ, it is not clear that this will ever be done. The problem is the small numerical factor in front of the integral in (1). Numerically,

$$\tau = 6 \times 10^{-6} \left[\frac{\int_0^{s_0} ds\, \rho_* s}{10^{10}\, \mathrm{M}_\odot/\mathrm{kpc}} \right], \tag{4}$$

where we have again assumed that the source is extragalactic. Figure 1 plots τ for two typical Galactic models. The model underlying the upper panel is the K-band luminosity model of Kent, Dame & Fazio (1991), which has been converted into a model of the stellar mass distribution by assuming a mass-to-light ratio $\Upsilon_K = 1$, which appears to be correct for the solar neighbourhood (§10.4.4 of Binney & Merrifield 1998). One sees that τ exceeds 10^{-6} only for lines of sight at fairly low latitudes.

Existing data indicate that the duration of a microlensing event is typically tens of days, so in an observing season one obtains at most a handful of statistically independent observations per line of sight. Hence, in a given area of the sky the number of lines of sight that must be monitored for of order years to distinguish τ from zero is $\sim 1/\tau \gtrsim 10^6$. Finding this number of extragalactic sources in each of a large number of patches of the sky is hard. Probably our best chance is offered by 'pixel lensing' towards nearby galaxies. Gould (1996) gives the necessary theory and I adopt his notation. A large rôle is played in this by the effective stellar flux F_*, which is defined in terms of the stellar luminosity function ϕ by

$$F_* = \int dF\, \phi(F) F^2 \Big/ \int dF\, \phi(F) F. \tag{5}$$

Per resolution element on a galaxy image, the rate at which microlensing events can be detected above a signal-to-noise threshold $Q_{\min}$ is

$$\Gamma = \frac{2\kappa \xi}{Q_{\min}^2} \tau \alpha F_*, \tag{6}$$

where $1 > \kappa, \xi$ are dimensionless functions, and αF_* is the rate at which the telescope detects photons from an object of flux F_*. In terms of the surface brightnesses of galaxy and sky S and S_{sky}, we have $\kappa \equiv (1+S_{sky}/S)^{-1} \simeq S/S_{sky}$ over most of a galactic image. The value of ξ depends on the degree to which the image is resolved into stars: if a significant part of the integral on the top of equation (5) comes from stars bright enough that lensing of them can be detected even at large impact parameter ($u \gtrsim 0.25$), then ξ is small, with $\xi \sim 1$ otherwise.

Gould shows that if one is interested only in measuring τ regardless of the masses of the lenses that generate it, the optimal observational strategy is to work in the regime $\xi \sim 1$ in which one detects only high-magnification events. We shall be in this regime provided

$$\frac{\kappa \alpha F_*^2}{\omega F_{psf}} < \frac{Q_{min}^2}{4\pi}, \tag{7}$$

where ω is the rate constant of a typical microlensing event and F_{psf} is the flux in a resolution element of the galactic image. How big will these quantities be in a typical case? Suppose we are using a diffraction-limited telescope of diameter D to study a galaxy of distance s at radius R_{25}, where the V-band surface brightness will be $\sim 24 \, \text{mag arcsec}^{-2}$. Gould gives the absolute magnitude corresponding to F_* as

$$M_{*I} = -4.84 + 3(V - I) \tag{8}$$

and estimates that from a star with $I = 20$ our telescope will collect photons at a rate $10(D/1\,\text{m})^2 \, \text{s}^{-1}$. Hence,

$$\alpha F_* = 10\Big(\frac{D}{1\,\text{m}}\Big)^2 10^{-0.4(30-20-4.84+3(V-I))}\Big(\frac{s}{10\,\text{Mpc}}\Big)^{-2}. \tag{9}$$

The I-band flux in the telescope's resolution element is

$$F_{psf} = F_0 10^{-0.4(24-(V-I))} \Big[0.206\Big(\frac{\lambda/1000\,\text{nm}}{D/1\,\text{m}}\Big)\Big]^2, \tag{10}$$

where F_0 is some universal constant. We can express F_* in terms of this same constant and M_{*I} thus

$$F_* = F_0 10^{-0.4(30-4.84+3(V-I))}\Big(\frac{s}{10\,\text{Mpc}}\Big)^{-2}. \tag{11}$$

When we substitute equations (9), (10) and (11) into equation (7) and assume (Table 4.4 of Binney & Merrifield, 1998)

$$\kappa \sim 10^{-0.4(4-(V-I))} \tag{12}$$

and $V - I \sim 1.25$ (de Jong 1995) the condition to be in the high-magnification regime becomes

$$10.4\Big(\frac{D}{1\,\text{m}}\Big)^4\Big(\frac{s}{10\,\text{Mpc}}\Big)^{-4}\Big(\frac{\lambda}{1000\,\text{nm}}\Big)^{-2}(\omega 1\text{week})^{-1} < \frac{Q_{min}^2}{4\pi}. \tag{13}$$

Gould finds that $Q_{\min} \sim 7$ is required for detection, so this condition is satisfied for $D \lesssim 4\,\text{m}$ and $s \sim 50\,\text{Mpc}$. By equations (6), (9) and (12) the event rate in this regime is

$$\begin{aligned}\Gamma &= \frac{2\tau}{Q_{\min}^2} 10^{-0.4(6.66+2(V-I))} \left(\frac{D}{1\,\text{m}}\right)^2 \left(\frac{s}{10\,\text{Mpc}}\right)^{-2} \text{s}^{-1} \\ &= \frac{262\tau}{Q_{\min}^2} \left(\frac{D}{1\,\text{m}}\right)^2 \left(\frac{s}{10\,\text{Mpc}}\right)^{-2} \text{week}^{-1}\end{aligned} \quad (14)$$

Hence, by directing a 4 m telescope along a line of sight with $\tau = 10^{-6}$, we should be able to detect three to four events per 10^6 resolution elements per week. In fact a somewhat higher event rate could be achieved if the telescope were in space, because the event rate is inversely proportional to the sky brightness, which will at least 2 mag fainter in space than the ground-based value I have assumed.

If the PSF of the telescope has FWHM of x arcsec, a galaxy of diameter y arcmin offers $3600(y/x)^2$ pixels to monitor. An L_* galaxy at a distance of 50 Mpc has $y \sim 1$, and will provide 10^6 pixels if $x \sim 0.06$. Thus, a diffraction-limited telescope of modest aperture monitoring galaxies within 50 Mpc for of order a year could map τ sufficiently extensively for it to be possible to recover ρ_* from a scheme such as (3).

Of course, the values of τ recovered from such a survey would include contributions from self-lensing within the target object in addition to the optical depth through the Milky Way. As Crotts (1992) pointed out in relation to pixel lensing of M31, variations in τ across the image of an highly inclined galaxy would help one to determine how much self lensing occurs in galactic halos rather than in disks or bulges. Consequently, the analysis of the data from the survey would proceed by modelling in some detail the distribution of τ in each target galaxy, with the Milky Way's contribution in that direction as a single number to be fitted to a considerable body of data.

3. Real data

Before we lobby NASA and ESA for a dedicated microlensing space telescope, we should ask what can be learned from the existing microlensing data. Three areas of the sky have been monitored: (i) towards the Galactic bulge; (ii) towards some spiral arms; and (iii) towards the Magellanic Clouds.

If one believes that the dark halo is comprised of elementary particles, only a very small optical depth is predicted towards the Clouds ($\sim 4 \times 10^{-8}$). At $1.2^{+0.4}_{-0.3} \times 10^{-7}$ the measured optical depth is about three times larger, but still deriving from only 17 events (Alcock et al. 2000a). The nature of these events is controversial. There is a powerful case that many of the lenses lie in the Clouds themselves (Kerins & Evans, 1999) because the only two lenses with reasonably securely determined distances (from finite-source effects) do lie in the Clouds. On the other hand, Ibata et al. (1999) may have detected a substantial population of old white dwarfs from their proper motions in the HDF. It is just possible that these objects provide the lenses for a significant number of the observed events. The arguments against a large column density in white dwarfs remain powerful, however (Gibson & Mould 1997).

The EROS collaboration (2000) has observed seven probable microlensing events in the directions of spiral arms, and from them derived an optical depth $4.5^{+2.4}_{-1.1} \times 10^{-7}$ that is compatible with conventional Galaxy models. Unfortunately, the error bar on this measurement is rather large.

3.1. The Galactic centre

Several hundred events have been detected by various groups along lines of sight towards the Galactic centre, and these data pose a fascinating puzzle. They are harder to interpret than the data for lines of sight to the Clouds because we do not know a priori where the sources are. However, if we confine ourselves to the data for red-clump stars, we can have quite a precise estimate of their distribution down each line of sight.

The red-clump stars must follow the general distribution of near infra-red light quite closely, because they are part of the population of evolved stars that are responsible for most of the Galaxy's near-IR luminosity. The DIRBE experiment aboard the COBE satellite mapped the Galaxy's IR surface brightness in several wavebands, and in the far-IR, where emission by dust is dominant. Spergel, Malhotra & Blitz (1996) used these data to estimate the effects of extinction on the near-IR data, and produce maps of what the Galaxy would look like in the near-IR bands in the absence of extinction.

Binney, Gerhard & Spergel (1997) used their Richardson–Lucy algorithm to deproject these corrected near-IR data under the assumption that the Galaxy has three specified perpendicular planes of mirror symmetry. The upper panel of Figure 2 shows the luminosity density that they recovered when projected perpendicular to the plane from $z = 225\,\mathrm{pc}$ upwards. Excluding the Galactic plane from the projection suppresses local maxima at $\sim 3\,\mathrm{kpc}$ along the y that are a notable feature of the density distribution in the plane (see Fig. 5 of Binney et al.). Binney et al argued that these maxima were artifacts resulting from the presence in the Galactic plane of spiral arms, which violate the assumed eight-fold symmetry.

Bissantz & Gerhard (2000) have recently deprojected the same data with an entirely different technique. Rather than using the Richardson–Lucy algorithm, they formulate the deprojection problem as a regularized likelihood maximization. They do not directly impose any symmetry on the model but have a term in the penalty function that discourages deviations from eight-fold symmetry. Another term in the penalty function encourages luminosity to lie along the spiral arms delineated by Ortiz & Lepine (1993).

The lower panel in Figure 2 shows that explicitly modelling the Galaxy's spiral structure in this way has the effect of making the bar longer and thinner than that recovered by Binney et al. (1997) – the axis ratio in the plane increases from 2:1 to 3:1. This change to the model bar enables the latter to reproduce an important datum that the Binney et al bar did not reproduce: the histograms from Stanek et al. (1994) that give for lines of sight at $l \simeq \pm 5°$ the number of clump stars at each apparent magnitude – see Figure 3. The ability of the model to reproduce these histograms to good accuracy strongly suggests that the model gives a faithful account of the distribution of red-clump stars. Hence,

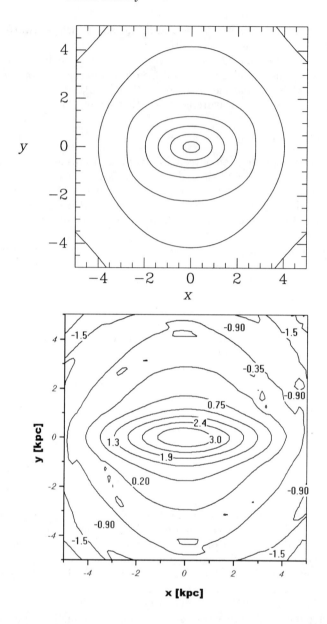

Figure 2. Top panel: the Galaxy in the L band projected perpendicular to the plane from $z = 225\,\mathrm{pc}$ upwards according to the model of Binney et al. (1997). Lower panel: the same view according to the model of Bissantz & Gerhard (2000). In both plots contours are logarithmically spaced. In the upper plot there are three contours per decade, while below contours are explicitly labelled.

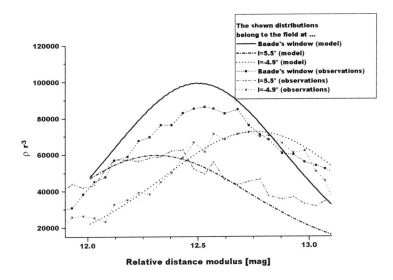

Figure 3. Predicted (curves) and measured (points) apparent-magnitude distribution of clump stars along three lines of sight through the bulge. The predictions are based on the model of Bissantz & Gerhard (2000).

when this model is used to predict the microlensing optical depth to red-clump stars, we may be confident that any discrepancy does not arise from an incorrect distribution of source objects.

Bissantz & Gerhard convert their luminosity model of the inner Galaxy into a mass model by adopting the constant near-IR mass-to-light ratio Υ of Englmaier & Gerhard (1999). This value of Υ was obtained by comparing the pattern of gas flow predicted by the Binney et al. model for some assumed Υ to the observed (l, v) diagrams for HI and CO; if Bissantz & Gerhard were to repeat this exercise with their new photometric model, they would surely obtain a very similar value of Υ. Once Υ has been chosen, one can calculate the microlensing optical depth for red-clump stars along any line of sight. In Baade's window they obtain $\tau = 1.24 \times 10^{-6}$, which is essentially identical to the value obtained by Bissantz et al. (1997) from the shorter, fatter bar of Binney et al. (1997), and significantly short of the values from the MACHO collaboration: $\tau = 3.9^{+1.8}_{-1.2} \times 10^{-6}$ (1σ) directly measured for bulge clump giants by Alcock et al. (1997) and $(3.9 \pm 0.6) \times 10^{-6}$ estimated by Alcock et al., (2000b) for bulge stars from a difference-imaging analysis of an inhomogeneous collection of sources. Evidently one cannot obtain agreement with the MACHO optical depth under the assumption that mass follows light.

Since the difference between the optical depth implied by constant Υ and the measured value is so large, it is natural to investigate an extreme model in which we ask, what is the minimum mass that is compatible with the red-clump optical depth attaining the MACHO value? Most of the red-clump sources lensed in

Baade's window $(l, b) = (1°, -4°)$ lie close to the Galaxy's z axis, so we simplify the calculation by considering a source that lies distance h from plane on the axis. Consider the contribution to τ from a band of mass M and radius r around the Galactic centre. If we assume that the band's surface density never increases with distance from the plane, then its mass will be minimized for a given optical depth when its surface density is constant and the line-of-sight to the source just cuts its edge. So we take the band's half-width to be $h(R_0 - r)/R_0$, which makes the band's surface density

$$\Sigma = \frac{M}{4\pi r h}\left(1 - \frac{r}{R_0}\right)^{-1}. \tag{15}$$

Substituting this into equation (1) we find the band's optical depth to be

$$\tau = \frac{GM}{c^2 h} \tag{16}$$

independent of radius (Kuijken 1997). This minimum mass estimate holds if the mass is widely distributed in radius rather than concentrated in a single band, because we can imagine a radially continuous mass distribution to be made up of a large number of bands, and we have shown that when the band is optimally configured, its optical depth depends only on its mass.

From equation (16) $(3.8 \pm 0.6) \times 10^{10}\,M_\odot$ is needed to produce the optical depth, to bulge sources that is implied by the latest MACHO results. A more realistic mass estimate is in excess of $8 \times 10^{10}\,M_\odot$ because realistically we must assume that the surface density of the band falls off smoothly with distance from the plane, and if this decline is exponential with the optimal scale-height $(h[1 - r/R_0])$, the band's mass must be e times that given by (16) for a given optical depth, while if the vertical density profile is Gaussian with optimal scale-height $(h[1 - r/R_0])$, equation (16) underestimates the band's mass by a factor $\sqrt{\pi e/2} \simeq 2.07$.

For comparison, the mass of the Galaxy interior to the Sun is of order $M = (220\,\mathrm{km\,s^{-1}})^2 \times 8\,\mathrm{kpc}/G \simeq 8.9 \times 10^{10}\,M_\odot$. Thus this naivest estimate of the mass interior to the Sun is just barely equal to the *minimum* mass required in a circular configuration to produce the reported optical depth, which suggests that an axisymmetric Galaxy is incompatible with the MACHO results This must be a tentative conclusion, however, until one has taken into account the effect on a body's circular-speed curve $v_c(r)$ of the body being strongly flatted. Binney, Bissantz & Gerhard (2000) show that it *is* possible to choose the Galaxy's radial density profile in such a way that the required optical depth is obtained without generating a value of v_c that conflicts with observation. However, such radial density profiles require more matter in the solar neighbourhood than observations of the Oort limit (Crézé et al. 1998; Holmberg & Flynn 2000) and the mass within 1.1 kpc of the plane (Kuijken & Gilmore 1991) imply. Consequently, we can safely conclude that an axisymmetric Galaxy cannot have as large an optical depth as that reported.

Can one achieve a higher optical depth within a give mass budget by making the bands elliptical rather than circular? Imagine deforming an initially circular band into an elliptical shape while holding constant the radius r at which the line of sight to our sources cuts the band. It is straightforward to show that if the column density through the band to the sources is to be independent of the band's eccentricity e, its mass $M(e)$ must satisfy

$$M(e) = M(0)\frac{1 - e^2 \cos^2 \phi}{\sqrt{1 - e^2}}, \qquad (17)$$

where ϕ is the angle between the band's major axis and the Sun–centre line. For $\phi < \pi/4$, $M(e)$ is a minimum with respect to e at

$$e_{\min} = \sqrt{2 - \sec^2 \phi}. \qquad (18)$$

Substituting equation (18) in equation (17) we find the minimum mass to be

$$M_{\min} = M(0)\sin 2\phi, \qquad (19)$$

and to require axis ratio $q_{\min} = \tan \phi$ (Zhao & Mao 1996).

For $\phi = 20°$, a value favoured by Binney et al. (1997), $q_{\min} = 0.36$ and $M_{\min}/M(0) = 0.64$; for $\phi = 15°$, we find $q_{\min} = 0.27$ and $M_{\min}/M(0) = 0.50$. Hence, making the bands elliptical realistically reduces the mass required to generate a given optical depth by at most 50%. In practice we cannot reduce our requirement for mass by so large a factor because the structure of the Galaxy's stellar bar is strongly constrained by both near-IR photometry (Blitz & Spergel 1991; Bissantz et al. 1997) and radio-frequency observations of gas that flows in the Galactic plane (Englmaier & Gerhard 1999; Fux 1999). Binney et al. (2000) estimate the possible reduction in mass by assuming that material at $R < 4\,\mathrm{kpc}$ forms a bar of optimal eccentricity whose long axis makes an angle of 20° with the Sun–centre line, while the Galaxy is axisymmetric at $R > 4\,\mathrm{kpc}$. They show that if the vertical structure of such a Galaxy is chosen to be optimal for lensing, then a radial density profile can be found that nowhere exceeds the observed value of v_c and is also compatible with the constraints on the density of matter at R_0. They argue, however, that such Galaxy models can be excluded for two main reasons. First, they predict values of v_c that are too small at $R \lesssim 350\,\mathrm{pc}$ because at small R they place significant mass high above the plane, where it contributes to τ but not v_c. Second, these models do not leave enough room in v_c for (i) departures from the optimal vertical profile and (ii) the presence of matter, such as interstellar gas and non-baryonic matter, that contributes to v_c but not τ.

Hence, even though we don't know much about the Galactic force-field, we know enough to exclude the measured optical depth in Baade's window! In fact, if we were to take seriously the prediction of simulations of the cosmological clustering of CDM that dark matter contributes substantially to the mass interior to the Sun (e.g., Navarro & Steinmetz 2000), our predicted optical depth for Baade's window would be significantly *less* than 10^{-6}, a factor of 4 or more below the measured value. Something is seriously wrong here.

References

Alcock, C. et al., 1997, ApJ, 479, 119
Alcock, C. et al., 2000a, ApJ, submitted (astro-ph/0001272)
Alcock, C., et al., 2000b, ApJ, submitted (astro-ph/0002510)
Binney, J.J., 1994, in Galactic and solar system optical astrometry ed. L. Morrison (Cambridge University Press), p. 141
Binney, J.J., Bissantz, N. & Gerhard, O.E., 2000, ApJ, submitted (astro-ph/0003330)
Binney, J.J. & Gerhard, O.E., 1996, MNRAS, 279, 1005
Binney, J.J., Gerhard, O.E., & Spergel, D.N., 1997, MNRAS, 288, 365
Binney, J.J. & Merrifield, M.R., 1998, Galactic Astronomy (Princeton University Press)
Bissantz, N., Englmaier, P., Binney, J.J. & Gerhard, O.E., 1997, MNRAS, 289, 651
Bissantz, N., Gerhard, O.E. 2000, in preparation
Blitz, L., Spergel, D.N., 1991, ApJ, 379, 631
Crézé, M., Chereul, E., Bienaymé, O. & Pichon, C., 1998, A&A, 329, 920
Crotts, A.P.S., 1992, ApJ, 399, L43
Dehnen, W. & Binney, J.J., 1996, in Formation of the Galactic Halo ... Inside and Out eds H. Morrison & A. Sarajedini, ASP Conf. Ser 92, p. 393
Dehnen, W. & Binney, J.J., 1998, MNRAS, 294, 429
de Jong, R.S., 1995, Spiral Galaxies (PhD thesis, University of Groningen)
Englmaier, P. & Gerhard, O.E., 1999, MNRAS, 304, 512
EROS Collaboration, 2000, A&A, submitted (astro-ph/0001083)
Fux, R., 1999, A&A, 345, 787
Gibson, B.K. & Mould, J.R., 1997, ApJ, 482, 98
Gould, A., 1996, ApJ, 470 201
Helmi, A., White, S.D.M., de Zeeuw, P.T. & Zhao, H., 1999, Nature, 402, 53
Holmberg J., Flynn C., 2000, MNRAS, 313, 209
Ibata, R.A., Richer, H.B., Gilliland, R.L. & Scott, D., 1999, ApJ, 524, L95
Johnston, K.V., Zhao, H., Spergel, D.N. & Hernquist, L., 1999, ApJ, 512, 771
Kent, S.M., Dame, T.M., Fazio, G., 1991, ApJ, 378, 131
Kerins, E.J. & Evans, N.W., 1999, ApJ, 517, 734
Kuijken K., 1997, ApJ, 486, L19
Kuijken, K. & Gilmore, G., 1991, ApJ, 367, L9
Navarro, J. & Steinmetz, M. 2000, ApJ, 528, 607
Ortiz, R. & Lepine, R.D., 1993, A&A, 279, 90
Spergel, D.N., Malhotra, S. & Blitz, L. 1996, in Spiral Galaxies in the Near-IR, eds Minniti, D. & Rix H.-W. (Springer: Heidelberg)
Stanek, K. Z., Mateo, M., Udalski, A., Szymanski, M., Kaluzny, J., Kubiak, M., 1994, ApJ, 429, L73

Zhao, H.-S. Mao, S., 1996, MNRAS, 283, 1197

Discussion

Sackett: You mentioned at the end of your talk that you were assuming that dark matter could not provide any microlensing. Do you think that there is evidence against compact disk dark matter given that the Kuijken & Gilmore constraints are all local and you need more lenses much closer to the Bulge?

Binney: The simple ring model I described shows that the mass distribution obtained from the L-band luminosity distribution is delivering good value for mass. So you don't gain anything by setting the mass-to-light ratio to zero + letting the lenses be provided by DM, whose distribution is only weakly constrained (by gas kinematics).

Gould: I think that pixel lensing against distant galaxies is pretty much hopeless. FAME will get k_z up to 500 pc. RR Lyraes seem to trace a spheroidal $r^{-3.5}$ distribution that could dominate the mass in the interior of the Galaxy.

Binney: At some angular resolution the pixel lensing proposal must be viable. Whether it will ever be cost-effective is another matter. The $r^{-3.5}$ distribution has to fail at some radius – it predicts infinite central mass. I believe the McWilliam & Rich metallicity distribution in Bulge fields shows that it is not dominant in Baade's window.

Bennett: The MACHO result toward the Bulge is centered at lower latitudes and higher longitudes than Baade's window. Also, I'd like to mention that MACHO will soon have a paper based upon Andrew Drake's image differencing analysis. We have 99 events and $\tau \approx 3.8 \times 10^{-6}$.

Binney: There is a strong vertical gradient in τ and we would clearly like to make field-by-field comparisons with the data. Unfortunately the MACHO papers do not give enough information for us to be able to do this. I'm delighted to hear that more information on more events is about to be released.

Bradbury: Nanotechnology would allow the construction of a large number of very large telescopes, so the "pie in the sky" is not unrealistic and should be examined further. In comparing observed results to theoretical models of the types and distributions of matter, it is better to trust the observations because the theorists are assuming that the universe is dead. If it isn't, the data used to build the theories may be wrong.

Graff: What is your opinion on whether the LMC can be driven far enough from virial equilibrium to cause enough microlensing?

Binney: I don't have a well informed opinion. My gut reaction is that the Clouds are near pericentre and their mutual orbit has been strongly disturbed by the Galaxy's tidal field. I can well imagine that this perturbation has caused the SMC to disturb strongly the LMC and get it out of vertical equilibrium. But gut reactions are no substitute for detailed modelling.

Microlensing 2000: A New Era of Microlensing Astrophysics
ASP Conference Series, Vol. 239, 2000
John Menzies and Penny D. Sackett, eds.

Galactic Bulge Microlensing Events with Clump Giants as Sources

P. Popowski[1], C. Alcock, R.A. Allsman, D.R. Alves, T.S. Axelrod,
A.C. Becker, D.P. Bennett, K.H. Cook, A.J. Drake, K.C. Freeman,
M. Geha, K. Griest, M.J. Lehner, S.L. Marshall, D. Minniti,
C.A. Nelson, B.A. Peterson, M.R. Pratt, P.J. Quinn, C.W. Stubbs,
W. Sutherland, A.B. Tomaney, T. Vandehei, D. Welch
(The MACHO Collaboration)

[1] *Institute of Geophysics and Planetary Physics, Lawrence Livermore National Laboratory*

Abstract. We present preliminary results of the analysis of five years of MACHO data on the Galactic bulge microlensing events with clump giants as sources. In particular, we discuss: (1) the selection of 'giant' events, (2) distribution of impact parameters, (3) distribution of event durations, (4) the concentration of long duration events in MACHO field 104 centered on $(l, b) = (3°.1, -3°.0)$. We report the preliminary average optical depth of $\tau = (2.0 \pm 0.4) \times 10^{-6}$ (internal) at $(l, b) = (3°.9, -3°.8)$. We discuss future work and prospects for building a coherent model of the Galaxy.

1. Introduction

The following short description of the most important observational studies of Galactic microlensing indicates that there is still an urgent need for a comprehensive analysis of the microlensing events toward the Galactic bulge. Udalski et al. (1994) found nine events in the first two years of the Optical Gravitational Lensing Experiment (OGLE) data. They set the lower limit on the optical depth to the Galactic bulge at $\tau = (3.3 \pm 1.2) \times 10^{-6}$. The uncertainties of this study are related to the detection efficiency analysis as well as small number statistics. Alcock et al. (1997) described a set of 45 events. The potential of this sample was not fully explored due to the use of sampling efficiencies only. The unbiased analysis was done only for 13 clump giants, which resulted in large uncertainties of the optical depth ($\tau = 3.9^{+1.8}_{-1.2} \times 10^{-6}$). Udalski et al. (2000) presented just a catalog of over 200 microlensing events from the last three seasons of the OGLE-II bulge observations. Unfortunately, no efficiency analysis has been done for those events so the information that can be extracted from this sample is very limited. Alcock et al. (2000a) performed Difference Image Analysis (DIA) of three seasons of bulge data in eight frequently sampled MACHO fields and found 99 events. They determined $\tau_{\rm bulge} = (3.2 \pm 0.5) \times 10^{-6}$. This was a major development in bulge microlensing. The DIA technique resulted in a substantial improvement in photometry, so this analysis was less vulnerable to uncertainties in the parameter determination. However, the results were obtained only for eight out of 94 MACHO bulge fields. Additionally, the detection efficiency estimate suffered from the fact that HST luminosity function was available for only one field. In summary, it is important to check the conclusions of Alcock et al. (2000a) with an independent set of events.

Blending is a major problem in any analysis of the microlensing data involving point spread function photometry. The bulge fields are crowded, so that the objects observed at a certain atmospheric seeing are blends of several stars. At the same time, typically only one star is lensed. In this general case, a determination of the event's parameters and the analysis of the detection efficiency of microlensing events is very involved and vulnerable to a number of possible systematic errors. If the sources are bright one can avoid these problems. First, a determination of parameters of the actual microlensing events becomes straightforward. Second, it is sufficient to estimate detection efficiency based on the sampling of the light curve alone. This eliminates the need of obtaining deep luminosity functions across the bulge fields. Red clump giants are among the brightest and most numerous stars in the bulge. Therefore, this analysis of the MACHO collaboration concentrates on the events where the lensed stars are clump giants.

2. Data

The MACHO Project observations were performed with the 1.27-m telescope at Mount Stromlo Observatory, Australia, from July 1992. Details of the telescope system are given by Hart et al. (1996) and of the camera system by Stubbs et al. (1993) and Marshall et al. (1994). Details of the MACHO imaging, data reduction and photometric calibration are described in Alcock et al. (1999). In total, we collected seven seasons (1993-1999) of data in the 94 Galactic bulge fields. The bulge data that are currently available for the analysis consist of five seasons (1993-1997) in 77 fields (Figure 1).

3. The Selection of Giant Events

The events with clump giants as sources have been selected from the sample of all events. The procedure that leads to a selection of microlensing events of general type consists of several steps. First, all the recognized objects in all fields are tested for any form of variability. Next, a microlensing light curve is fitted to all stars showing any variation and the objects that meet very loose selection criteria (cuts) enter the next phase. Here, this selection returns almost 43000 candidates. These candidates undergo more scrutiny and are subject to more stringent cuts, most of which test for a signal-to-noise of the different parts of the light curve. Here, this last procedure narrows the list of candidate events to ~ 280. The question, which of those sources are clump giants, is investigated through the analysis of the global properties of the color-magnitude diagram (CMD) in the Galactic bulge. The clump giant selection is based on four assumptions:
(1) the clump in the Baade's Window is representative,
(2) the OGLE-II (e.g. Paczyński et al. 1999) and MACHO photometry are

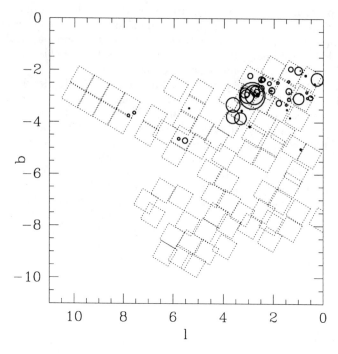

Figure 1. Location of the 77 MACHO fields [excluding three high-longitude fields at $(l,b) \sim (18,-2)$] and spatial distribution of events with clump giants as sources. The sizes of points are proportional to the Einstein ring crossing times with the longest event lasting ~ 150 days.

consistent,
(3) the intrinsic colors follow the relation: $(V-R)_0 = 0.5(V-I)_0$,
(4) the extinction toward the bulge follows the relation: $A_V = 5.0E(V-R)$.
Using the accurately measured extinction towards Baade's Window (Stanek 1996 with zero point corrected according to Gould et al. 1998 and Alcock et al. 1998) allows one to locate *bulge* clump giants on the intrinsic color – absolute magnitude diagram. Such a diagram can be then used to predict the positions of clump giants on the color – apparent magnitude diagram for fields with different extinctions. One obtains the following average values and color ranges: $\left\langle I_0^{\rm BW} \right\rangle = 14.35$, $\left\langle (V-I)_0^{\rm BW} \right\rangle = 1.1$, $(V-I)_0^{\rm BW} \in (0.9, 1.3)$ [which corresponds to $(V-R)_0^{\rm BW} \in (0.45, 0.65)$]. Combination of average I_0 and $(V-I)_0$ range allows one to determine a central V_0 of the clump for a given color. For example, for $(V-I)_0 = 0.9$ one obtains $V_0 = 15.25$, and for $(V-I)_0 = 1.3$ one gets $V_0 = 15.65$. We assume that the actual clump giants scatter in V around this central value, but by not more than 0.6 mag toward both fainter and brighter V_0. This defines the parallelogram-shaped box in the upper left corner of Figure 2.

Bulge Microlensing with Clump Giant Sources

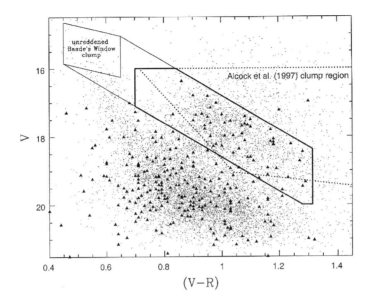

Figure 2. The region surrounded by a bold line is our clump region. For comparison, we plot with a dotted line the clump region from Alcock et al. (1997). Both selections return very similar events.

With the assumption that the clump populations in the whole bulge have the same properties as the ones in Baade's Window, the parallelogram described above can be shifted by the reddening vector to mark the expected locations of clump giants in different fields. The solid lines are the boundaries of the region where one could find the clump giants of fields with different extinctions. There are a few more V- and $(V-R)$ cuts that determine the final shape of the clump region. The clump regions from Figure 2 contains 52 unique clump events. There are 6 identified binary lenses among these events.

Note that several assumptions that went into creating this region should be carefully reviewed. In particular, the assumption that $(V-R)_0 = 0.5(V-I)_0$ is only approximately true, the clump region is rather sensitive to color, the assumed spread in V magnitudes can be either bigger or smaller or asymmetric around the central value, clump giants in different fields may have different characteristics. Taking into account that all of the above might have gone wrong, the obvious success of the outlined procedure (see Figure 2) is very encouraging.

4. Distribution of Impact Parameters

In Figure 3, we plot the cumulative distribution of the impact parameter u_{min} (solid line). The impact parameter was obtained from the maximum amplifica-

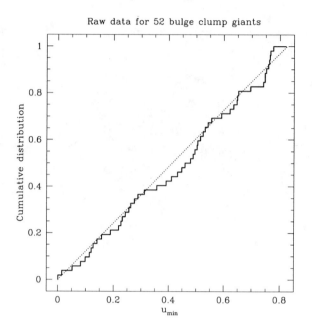

Figure 3. Cumulative distribution of impact parameters for all the candidates including six binaries.

tion, A_{max}, according to the formula:

$$u_{min} = \sqrt{-2 + \frac{2 A_{max}}{\sqrt{A_{max}^2 - 1}}}. \qquad (1)$$

No efficiency correction was applied. The dashed line is the expected theoretical distribution if the minimum recorded A_{max} equals 1.5. The agreement is beautiful.

5. Distribution of Event Durations

We note that the first accurate description of the bulge event durations was given by Dr. Seuss (1960): "We see them come. We see them go. Some are fast. And some are slow." The left panel of Figure 4 shows the number of events as a function of event duration uncorrected for detection efficiencies. The right panel presents a contribution of events with particular duration to the optical depth. About 40 % of the optical depth is in the events longer than 50 days ($t_E > 50$). This is at odds with standard models of the Galactic structure and kinematics.

6. Concentration of Long Events in Field 104

Ten clump giant events out of 52 are in the MACHO field 104. There is a high concentration of long-duration events in this field (five out of 10 events longer than 50 days are in 104, including the longest two). We investigate how statistically significant is this concentration. Ideally, one would like to account for the change in the detection of efficiency of events with different durations in different fields. However, reliable efficiencies for individual fields are not available at this point. Nevertheless, it should be possible to place a lower limit on the significance of this difference. The efficiency for detecting long events should be similar in most fields, because this does not depend strongly on the sampling pattern. The detection of short events will be lower in sparsely sampled fields. Therefore, the number of short events in some of the fields used for comparison may be relatively too small with respect to the frequently-sampled field 104, but this is only going to lower the significance of the t_E distribution difference. In conclusion, the analysis of event durations *uncorrected* for efficiencies should provide a lower limit on the difference between field 104 and all the remaining clump giant fields. We use the Wilcoxon number-of-element-inversions statistic to test this hypothesis. First, we separate events into two samples: events in field 104 and all the remaining ones. Second, we order the event in the combined sample from the shortest to the longest. Then we count how many times one would have to exchange the events from field 104 with the others to have all the 104 field events at the beginning of the list. If N_1 and N_2 designate the numbers of elements in the first and second sample, respectively, then for $N_1 \geq 4$, $N_2 \geq 4$, and $(N_1 + N_2) \geq 20$, the Wilcoxon statistic is approximately Gaussian distributed with an average of $N_1 N_2/2$ and a dispersion σ of $\sqrt{N_1 N_2 (N_1 + N_2 + 1)/12}$. The Wilcoxon statistic is equal to 320, whereas the expected number is 210 with an error of about 43. Therefore the events in field 104 differ (are longer) by 2.55σ from the other fields. That is, the probability that events in field 104 and other fields originate from the same parent population is of order of 0.011.

7. The Optical Depth

We use the following estimator of the optical depth

$$\tau = \frac{\pi}{2NT} \sum_{\text{all events}} \frac{t_E}{\epsilon(t_E)}, \quad (2)$$

where N is the number of observed stars (here about 2.1 million clump giants), T is the total exposure (here about 2000 days) and $\epsilon(t_E)$ is an efficiency for detecting an event with a given t_E. The sampling efficiencies were obtained with the pipeline that has been previously applied to the LMC data (for a description see Alcock et al. 2000b). In brief, artificial light curves with different parameters have been added to 1% of all clump giants in our 77 fields and the analysis used to select real events was applied to this set. For a given duration of the artificial event, the efficiency was computed as a number of recovered events divided by a number of input events. The efficiencies used in this analysis are global efficiencies averaged over clump giants in all 77 fields. The optical depth

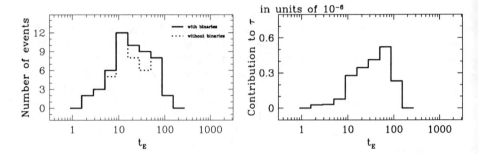

Figure 4. The left panel shows the histogram of the distribution of Einstein crossing times. The right panel presents the contribution of events with different durations to the total optical depth.

is reported at the central position that is an average of positions of 1% of the analyzed clump giants.
We obtain:

$$\tau = (2.0 \pm 0.4) \times 10^{-6} \quad \text{at} \quad (l, b) = (3°\!.9, -3°\!.8). \tag{3}$$

with the error computed according to the formula given by Han & Gould (1995). We caution that this result is only preliminary. The details of the analysis as well as full discussion of the statistical and possible systematic errors (which may be a fair fraction of the statistical error) will be given in Alcock et al. (2000c).

8. Future work and conclusions

Microlensing gives two major types of constraint: the optical depth spatial distribution and the distribution of event durations. The first type of information measures the mass density along different lines of sight, whereas the second one is related to the mass function of the lenses and the kinematics of involved populations. These suggest two major routes of attack. The structure of the Galaxy (e.g. disk scale length and height, bar shape and inclination) can be constrained by analyzing the average and gradient of the optical depth. This is most easily achieved with a maximum likelihood technique (e.g. Gyuk 1999), which allows one to extract information from fields with and without any events. Additionally, it is possible to constrain the location of clump sources comparing reddening-free indices of lensed and unlensed stars (Stanek 1995). The difference can be up to 0.2 mag depending on the bulge/bar model, and our sample allows one to find the difference with an accuracy of 0.07 mag. The microlensing data in the bulge should be combined with all the other types of information, e.g. kinematic and density distributions of RR Lyrae stars, rotation curve, star counts, motions of the Milky Way's satellites, etc.

Because the distribution of event durations depends on both the mass function of the lenses as well as the kinematics of the observer, sources and deflectors, an unambiguous solution of the entire system requires hundreds of events.

However, Han & Gould (1996) showed that for a sample of ~ 50 events, the errors in mass function reconstruction are small if the kinematics are assumed to be known. The kinematics are not exactly under control but it is possible to construct a plausible model consistent with observational constraints. Determination of the mass function of the microlenses toward the Galactic bulge would be a very exciting development. As a matter of fact, microlensing is currently the only method that could enable one to find a mass function of objects as distant as several kpc.

The conclusions from the analysis described above are the following: it is possible to select an unbiased (when $u_{\min}$ is concerned) sample of clump events in a universal way based only on V and $(V - R)$ (or more generally, an event's position on the CMD). Ten out of 52 clump events have durations > 50 days, which implies that ~ 40 % of the optical depth is in the long duration events. This is surprising, because long events are most likely to be the result of disk-disk lensing (Kiraga & Paczyński 1994). However, it is widely believed that clump giants trace the bar rather than the inner disk (Stanek et al. 1994). Events in the area of field 104 centered at $(l, b) = (3°.1, -3°.0)$ have longer durations than the events in other fields. The explanation can be as exotic as a cluster of remnants along the line of sight or some conspiracy of the bar orbits. The optical depth averaged over the clump giants in 77 fields is $\tau = (2.0 \pm 0.4) \times 10^{-6}$ at $(3°.9, -3°.8)$. Allowing for the optical depth gradient of $\sim 0.5 \times 10^{-6}/\deg$, this optical depth is lower but consistent both with the Alcock et al. (1997) clump result of $3.9^{+1.8}_{-1.2} \times 10^{-6}$ at $(l, b) = (2°.55, -3°.64)$ as well as the DIA result of $(3.2 \pm 0.5) \times 10^{-6}$ at $(l, b) = (2°.68, -3°.35)$ (Alcock et al. 2000a). The new optical depth is still rather high but in the range accessible to some models of Galactic structure.

Acknowledgments. This work was performed under the auspices of the U.S. Department of Energy by the University of California Lawrence Livermore National Laboratory under contract No. W-7405-Eng-48.

References

Alcock, C., et al. 1997, ApJ, 479, 119

Alcock, C., et al. 1998, ApJ, 494, 396

Alcock, C., et al. 1999, PASP, 111, 1539

Alcock, C., et al. 2000a, ApJ, 541, 000 (astro-ph/0002510)

Alcock, C., et al. 2000b, submitted to ApJ (astro-ph/0003392)

Alcock, C., et al. 2000c, in preparation

Gould, A., Popowski, P., & Terndrup, D.T. 1998, ApJ, 492, 778

Gyuk, G. 1999, ApJ, 510, 205

Han, C, & Gould, A. 1995, ApJ, 449, 521

Han, C, & Gould, A. 1996, ApJ, 467, 540

Hart, J., et al. 1996, PASP, 108, 220

Kiraga, M., & Paczyński, B. 1994, ApJ, 430, L101

Marshall, S.L., et al. 1994, in IAU Symp. 161, Astronomy From Wide Field Imaging, ed. H.T. MacGillivray, et al., (Dordrecht: Kluwer), p. 67

Paczyński, B., et al. 1999, Acta Astron., 49, 319
Seuss, Dr. 1960, in "One fish, two fish, red fish, blue fish", (New York: Random House)
Stanek, K.Z., et al. 1994, ApJ, 429, L73
Stanek, K.Z. 1995, ApJ, 441, L29
Stanek, K.Z. 1996, ApJ, 460, L37
Stubbs, C.W., et al. 1993, in Proceedings of the SPIE, Charge Coupled Devices and Solid State Optical Sensors III, ed. M. Blouke, 1900.
Udalski, A., et al. 1994, Acta Astron., 44, 165
Udalski, A., et al. 2000, Acta Astron., submitted (astro-ph/0002418)

Discussion

Glicenstein: Can you give more details about the reconstruction process?

Popowski: Reconstruction is one of the possible ways to recreate the underlying distribution of event durations. It is done in the following way:
1. At every $t_{e,i}$ location I center a normalized-to-1 Gaussian distribution with a dispersion which depends on the distances to neighbouring $t_{e,k}$ – the smaller the distance, the narrower the Gaussian.
2. The multi-peak distribution resulting from the above procedure is smoothed to produce the final reconstruction. The amount of smoothing determines how featureless the final distribution is. The minimum amount of smoothing will produce very spiky distributions which trace the cumulative distribution of actual events very nicely. More smoothed distributions will miss some small features of the actual distribution, but will look better.

Gaudi: 1. Does MACHO plan on doing the full efficiency analysis?
2. Are the event achromatic?
3. Didn't Mao & Paczyński prove that it was impossible to determine the mass function from bulge microlensing?

Popowski: 1. Yes, we still have about 2 years of funding, and hopefully will be able to do the full efficiency analysis.
2. Yes, the events in this clump sample are achromatic.
3. It may be hard to determine the character of the mass function. However, it should be possible to ascertain the parameters of the mass function if one assumes its functional form.

Zinnecker: The stellar mass function derived from the distribution of your event durations is quite steep (a power law almost as steep as the Salpeter IMF), certainly steeper than the observed low-mass IMF in the solar neighbourhood. Is this result robust and therefore significant? Also, your duration distribution for the clump giant events looked more log-normal than power-law, suggesting the question whether the corresponding mass function might be better fitted by a log-normal function instead of a power law?

Popowski: I have to stress that my statement about a steep mass function is completely preliminary. I rather wanted to demonstrate the feasibility of such a calculation than to give a specific answer (this has to wait for reliable detection efficiencies). I fitted a log-normal function to the t_e distribution and it is not a good fit. I agree, however, that one should not consider just a power-law type mass function.

Han: You said that long-t_e events might be explained by disk self-lensing. However, you already carefully selected only bulge clump giants. Then, long-t_e events cannot be explained by disk self-lensing. What do you think?

Popowski: I selected sources consistent with being bulge clump giants. However, it is not obvious that there are only bulge clump giants in this sample, so it may be too early to reject disk-disk microlensing as an explanation of long events seen toward the bulge.

Glicenstein: How can you be sure that you select only clump giants inside your cuts.

Popowski: I cannot be sure that they are only clump giants. But the main idea behind using only clump giants is to work with unblended events. As long as the selected events are unblended (and I showed you the impact parameter cumulative distribution indicating that this is the case) it is OK, even if technically speaking the sources are not clump giants.

Bennett: I just wanted to point out that the microlensing parallaxes seem to indicate that most of the long timescale events don't come from disk sources.

A Galactic Bar to Beyond the Solar Circle and its Relevance for Microlensing

Michael Feast

Department of Astronomy, University of Cape Town, Rondebosch, 7701, South Africa.

Patricia Whitelock

South African Astronomical Observatory, PO Box 9, Observatory, 7935, South Africa.

Abstract. The Galactic kinematics of Mira variables have been studied using infrared photometry, radial velocities, and Hipparcos parallaxes and proper motions. For Miras in the period range 145 to 200 days (probably corresponding to [Fe/H] in the range –0.8 to –1.3) the major axes of the stellar orbits are concentrated in the first quadrant of Galactic longitude. This is interpreted as a continuation of the bar-like structure of the Galactic Bulge out to the solar circle and beyond.

1. Introduction

A somewhat unexpected discovery of microlensing experiments towards the Galactic Bulge was the need to take into account a bar-like structure there (see, e.g. Paczyński 1996). The precise nature of this bar – its composition, extent and evolution – is known only sketchily. This is an area where progress is likely to come from the combination of microlensing results with other types of observations. In the present paper we summarize the results of an analysis of the Galactic kinematics of Mira variables based on infrared photometry, radial velocities, and Hipparcos parallaxes and proper motions (Whitelock, Feast & Marang 2000, Whitelock & Feast 2000, Feast & Whitelock 2000b). This provides evidence for a rather extended bar-like structure in the Galaxy.

2. Background

Mira variables have a number of properties that makes them particularly interesting for Galactic structure studies[1].

Bolometrically and in the near infrared, e.g. at K (2.2μm), they are the brightest members of the old populations in which they are found (see for instance Fig. 1 of Feast & Whitelock 1987). These variables therefore define the

---

[1]Note that the present paper refers entirely to oxygen-rich Miras.

tip of the AGB in these populations. Because of their brightness they can be studied with relative ease to large distances.

Observations in the LMC show that Miras follow a well defined, narrow period–luminosity (PL) relation at K or in M_{bol} (Feast et al. 1989). Thus Mira distances can be estimated from infrared photometry with good accuracy.

It has long been known that the kinematics of Galactic Miras are a function of period (see, e.g. Fig. 12 of Feast 1963). Their asymmetric drift gets numerically larger as one moves from the longer period Miras to the shorter period ones with a maximum of $\sim 100\,\mathrm{km\,s^{-1}}$ at a period of about 175 days. This is intermediate between the value usually quoted for the thick disc ($30-40\,\mathrm{km\,s^{-1}}$, e.g. Freeman 1987) and that appropriate to halo objects ($\sim 230\,\mathrm{km\,s^{-1}}$). This apparently intermediate population has generally been neglected in discussions of Galactic structure and kinematics. Judging from Miras in globular clusters this corresponds to a metallicity, [Fe/H], range from about –0.8 to about –1.3. As one moves to Miras of even shorter periods the numerical value of the asymmetric drift decreases, making the shortest period Miras similar in kinematics to Miras of much longer period. The relationship of this shortest period group to the other Miras was long a puzzle.

Another important property of the Miras is shown by those in globular clusters. Miras occur in relatively metal rich clusters. If there are more than one Mira in any cluster, their periods are all close together. Such clusters also contain semiregular (SR) variables. In any one cluster these have a range of periods, and in the period–luminosity plane they lie on an evolutionary track which terminates on the Mira PL relation (see Whitelock 1986, Feast 1989). The clusters containing Miras show that for these stars there is a good relation between period and metallicity (see Fig. 1 of Feast & Whitelock 2000a). Since the change of mass with metallicity is presumably small for these stars, the period–metallicity relation is primarily indicating a relation between metallicity and stellar radius at the tip of the AGB.

These various properties have been important in using Miras to study the Galactic Bulge itself. The work of Lloyd Evans (1976) and others showed that there is a wide distribution of periods amongst Miras in the Bulge. This, then, implies a wide range in metallicities. Indeed the metallicity distribution inferred from the Mira period distribution (Feast & Whitelock 2000a) is similar to that derived by Sadler et al. (1996) from Bulge K-type giants. The PL relation allows one to estimate the distances to individual stars in the Bulge (see, e.g. Glass et al. 1995). In this way Whitelock & Catchpole (1992) found that Bulge Miras at positive Galactic longitudes are nearer to us than those at negative longitudes. This was one of the first pieces of evidence for a barred structure of the Bulge.

3. Present Work

Our present study consists of extensive near infrared, $JHKL$, photometry of Galactic Mira-like variables[2], made at SAAO Sutherland, combined with Hip-

---

[2]The sample contains a few stars classified as semiregular (SR) variables which have Mira-like properties (see Whitelock et al. 2000).

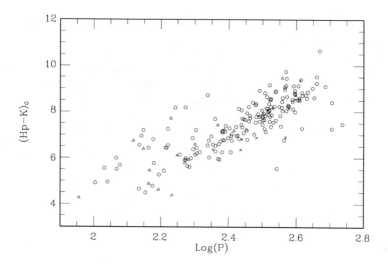

Figure 1. The period-colour relation for oxygen-rich Mira-like variables. Stars catalogued as Miras and semi-regulars are shown as circles and triangles, respectively. S-type stars are shown as crosses

parcos astrometric and photometric data and with radial velocities from the literature.

Combining the infrared observations with the Hipparcos photometry allows us to see clearly the cause of the anomalous result for the asymmetric drift of the shortest period Miras discussed above. In the $(Hp - K)_0 - \log P$ plane (where Hp is the mean Hipparcos magnitude) there are two near parallel sequences (see Fig. 1): a main (blue) sequence and a less populated (redder) sequence which is largely confined to the shorter periods and dominates at the very shortest periods. In the period range where the two sequences overlap their kinematics are quite different. The red sequence stars have kinematics that associate them with much longer period Miras. It is tempting therefore to identify at least some of the stars on this red sequence as analogous to the SRs in metal-rich globular clusters, which were discussed above, but at somewhat higher luminosities, and to suggest that they are evolving into longer period Miras. If this is the case we would expect the red sequence stars to be somewhat brighter than the blue sequence ones at a given period. This is because evolutionary tracks (at least in metal-rich globular clusters) lie above the Mira PL relation in the period–luminosity plane. Indeed the Hipparcos parallaxes do indicate that at a given period the red sequence stars are brighter than the blue sequence ones. It is also possible that some of the red sequence stars are pulsating in a higher mode than the blue sequence stars. The following discussion refers only to the main (blue) sequence stars. These stars may also be recognized from their infrared colours and these show that the Miras that have been found in globular clusters and in the LMC belong to the blue sequence.

Hipparcos parallaxes of blue sequence stars were used to derive the zero-point of the PL relation. Using this calibration one can determine a distance modulus for the LMC. This is found to be 18.64 ± 0.14 mag, or slightly greater if a correction is necessary for a metallicity difference between the local and LMC Miras. This modulus agrees well with that determined from Cepheids and some other distance indicators (see the summary in Feast 1999).

We have studied the Galactic kinematics of the blue sequence Miras using Hipparcos proper motions and radial velocities from the literature. These radial velocities include radio observations of OH and other molecules. Distances were derived from the mean K magnitudes and the Hipparcos-calibrated Mira PL relation. However, the results are rather insensitive to the distance scale adopted.

The three components of the space motion of each star were derived in non-rotating cylindrical co-ordinates centred at the Galactic centre. These are V_R radially outwards and parallel to the plane, V_θ at right angles to V_R in the direction of Galactic rotation, and w in the direction of the North Galactic Pole. The data were analysed by dividing the Miras into six (blue sequence) groups, according to period.

The results for V_θ confirm previous work. In the longest period group (mean period 453 days), $V_\theta = 223 \pm 4\,\mathrm{km\,s^{-1}}$. This is only slightly smaller than the circular velocity implied by the kinematics of Cepheids ($231\,\mathrm{km\,s^{-1}}$, Feast & Whitelock 1987). The value of V_θ drops as the period decreases, reaching $147 \pm 14\,\mathrm{km\,s^{-1}}$ in the group of mean period 175 days (17 stars). In this calculation one Mira in the period range 145 to 200 days (S Car) has been omitted, as it is moving on a highly eccentric retrograde orbit.

The main surprise of the analysis was in the values of V_R. This is $+67 \pm 17\,\mathrm{km\,s^{-1}}$ in the group of mean period 175 days, whilst for the other groups there is a small net outward velocity.

There seem to be three possible explanations of this large mean outward radial motion in the 145 to 200 day group.

If the Galaxy is axially symmetric the result would imply that there is a general, axially symmetric, outward, radial motion of Miras in the period range 145 - 200 days. This seems unlikely.

Another alternative is that Miras in this period range belong to some galactic interloper. Whilst this cannot be ruled out we might ask why there are, in that case, no local stars of this type belonging to our Galaxy (such stars exist in the Galactic Bulge itself and in Galactic globular clusters). Also the mean motion perpendicular to the plane is small for this period group ($w = -12 \pm 12\,\mathrm{km\,s^{-1}}$) indicating that any interloper must be moving very closely parallel to the Galactic plane.

The most likely explanation would seem to be that there is a Galactic asymmetry leading to a deficit of incoming orbits in the solar neighbourhood for Miras in this period range. The values of V_θ and V_R for this group then show that the major axes of their Galactic orbits are concentrated in the first quadrant of galactic longitude. All workers place the major axis of the Bulge-bar in this quadrant. A simple first-order calculation shows that the mean orbit has a major axis making an angle of 17^{+11}_{-4} degrees with the Sun–Centre line. This agrees with the position angle (16 degrees) of the Bulge bar derived by Binney et al. (1991) from gas dynamics in the central region. It is also consistent with a number of other estimates which are in the 20 to 30 degree range, though values near 45 degrees derived by some workers agree less well. Rough estimates suggest

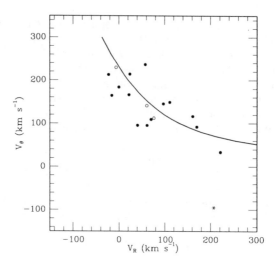

Figure 2. A plot of V_θ against V_R for stars with periods in the range 145 to 200 days. The curve is the relation expected in a simple model in which the major axes of the orbits of all the stars are aligned at 17° to the Sun–Centre line. Stars for which any velocity component has a standard error in excess of 20 km s^{-1} are represented by open symbols. The asterisk denotes S Car.

that the orbits of some of these short period Miras are sufficiently eccentric to penetrate into the Bulge region itself at perigalacticum.

Figure 2 shows a plot of V_θ against V_R for stars in the 145 to 200 day group. The curve is the relation that would be expected, in a simple model, if the major axes of the orbits of all the stars were aligned at an angle of 17 degrees to the Sun–Centre line.

The number of Miras in the short period group is small, 17 stars (omitting S Car for reasons mentioned above). However, the lifetime of a Mira is short ($\sim 2 \times 10^5$ years). Thus these stars are tracers of a much larger population. This can be seen from the metal-rich globular clusters, where one such cluster may contain just one Mira. It would be particularly interesting to identify other objects associated with the 145 to 200 day Miras. The globular cluster results suggest that such objects would have metallicities in the range $-0.8 \geq [Fe/H] \geq -1.3$ and the age of globular clusters in this metallicity range. However, the situation may be complex. Minniti et al. (1997) suggest that the RR Lyrae variables in the Bulge do not belong to a bar-like distribution, despite the fact that the ones in the NGC 6522 field, at least, have $[Fe/H] \simeq -1$ (Walker & Terndrup 1991). It may be that choosing Miras by period allows one to sort out a more homogeneous population than is possible in other ways.

So far as microlensing is concerned this work would seem relevant for at least two reasons.

1. It gives hope of defining further the composition and general character of the Galactic bar.

2. It warns that one should not necessarily consider Galactic structure and kinematics as "knowns" into which microlensing studies must be fitted. There may well be surprises; and microlensing data may make an important contribution to uncovering them.

Acknowledgments. This paper is based on observations made with the Hipparcos satellite and at the South African Astronomical Observatory (SAAO).

References

Binney, J.J., Gerhard, O.E., Stark, A.A., Bally, J., & Uchida, K.I. 1991, MNRAS, 252, 210
Feast, M.W. 1963, MNRAS, 125, 367
Feast, M.W. 1989, in: The Use of Pulsating Stars in Fundamental Problems of Astronomy, ed. E.G. Schmidt, Cambridge University Press, p. 205
Feast, M.W. 1999, PASP, 111, 775
Feast, M.W., Glass, I.S., Whitelock, P.A. & Catchpole, R.M. 1989, MNRAS, 241, 375
Feast, M.W. & Whitelock, P.A. 1987, in: Late Stages of Stellar Evolution, eds. S. Kwok & S.R. Pottasch, Reidel, p. 33
Feast, M.W. & Whitelock, P.A. 2000a, in: The Chemical Evolution of the Milky Way, eds. F. Giovannelli & F. Matteucci, Kluwer, in press (astro-ph/9911393)
Feast, M.W. & Whitelock, P.A. 2000b, MNRAS, submitted
Freeman, K.C. 1987, ARA&A, 25, 603
Glass, I.S., Whitelock, P.A., Catchpole, R.M., & Feast, M.W. 1995, MNRAS, 273, 383
Lloyd Evans, T. 1976, MNRAS, 174, 169
Minniti, D. 1997, in: Variable Stars and the Astrophysical Returns of Microlensing Surveys, eds. R. Ferlet, J.P. Maillard & B. Raban, Editions Frontières, p. 257
Paczyński, B. 1996, ARA&A, 34, 419
Sadler, E.M., Rich, R.M. & Terndrup, D.M. 1996, AJ, 112, 171
Walker, A.R. & Terndrup, D.M. 1991, ApJ, 378, 119
Whitelock, P.A. 1986, MNRAS, 219, 525
Whitelock, P.A. & Catchpole, R.M. 1992, in: The Center, Bulge and Disk of the Milky Way, ed. L. Blitz, Kluwer, p. 103
Whitelock, P.A. & Feast, M.W. 2000, MNRAS, submitted
Whitelock, P.A., Feast, M.W. & Marang, F. 2000, MNRAS, submitted

Discussion

Binney: It would be fascinating to see your Miras placed on the (U,V) plane as obtained from the Hipparcos catalogue by Dehnen (AJ115, 1998).

Feast: They lie on the outer edge of the Dehnen distribution and occupy a very limited locus in the (U,V) or (V_R, V_θ) plane.

Gould: Do metal poor stars become carbon Miras?

Whitelock: We do not find carbon Miras in old metal poor systems such as the Galactic globular clusters. But in younger systems, e.g. the Magellanic Clouds, they are very common. The Mira population in the SMC is almost entirely carbon-rich.

New EROS2 Results towards the Galactic Disk

J. -F. Glicenstein (EROS collaboration)

DSM/DAPNIA/SPP, CEA Saclay, F-91191 Gif-sur-Yvette

Abstract. EROS2 has taken 3 years of data towards the Galactic Center and four directions in the Galactic Disk ("Spiral Arms"). The average optical depth $\bar{\tau} = 0.45^{+0.25}_{-0.17} \times 10^{-6}$ found towards the Spiral Arms is consistent with the expectation from a simple Galactic model. The sensitivity of the measurements to the parameters of the Galactic model is discussed.

1. Introduction

The main goal of EROS is to search for massive compact celestial objects in the Galactic halo with microlensing, following the suggestion by Paczyński (1986). EROS started in 1990 and was upgraded in 1996 with two new CCD cameras, a new 1-m class telescope and a faster acquisition. A program of Galactic structure study was started in 1996. The observed optical depth measured towards the Galactic Center by the OGLE (Udalski et al 1994) and MACHO (Alcock et al 2000) collaborations is large, but not well understood (Binney 2000). The EROS group decided to monitor sources distributed over a large area, towards the Galactic Center and four directions in the Galactic Disk. This observing strategy should help disentangle the contributions to the optical depth from disk and bulge lenses. This paper reports the analysis of the three first years of data taking towards the Galactic Disk, which is now being submitted (Derue et al 2000). The analysis of the two first years has already been published (Derue et al 1999).

2. EROS observations

The four directions monitored towards the Galactic Disk correspond to a total area of 29 square degrees. We refer to them as γ Nor, θ Mus, β and γ Sct. The average coordinates of the fields and the number of stars monitored are given in Table 1. The EROS catalog of sources from the Galactic disk fields contains 9.1 million stars. These stars have been monitored since July 1996. The data presented in this paper span a period of three years between July 1996 and November 1998, except for θ Mus, which has been monitored only since January 1997. There are roughly 100 measurements in both EROS passbands for each star.

The distance to the source stars is poorly known in our Galactic disk fields. This is an important difference from the other targets monitored by the microlensing surveys (Magellanic Clouds and Galactic Center fields) and a major

Table 1. Fields observed by EROS towards the Galactic Disk.

Field	av. longitude (degrees)	av. latitude (degrees)	# of stars (millions)
β Sct	27	-2.5	2.1
γ Sct	18.6	-2.6	1.8
γ Nor	331.2	-2.7	3.0
θ Mus	306.4	-1.8	2.2

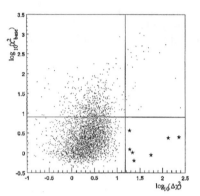 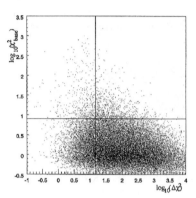

Figure 1. Distribution of $\Delta\chi^2$ versus χ^2_{base} for data (left) and simulated light curves (right). $\Delta\chi^2$ and χ^2_{base} are defined in the text. The cuts on these variables are shown by straight lines. The location of the seven candidates selected by the microlensing analysis is shown by stars.

source of systematic error in the optical depth prediction. Different studies using synthetic color-magnitude diagrams (Mansoux 1997) and a dedicated Galactic Cepheid search (Derue 1999) indicate distances in the range 5–8 kpc. An average distance of 7 kpc has been used in this paper.

3. Microlensing candidates selection

The microlensing candidate selection is described in detail in (Derue 1999). The candidates are first selected by requiring the detection of a single "bump", simultaneously in time in both EROS passbands. Events with low signal-to-noise ratio are then rejected by demanding a significant improvement of a microlensing fit (which gives a chi-square of χ^2_{ml} for N_{dof} degrees of freedom) over a constant flux fit (with a chi-square of χ^2_{cst}). A last cut on the stability of the ligth curve outside the "bump" rejects a fraction of the remaining variable stars.

These last cuts are illustrated on Figure 1, which shows $\Delta\chi^2$ versus χ^2_{base}. $\Delta\chi^2$ is defined by

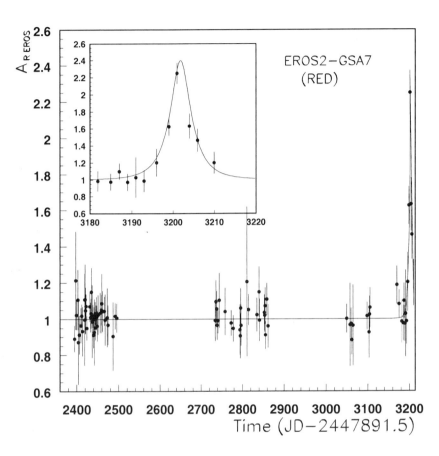

Figure 2. Light curve of the EROS candidate GSA7 in the EROS red passband.

$$\Delta\chi^2 = \frac{1}{\sqrt{2N_{dof}}} \frac{\chi^2_{cst} - \chi^2_{ml}}{\chi^2_{ml}/N_{dof}} \quad (1)$$

and χ^2_{base} is the chi-square of the part of the light-curve of the candidate outside the "bump".

Seven microlensing candidates have been found by EROS in the three years data set. The characteristics of these candidates are reported on Table 2. The light curve of candidate GSA7, which has a small duration of 6.3 ± 0.5 days is plotted on Figure 2. Five of the candidates are found in the direction of γ Sct, and two are found towards γ Nor.

The efficiency of the sampling and of the selection cuts has been estimated using simulated amplifications on a representative sample of observed light curves. Blending was not taken into account. The average detection effi-

Table 2. Characteristics of the 7 microlensing candidates and contribution to the optical depth.

Candidate	GSA1	GSA2	GSA3	GSA4
field	$\gamma\,Sct$	$\gamma\,Nor$	$\gamma\,Nor$	$\gamma\,Sct$
α(h:m:s)	18:29:09.0	16:11:50.2	16:16:26.7	18:32:26.0
δ(d:m:s)	-14:15:09	-52:56:49	-54:37:49	-12:56:04
V_{EROS} - R_{EROS}	20.7 - 17.7	19.4 - 17.8	18.6 - 17.5	17.9 - 17.1
t_E (days)	73.5 ± 1.4	98.3 ± 0.9	70.0 ± 2.0	23.9 ± 1.1
Max. magnification	26.5 ± 0.6	3.05 ± 0.02	1.89 ± 0.01	1.72 ± 0.02
contribution to τ ($\times 10^6$)	0.51	0.15	0.12	0.30

GSA5	GSA6	GSA7
$\gamma\,Sct$	$\gamma\,Sct$	$\gamma\,Sct$
18:32:12.0	18:33:56.7	18:34:10.0
-12:55:16	-14:33:52	-14:03:40
19.9 - 17.9	18.5 - 17.2	18.7 - 17.5
59.0 ± 5.5	37.9 ± 5.0	6.20 ± 0.50
1.71 ± 0.03	1.35 ± 0.02	2.70 ± 0.30
0.44	0.35	0.22

ciency of a microlensing event with an Einstein radius crossing time $t_E = 75$ days is $\sim 30\,\%$ for the γ Nor and γ Sct directions and $\sim 10\,\%$ for the other directions.

4. Discussion

The number of events for each direction and the distribution of observed Einstein radius crossing times are the two useful pieces of information that can be extracted from the microlensing candidates. These observables can be compared to models.

4.1. Galactic model

We use a simple Galactic model, with three components, a central bulge, a disk and a dark halo (which has only a small influence). The predicted optical depth depends on the assumed number density of the lenses. The Einstein radius crossing time distribution depends both on the number density and on the velocity distribution. The number density is in turn calculated from the mass density and the mass function of the lenses.

Number densities

- The density distribution for the bulge – a barlike triaxial model – is taken from (Dwek et al 1995) model G2 (in Cartesian coordinates):

$$\rho_{Bulge} = \frac{M_{Bulge}}{8\pi abc} e^{-r^2/2}, \quad r^4 = \left[\left(\frac{x}{a}\right)^2 + \left(\frac{y}{b}\right)^2\right]^2 + \frac{z^4}{c^4},$$

where $M_{Bulge} = 2.1 \times 10^{10} M_\odot$ is the bulge mass, and a=1.49, b=0.58, c=0.40 kpc are the length scale factors. The bar major axis is inclined at an angle of 15° with respect to the Sun-Galactic Center line.

- The matter distribution in the disk is modeled in cylindrical coordinates by a double exponential:

$$\rho_{thin}(R,z) = \frac{\Sigma_{thin}}{2H_{thin}} \exp\left(\frac{-(R-R_\odot)}{R_{thin}}\right) \exp\left(\frac{-|z|}{H_{thin}}\right) ,$$

$\Sigma_{thin} = 50 M_\odot \text{pc}^{-2}$ is the column density of the disk at the Sun position, $H_{thin} = 0.325$kpc is the height scale and $R_{thin} = 3.5$kpc is the length scale of the Disk.

- We use a standard isotropic and isothermal halo with a density distribution given in spherical coordinates by :

$$\rho(r) = \rho_{h\odot} \frac{R_\odot^2 + R_c^2}{r^2 + R_c^2} ,$$

where $\rho_{h\odot} = 0.008 \ M_\odot \text{pc}^{-3}$ is the local halo density, $R_\odot = 8.5$ kpc is the distance between the Sun and the Galactic Center, and $R_c = 5$ kpc is the Halo "core radius".

- The mass function of lenses in the Bulge is taken from (Richer & Fahlman 1992). The disk lens mass function is taken from (Gould, Bahcall & Flynn 1997). Lenses belonging to the (non-rotating) halo have the same mass $(0.5 M_\odot)$.

Velocities

- The velocities transverse to the line of sight of bulge stars follow a Boltzmann distribution with a dispersion of ~ 110km/s.

- The velocity dispersions of the disk population are expected to be $(\sigma(V_R) = 40, \ \sigma(V_\theta) = 30, \ \sigma(V_z) = 20)$ (in km/s).

- The velocities of halo lenses follow a Boltzmann distribution with a dispersion of ~ 150km/s.

The variation of the predicted optical depths and event durations with the Galactic model parameters has been studied. Details can be found in Derue (1999). The most sensitive parameters are the scale length parameter a of the bar and the column density Σ_{thin}.

4.2. Einstein radius crossing time distribution

The distribution of observed Einstein radius crossing time t_E is displayed on Figure 3 separately for γ Sct and γ Nor. Figure 3 also shows the expected distributions for disk and bulge lenses. Bulge lenses contribute only towards γ Sct. The average t_E expected for disk lenses is

$$< t_E >_{expected} = 65 \text{ days} \qquad (2)$$

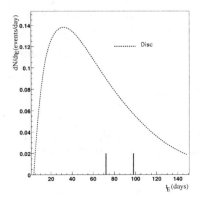 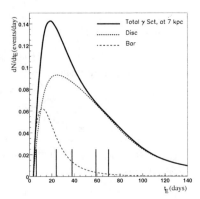

Figure 3. Measured Einstein radius crossing time distributions for γ Nor (left) and γ Sct (right). The measurements are compared to the predictions of the Galactic model of section 4.1, with the disk and the bar contributions plotted separately.

and the measured values are

$$< t_E >_{\gamma Sct} = 39.8 \pm 11.8 \text{ days} \tag{3}$$
$$< t_E >_{\gamma Nor} = 84 \pm 14 \text{ days} \tag{4}$$

The measured values are in good agreement with the expectation from the model.

4.3. Optical depth

For a given target, the optical depth (or a limit when no event was observed) has been computed using the expression

$$\tau = \frac{\pi}{2 N_{obs} T_{obs}} \sum_{events} \frac{t_E}{\epsilon(t_E)} \tag{5}$$

where N_{obs} is the number of observed stars, T_{obs} is the duration of the search period of 1170 days, except for θ Mus for which it is 990 days, and $\epsilon(t_E)$ is the average detection efficiency. The contribution of the candidates to the optical depth is given in Table 2.

The measured optical depth, averaged over the four directions is:

$$\bar{\tau} = 0.45^{+0.25}_{-0.17} \times 10^{-6} \tag{6}$$

The 1σ error bar reported here is statistical only. This is in agreement with the expectation from the model, (assuming as before a distance to the sources of 7 kpc):

$$\tau_{expected} = 0.55 \times 10^{-6} \tag{7}$$

The value of the measured optical depth towards the four directions is given in Table 3. A 95% confidence limit was computed for the θ Mus and β Nor

Table 3. Measured and predicted optical depth for the four directions monitored by EROS towards the Galactic disk.

Field	measured τ	expected τ
β Sct	≤ 1.08 (95% C.L)	0.57
γ Sct	1.82 (+1.33-0.85)	0.65
γ Nor	0.27 (+0.36-0.17)	0.49
θ Mus	≤ 0.6 (95% C.L)	0.34

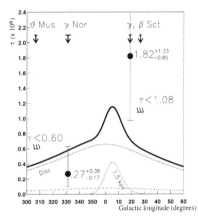 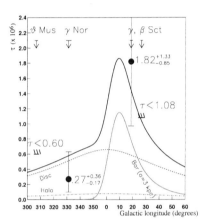

Figure 4. Microlensing optical depth towards the Galactic disk. The four directions monitored by EROS are indicated by arrows. The measured values are compared to the Galactic models of section 4.1 with a bar length parameter $a = 1.5$ kpc (left) and $a = 3$ kpc (right). The contribution of the bar, the disk and the halo are shown separately.

directions assuming a mean event Einstein radius crossing time of 50 days. The measured τ agrees well with the expectation from the model. There is a hint of an excess optical depth towards γ Sct, which is however not significant.

It is an interesting exercise to see what can be changed in the model to accommodate this possible excess optical depth towards γ Sct.

- An increase of the bar length parameter enhances the asymmetric contribution to the optical depth. Changing this parameter from $a = 1.5$ kpc to $a = 3$ kpc (see discussion in Gerhard 2000) leads to an optical depth towards γ Sct $\tau(\gamma\ Sct) = 1.40 \times 10^{-6}$. This is illustrated on Figure 4.

- The optical depth is very sensitive to the poorly known distance distribution of the monitored source stars. For example, changing the $\gamma\ Sct$ source star distances from 7 to 9 kpc increases the expected optical depth from 0.75 to 1.3×10^{-6}.

5. Conclusion

We have searched for microlensing events with durations ranging from a few days to a few months in four Galactic Disk fields lying 18° to 55° from the Galactic Center. Seven events that can be interpreted as microlensing effects due to massive compact objects were found. The average Einstein radius crossing time observed towards γ Nor and γ Sct is in agreement with the predicted value for disk lenses. The average optical depth measured towards the four directions is $\bar{\tau} = 0.45^{+0.25}_{-0.17} \times 10^{-6}$. Assuming the sources to be 7 kpc away, the expected optical depths from a simple Galactic model is 0.55×10^{-6}, in agreement with our measurement.

The predicted value for the optical depth is very sensitive to parameters such as the bar length scale. However, the knowledge of the location of the sources and the statistics will have to be improved before any conclusion can be drawn on Galactic structure.

References

Alcock, C., et al. (MACHO collaboration) 2000, ApJ, submitted (astro-ph/0002510)
Binney, J. 2000, astro-ph/0004362
Derue, F. et al. (EROS collaboration) 1999, A&A, 351, 87
Derue, F. et al. (EROS collaboration) 2000, A&A, submitted (astro-ph/0001083)
Derue F. 1999, PhD thesis, Université Paris XI, LAL 99-14
Dwek, E., Arendt, R.G, Hauser, M.G., et al. 1995, ApJ, 445,716
Gerhard, O.E. 2000, in Galaxy Dynamics, ASP Conf. Series, eds D. Merritt, M. Valluri (astro-ph/9902247)
Gould, A., Bahcall, J., Flynn, C. 1997, ApJ, 482, 913
Mansoux B. 1997, PhD thesis, Université Paris VII, LAL 97-19
Paczyński, B. 1986, ApJ, 304, 1
Richer, H.B., Fahlman G.G. 1992, Nature, 358, 383
Udalski, A., et al. (OGLE collaboration) 1994, Acta Astronomica, 44, 165

Discussion

Bennett: MACHO has 3 fields in the direction of your highest optical depth spiral arm field. We have $\gtrsim 10$ events in these fields, so we expect a high optical depth. Also, do you know if your sample has any overlap with the MACHO Alert sample?

Glicenstein: To my knowledge, we see none of the MACHO alerts in our data. They all fall in dead CCDs or are not in the fields that we monitor.

Gaudi: The event you fit with a binary source – did you try fitting it with a binary lens?

Glicenstein: As far as I know, we did not try to fit this event by a binary lens event. Since wiggles are seen on the residuals of the point-source point-lens fit, it seems to be more logical to try a binary source. This works quite well: the $\chi^2 \sim 550/440$ compared to $877/439$ for point-source point-lens.

Wambsganss: You said that in your analysis you only look for the simplest events, i.e. point-source point-lens. Do you plan to do a refined analysis in order to look for binary lens events as well?

Glicenstein: The selection of the events is optimized for point-source point-lens events. Nevertheless, a large fraction of binaries should pass the cuts. An instance of when it might fail the cut is when the binary lens event has two independent bumps.

Bradbury: Does the screening process eliminate events where the initial brightness is greater than the final brightness after the brightening event.

Glicenstein: To the best of my knowledge, yes.

Popowski: You have very few events and your efficiencies are not full efficiencies. It seems that the error on your optical depth is underestimated.

Glicenstein: I agree with this comment. The error was calculated as follows:
1) It is a statistical only error.
2) The error is estimated from Poisson error bars on the number of events located in each bin. Furthermore, these are not classical 68% confidence limit intervals but Bayesian intervals, so that it underestimates somewhat the size of the error bar. The errors will be re-examined.

Han: Why do you not have a continuous source star distribution?

Glicenstein: We have a continuous distribution of sources, with a large spread in distances. It may be possible in some fields to select the sources from a given spiral arm.

Evidence for Isolated Black Hole Stellar Remnants from Microlensing Parallax Events

D. Bennett, for the MACHO/GMAN and MPS Collaborations

Physics Department, University of Notre Dame, 225 Nieuwland Science Hall, Notre Dame, IN 46556, USA

Abstract. We present the analysis of microlensing parallax events discovered by the MACHO/GMAN and Microlensing Planet Search (MPS) Collaborations. We find two events towards the Galactic bulge which are likely to be caused by black hole lenses: MACHO-96-BLG-5 and MACHO-98-BLG-6. The microlensing parallax constraint on the lens masses and the observed upper limit on the lens brightness imply that these lenses are likely to be black holes of $\sim 6 M_\odot$.

A New Component of the Galaxy as the Origin of the LMC Microlensing Events

Evalyn Gates

Adler Planetarium & Astronomy Museum, Chicago, IL 60605
Department of Astronomy & Astrophysics, University of Chicago, Chicago, IL 60637

Geza Gyuk

Physics Department, University of California, San Diego, CA 92093

Abstract. We suggest a new component of the Milky Way that can account for both the optical depth and the event durations implied by microlensing searches targeting the Large Magellanic Cloud. This component, which represents less than 4% of the total dark matter halo mass, consists of mainly old white dwarf stars in a distribution that extends beyond the thin and thick disks, but lies well within the dark matter halo. It is consistent with recent evidence for a significant population of faint white dwarfs detected in a proper motion study of the Hubble Deep Field that cannot be accounted for by stars in the disk or spheroid. Further, it evades all of the current observational constraints that restrict a halo population of white dwarfs.

1. Introduction

The past few years have yielded exciting new data from observational teams searching for evidence of MACHOs (dark compact objects) in the halo of our galaxy, data which have raised many new questions. The microlensing optical depth toward the Large Magellanic Cloud (LMC) obtained by the MACHO collaboration, $\tau_{\rm LMC} = (1.2^{+0.4}_{-0.3}) \times 10^{-7}$ (Cook et al. 2000), is consistent with a significant fraction of the Galactic halo ($\sim 20\%$) in the form of MACHOs. The duration of these events also indicates that, under the assumption of a spherical isothermal halo distribution for the MACHOs, the average MACHO mass is $\sim 0.5 M_\odot$ (with large statistical uncertainties).

Such masses suggest several candidates for the lenses including faint halo stars, white dwarfs, and black holes. Direct searches for faint halo stars have placed severe limits on their contribution to the halo, requiring it to be less than about 3% (Flynn, Gould, & Bahcall 1996; Graff & Freese 1996). Primordial black hole candidates require a fine tuning of the initial density perturbations, and details of QCD phase transition black hole formation remain to be worked out in order to assess their viability and mass function (Jedamzik & Niemeyer

1999). Thus in the standard halo model interpretation, white dwarfs appear to be the remaining strong candidate for the lenses.

However, other possible interpretations of the MACHO results have been proposed in order to avoid the difficulties associated with a large halo population of white dwarfs. The mass estimate for the lenses depends upon the assumed phase space distribution of MACHOs. There are large uncertainties in the halo model parameters, including the distribution and velocity structure of the dark matter, and attempts have been made to exploit these uncertainties in order to obtain mass estimates from the current data that are consistent with lenses in the substellar regime (e.g. brown dwarfs). Previous work by the authors and others have examined a wide range of halo models, including flattened halos, halos with a bulk rotational component to the velocity structure (Gyuk & Gates 1998) and halos with anisotropic velocity dispersions (Gyuk, Evans, & Gates 1998). These analyses have shown that for any reasonable (smoothly varying) phase space distribution of the lenses, the implied lens mass is still much larger than the hydrogen burning limit, and thus one cannot appeal to modeling uncertainties in order to invoke brown dwarfs as candidates for the MACHOs.

Other work has explored the possibility that the lenses are not in the halo of the Milky Way. LMC self-lensing was originally suggested by Sahu (1994), but recent work by Gyuk, Dalal & Griest (1999) has argued that this is unlikely to be able to account for the current data. More recently, an unvirialized component of the LMC has been proposed in order to produce sufficient LMC self-lensing (cf Graff 2000). Zaritsky & Lin (1997) and Zhao (1998) have suggested an intervening population of stars toward the LMC (tidal debris or a dwarf galaxy) could be responsible for the microlensing events. Again this suggestion has been subject to much debate, and a recent paper by Gould (1999) argues strongly against such a scenario. Galactic models in which dark extensions of known populations, such as a heavy spheroid or thick disk, could be the source of the lenses were explored by Gates et al. (1998).

However, recent results from a proper motion study of the Hubble Deep Field (HDF) (Ibata et al. 1999), and from a comparison of the north and south HDF images (Mendez & Minniti 1999) have added a new piece to the puzzle. These studies provide further evidence that there may be a previously undetected population of old white dwarfs in the galaxy. Using new models for white dwarf cooling (Hansen 1999), the sources detected by Ibata et al. are consistent with 2-5 old ($> 12 Gyr$), $0.5 M_\odot$ white dwarfs with kinematics roughly consistent with that expected for halo objects. While earlier searches for the local counterparts did not find any such objects (Flynn et al. 2000), recent surveys have turned up several candidates (Hodgkin et al. 2000; Ibata et al. 2000). Overall, these data strengthen the interpretation of the MACHO lenses as white dwarfs in our own galaxy, and thus make alternative scenarios, such as LMC self-lensing, less appealing.

While these new data are still somewhat preliminary, they do raise the intriguing possibility that there is a previously undetected population of white dwarf stars that are not part of the disk or (known) spheroid. Along with the MACHO data and the inability of modeling to significantly change the mass estimates, these new results, seem to be relentlessly pointing to white dwarfs as the lenses. So, is the halo of our galaxy filled with white dwarfs? A standard halo interpretation of these data would say yes – a significant fraction of the Galactic halo must be in white dwarfs. However, such a scenario faces serious challenges from many directions.

2. White Dwarfs in the Halo?

When considering the possibility that a large fraction (or all) of the Galactic halo might be in the form of white dwarfs, it is extremely important to recall the evidence for Galactic dark matter, including estimates of the total mass of the Milky Way. A recent analysis of satellite radial and proper motions by Wilkinson and Evans (1999) found a total mass of the Galaxy $M_{TOT} \sim 2 \times 10^{12} M_\odot$, in good agreement with other recent estimates (Kochanek 1996; Zaritsky 1998). Wilkinson and Evans also find that the halo extends to at least 100kpc, and possibly much further to 150 or 200 kpc. Thus the total mass in a white dwarf population that comprises a significant fraction of the halo would be of order $10^{12} M_\odot$, a number which already severely strains the baryon budget of the Universe.

Models which propose such a population must also account for the mass in the progenitor population of stars and in the metal enriched gas produced during the formation of the white dwarfs. Combined with the above mass estimate for the Galactic halo, such considerations provide serious challenges for these models. For example, consider a white dwarf halo which is comprised of at least 50% white dwarfs. The total mass in white dwarfs today is thus of order $10^{12} M_\odot$. The efficiency $\epsilon(m)$ for producing a white dwarf from a progenitor star of mass m is likely to be 0.25 or smaller, depending on the progenitor mass (with an upper limit of $\epsilon = 0.5$ for progenitor stars of $1 M_\odot$) (Adams & Laughlin 1996). Thus, for a white dwarf halo mass of M_{wd}, we expect a mass in the progenitor population $M_{Pstars} \geq 4 M_{wd}$ and a mass in processed, metal rich gas $M_{zgas} \geq 3 M_{wd}$. A halo of mass $M_{TOT} = 2 \times 10^{12} M_\odot$, half of which is in white dwarfs, requires a progenitor mass of $M_{Pstars} \geq 4 \times 10^{12} M_\odot$. This in turn requires an extremely efficient early burst of star formation, through which essentially all of the baryons in the Universe are processed.

From a cosmological point of view, we can consider the contribution of the white dwarfs and the progenitor population to the matter density of the Universe. The Milky Way has a mass to light ratio $M/L \sim 100$ or greater (Zaritsky 1998). If we assume that this is a typical value for all galaxies, then galaxies contribute $\Omega_g \gtrsim 100/1200h = 0.08 h^{-1}$. Comparing this with $\Omega_b h^2 = 0.019 \pm 0.0024$ (95%cl, Burles et al. 1999), we find that a 50% white dwarf halo exceeds the baryon budget ($\frac{\Omega_{MACHOs}}{\Omega_b} \sim 2h$) even before considering the effects of processing most of the baryons through an early star phase. A 20% white dwarf halo is also difficult to reconcile with the above estimate of Ω_b, since the contributions of the progenitor stars will exceed Ω_b.

Many authors have explored the implications of a halo filled with white dwarfs. These analyses, combined with the estimates of the total mass in the halo, make the possibility of a white dwarf halo even less tenable.

First, the initial mass function (IMF) of the progenitor stars must be markedly different than the disk IMF. Limits on the IMF arise from both low and high mass stars. Low mass stars ($< 1 M_\odot$) would still be burning hydrogen today and should be visible. High mass stars ($> 8M_\odot$) would have evolved into Type II supernovae, ejecting heavy metals back into the interstellar medium [1]. From limits on red dwarfs in the halo and the Galactic metallicity, Adams & Laughlin (1996) and Chabrier et al (1996) find that the IMF must be sharply peaked about a progenitor star mass of $m \sim 2M_\odot$. Adams & Laughlin conclude that even with the above IMF, the white dwarf contribution to the halo is limited to less than 25% (with 50% being an extreme upper limit).

Next, the metal enriched gas that is produced when these stars become white dwarfs will pollute the remaining unprocessed gas, leading to high metallicities predicted for the Galactic disk and the interstellar medium (into which much of this gas must be blown out since the total mass in processed gas is much larger than the mass of the disk)(Fields, Matthews, & Schramm 1997). Gibson & Mould (1997) have estimated that the expected amount of C, N and O produced by the above IMFs, where the white dwarf mass is about $4 \times 10^{11} M_\odot$ (for a 20% MACHO halo), would be difficult to reconcile with that in pop II white dwarfs.

White dwarfs in the halo would also produce heavy metals via Type Ia supernovae. Canal, Isern and Ruiz-Lapuente (1997) use this to limit the halo fraction in white dwarfs to less than $5 - 10\%$ (or a total mass in white dwarfs of $5 - 10 \times 10^{10} M_\odot$). In addition, deep galaxy counts limit the fraction of the halo in white dwarfs to less than about 10%, since the brightly burning progenitor stars would be visible (Charlot & Silk 1995). Finally, it is worth mentioning that an all white dwarf halo would rule out the existence of other dark matter in the Universe (for example cold dark matter) (Gates & Turner 1994), opening the door to a host of problems with large scale structure formation.

3. A New Component of the Galaxy

Given the evidence for a previously undetected population of white dwarfs and the severe constraints on a halo population consistent with this evidence we propose a new component of the Galaxy. Such a component was first considered by the authors (Gyuk & Gates 1999) in the context of attempting to lower the mass estimates for the MACHO lenses, and in Gates et al. (1998) in considering dark extensions to know components.

This new component is essentially a very thick (scale height > 2 kpc) population of (mostly) old white dwarf stars. It is distinct from known Galactic populations, in both distribution and age. This "extended protodisk" extends beyond the thin and thick disk populations, but lies well within the halo. While the details of the distribution cannot be determined without significantly more data, the general features of this proposed model can be illustrated with the following example:

Consider an exponential disk with a volume density given by

---

[1] However, Venkatesan, Olinto & Truran 1999 have argued that the bounds on high mass stars are significantly relaxed for progenitor stars with very low ($10^{-4} Z_\odot$) or zero metallicity, which might relax this constraint somewhat.

$$\rho(r,z) = \frac{\Sigma_0}{2h_z} exp((r_0 - r)/r_d) sech^2(z/h_z) \qquad (1)$$

where $r_d = 4.0$kpc is the scale length and $h_z = 2.5$kpc is the scale height. We assume standard values for the position and circular velocity of the Sun, $r_0 = 8.0$kpc and $v_c = 220$km/s.

We also assume a velocity structure, which includes a rotational component $\tilde{v}_\phi \sim 130 - 170$km/s, of the form

$$f = \frac{\rho(r,\phi,z)}{m} \frac{1}{\sqrt{(2\pi)^3}\sigma_r \sigma_\phi \sigma_z} e^{-\left[\frac{v_r^2}{2\sigma_r^2} + \frac{(v_\phi - \tilde{v}_\phi)^2}{2\sigma_\phi^2} + \frac{v_z^2}{2\sigma_z^2}\right]} \qquad (2)$$

where the velocity ellipsoid varies as

$$\sigma_z^2 = 2\pi G \rho_0 h_z^2 = \pi G \Sigma_0 h_z. \qquad (3)$$

$$\sigma_r^2 \approx 2\sigma_z^2$$
$$\sigma_\phi^2 \approx \sigma_z^2.$$

For details on varying these model parameters as well as different parameterizations of the generic extended protodisk model see Gyuk & Gates (1998, 1999).

We can also consider a spheroid-like distribution for this component. Dynamical estimates for the mass of the spheroid are considerably larger than the luminous mass, although recent studies of the mass function of the spheroid indicate that the known spheroid population is unlikely to be able to account for the microlensing events (Gould, Flynn, & Bahcall 1998). Thus a spheroidal distribution would again correspond to a previously undetected component. For such a distribution the total mass is constrained in order not to conflict with the inner rotation curve of the galaxy, which limits LMC optical depths $\tau \lesssim 1.2 \times 10^{-7}$.

The extended protodisk supports approximately half of the local rotation speed, with the remainder coming from the thin disk and dark (non-MACHO) halo (see e.g. Figure 1). The dark halo in these models has a large core radius (> 7kpc) and an asymptotic rotation speed of ≈ 180kpc. The total mass in the Galaxy out to 50 kpc is $\approx 4.6 \times 10^{11} M_\odot$. For a total mass in the white dwarf extended protodisk of $M_{wd} = 7 \times 10^{10} M_\odot$, we find:

- The optical depth toward the LMC generated by this component is $\tau \sim 1 - 1.5 \times 10^{-7}$;

- The lens mass estimates for the current MACHO event durations is $m \sim 0.5 M_\odot$, consistent with white dwarf masses;

- We expect to see roughly twice as many white dwarfs in the HDF-South compared to HDF-North, similar to the halo models;

- Simulations of the proper motions for white dwarfs in this model are broadly consistent with the observations of Ibata et al. (1999).

The main feature of this model, however, is that it has a much lower total mass in white dwarfs than halo models. As outlined above, it is consistent with both the MACHO data and the HDF studies for a total mass in white dwarfs of $M_{wd} \lesssim 7 \times 10^{10} M_\odot$. This is approximately half of the mass that would be required for a halo distribution of MACHOs which would produce the same optical depth. Basically, this reduction can be understood because most microlensing is due to lenses within about 20 kpc of the Sun for either configuration. The extended protodisk has less mass beyond that distance than a halo.

The smaller mass in white dwarfs today also implies a smaller total mass in the progenitor population. For our above example the progenitor mass is $M_{Pstars} \sim 3 \times 10^{11} M_\odot$, a crucial factor of 5 less than that for a 20% white dwarf halo, assuming the same IMF in both cases.

There are several predictions of this model that can eventually allow it to be distinguished from a standard halo white dwarf population. First, the LMC optical depth cannot be much greater than about 1.5×10^{-7}. Second, because the lenses are concentrated closer to the plane of the galaxy, the typical lens-observer distance will be smaller (of order 5 kpc). This in turn implies an increase in the expected number of parallax events (Gyuk & Gates 1999). Finally, the ratio of optical depths toward the Small and Large Magellanic Clouds is expected to be of order $\tau_{SMC}/\tau_{LMC} \sim 0.8$, in contrast with $\tau_{SMC}/\tau_{LMC} \sim 1.5$ (Sackett & Gould 1993) predicted for a standard halo.

The distribution of event durations and a detailed comparison of the distribution (in direction and magnitude) of the observed proper motions will also differ for a halo vs. extended protodisk white dwarf population, but these seem unlikely to be able to differentiate between models without significantly more data.

4. Model Implications

Because the total mass in the white dwarf population today is significantly lower in this model, some of the constraints on a halo white dwarf population can be avoided, including those which consider the progenitor population and the ejected metal enriched gas. The total mass in this new component represents less than about 4% of a total halo mass of $2 \times 10^{12} M_\odot$. Since essentially all of the current constraints on white dwarf halos which limit the halo mass fraction in white dwarfs do so at only the 10% level, these constraints can be satisfied by our model. This includes the Type Ia supernovae constraints which are dependent on the mass in white dwarfs today, and cannot be evaded by scenarios which involve somehow hiding the metal enriched gas produced by the progenitor stars.

However, there remains much work to be done to more carefully consider the implications of this new component. First, we still require an IMF which differs significantly from the disk IMF. Assuming a log-normal distribution, Adams & Laughlin (1996) and Chabrier et al. (1996) used conservative constraints to limit the mass fraction of the high and low mass end of the progenitor IMF for a halo white dwarf population. While the lower total mass in our progenitor population will relax the constraints (which are based on the metallicity of the Galactic disk) somewhat at the high mass end, the mass fraction of low mass ($m < 1 M_\odot$) stars is constrained by number counts of faint low mass stars locally. The extended protodisk has an increased local density relative to a halo distribution, but a lower total mass, resulting in a constraint similar to that for a halo. Thus we expect to require a fairly sharp low mass drop-off in the progenitor IMF. The implications of such an IMF, including the lower fraction of primordial baryons which is processed through this early population, need to be examined in greater depth.

This new component also provides some intriguing hints for cosmology. When did this component form and how is it related to galaxy formation scenarios? Can this early starburst population help us to trace the baryons in the Universe from their primordial state to the present, where we find most of the baryons in the intracluster medium?

5. Conclusions

We have argued that the microlensing data toward the LMC, combined with observations of white dwarf stars in a proper motion study of the HDF indicate the presence of a new component of the galaxy. This component can be generally described as an extended distribution that extends at least 2 kpc above the Galactic plane, but resides well within the dark matter halo. It is consistent with all data and observations of the structure and kinematics of the galaxy, and significantly alleviates the considerable problems with a halo population of white dwarf stars that is consistent with microlensing data. Much work remains to carefully consider the implications of such a component, in particular the formation and evolution of the early population of progenitor stars (and resulting metal enriched gas) that produced this component. However, the significantly lower mass in the progenitor population as compared to that for a halo population of white dwarfs will allow a reasonable fraction of the baryonic mass of the Universe to remain in gas that has not been processed through these very early stars. Moreover, this component may be a more reasonable distribution for the remains of an early starburst population, in which one would expect a more condensed distribution than that of the halo.

Acknowledgments. This work was supported in part by the DOE (at Chicago and Fermilab) and by the NASA (through grant NAG5-2788 at Fermilab).

References

Adams, F. & Laughlin, G. 1996 ApJ, 46, 586

Burles, S., Nollett, K.M., Truran, J.N. & Turner, M.S. 1999, Phys.Rev.Lett, 82, 4176

Canal, R., Isern, J., and Ruiz-Lapuente, P. 1997, ApJ, 488, L35

Chabrier, G., Segretain, L. & Mera, D. 1996, ApJ, 468, L21

Charlot, S. and Silk, J. 1995, ApJ, 445, 124

Cook, K. and the MACHO Collaboration 2000, these proceedings

Fields, B.D., Matthews, G. & Schramm, D. 1997, ApJ, 483, 625

Flynn, C., Gould, A. & Bahcall, J. 1996, ApJ, 466, L55

Flynn, C., Sommer-Larsen, J., Fuchs, B., Graff, D., & Salim, S. 2000 (astro-ph 9912264)

Gates, E., Gyuk, G., Holder, G. & Turner, M.S. 1998, ApJ, 500, L145

Gates, E., Gyuk, G., & Turner, M.S. 1996, Phys.Rev.D, 53, 4138

Gates, E. & Turner, M.S. 1994, Phys.Rev.Lett, 72, 2520

Gibson, B.K. & Mould, J.R. 1997, ApJ, 482, 98

Gould, A. 1999, astro-ph/9902374

Gould, A., Flynn, C. & Bahcall, J.N. 1998, ApJ, 503, 798

Graff, D. & Freese, K. 1996, ApJ, 456, L49

Graff, D. 2000, these proceedings

Gyuk, G., Dalal, & Griest, K. 1999, astro-ph/9907338

Gyuk, G., Evans, W. & Gates, E. 1998, ApJ, 502, L29

Gyuk, G. & Gates, E. 1999, MNRAS, 304, 281

Gyuk, G. & Gates, E. 1998, MNRAS, 294, 682

Hansen, B. 1999, ApJ, 20, 680

Hodgkin, S., Oppenheimer, B., Hambly, N., Jameson, R., Smartt, S., & Steele, I. 2000, Nature, 403, 57

Ibata R., Richer H., Gilliland, R., & Scott, D. 1999, ApJ, 524, 1

Ibata, R., Irwin, M., Bienayme, O., Scholz, R., & Guibert, J. 2000, ApJ, 532, L41

Jedamzik, K. & Niemeyer, J.C. 1999, Phys.Rev.D, 59, 124014

Kochanek, C. 1996, ApJ, 457, 228

Mendez, R.A. & Minniti, D. 1999, ApJ, in press (astro-ph/9908330)

Sackett, P. & Gould, A. 1993, ApJ, 419, 648

Sahu, K. 1994, Nature, 370, 275

Wilkinson & Evans 1999, MNRAS, in press (astro-ph/9906197)

Venkatesan, A., Olinto, A.V. & Truran, J.T. 1999, ApJ, 516, 863

Zaritsky, D. 1998, in The Third Stromlo Symposium, ASP Conf. Ser. 165, eds. B.K. Gibson, T.S. Axelrod & M.E. Putnam, p. 34

Zaritsky, D. & Lin, D.N.C. 1997, AJ, 114, 2545

Zhao, H.-S. 1998, MNRAS, 294, 139

Discussion

Gould: The IMF must be different not only from the disk but also from the globular clusters, since these do not have many disk WDs.

Gates: The required IMF will be distinctly different from the observed IMF (in either the disk or globular clusters). I'm assuming that this population forms from zero-metallicity gas, which could lead to an IMF which might be very different from those observed.

Hansen: Was the progenitor population in a much thinner disk? I would guess that, if they were all in 4 $M_\odot$ stars initially, losing most of the mass should "fatten" the disk.

Gates: The "extended protodisk" forms from the merger of smaller fragments, and the answer to this question depends at least in part on when the progenitor star mass loss occurs relative to the merging history. I expect some of this mass to be lost to the system and some to remain to collapse to form the thin disk.

Graff: 1) I think your model would be strengthened if you relaxed the high-mass cutoff of the IMF. High-mass stars are no worse polluters than mid-mass stars. Thus, you only need one bit of new physics to constrain the low-mass cutoff of the IMF, not two (low mass and high mass).
2) Chemical pollution arguments limit $\Omega_{MACHO} \lesssim 0.1\Omega_B$, much tighter than your limits. Could you comment on this?

Gates: 1) I would agree that we are not so tightly constrained at the high-mass (M $\gtrsim$ 8 $M_\odot$) end as at the low-mass (M $\lesssim$ 1 $M_\odot$) end of the IMF, which is why I'm considering a log-normal distribution with a (high-mass end) tail. There are two reasons for this: first, the lower total mass in my model relative to a halo population of lenses and secondly, recent work suggesting that zero- (or close to zero) metallicity high-mass stars may eject fewer of their metals.
2) Some caution is indicated in attempting to put such tight constraints on Ω_{MACHO} from Lyα systems. The total amount of metals seen in the Lyα forest, damped Lyα systems and Ly-break galaxies is only $\sim$ 10% of what is produced by stellar activity directly observed at z $\gtrsim$ 3. This implies that the observed metallicity in these systems is not a good measure of early stellar activity.

The Local Group

Eva K. Grebel[1,2,3]

[1] *University of Washington, Department of Astronomy, Box 351580, Seattle, WA 98195-1580, USA*

[2] *Hubble Fellow*

[3] *Max-Planck-Institut für Astronomie, Königstuhl 17, D-69117 Heidelberg, Germany*

Abstract. Local Group galaxies such as the Milky Way, the Magellanic Clouds and M31 are being used by a number of international collaborations to search for microlensing events. Type and number of detections place constraints on dark matter and the stellar populations within and along the line of sight to these galaxies. In this review I briefly discuss the stellar populations, evolutionary histories, and other properties of different types of Local Group galaxies as well as constraints on the dark matter content of these galaxies. Particular emphasis is placed on the dwarf companions of the spiral galaxies in the Local Group.

1. Introduction

The "Local Group" is the small group of galaxies around the Milky Way and M31. The size of the Local Group is not well known, and its galaxy census is incomplete for low-surface-brightness galaxies. Recent studies suggest that the radius of the zero-velocity surface of the Local Group is ~ 1.2 Mpc (Courteau & van den Bergh 1999) when a spherical potential is assumed. Within this radius 35 galaxies have been detected (see Grebel 2000a for a list). Since information about orbits is lacking it is unknown which ones of the more distant galaxies within and just outside of the adopted Local Group boundaries are actually bound to the Local Group. Many faint Local Group galaxies were only discovered in recent years, and searches are continuing. Hierarchical cold dark matter (CDM) models predict about 10 times more dark matter halos than the number of known Local Group satellites (e.g. Klypin et al. 1999). Compact high-velocity clouds (Braun & Burton 1999), which appear to be dark-matter-dominated with total estimated masses of a few 10^8 M$_\odot$, may be good candidates for the "missing" satellites.

The Local Group comprises galaxies with a variety of different morphological types, a range of masses, ages, and metallicities, and differing degrees of isolation. Their proximity makes these galaxies ideal targets for detailed studies of their star formation histories from their resolved stellar populations and of galaxy evolution in general. Furthermore, Local Group galaxies provide a convenient set of targets for studies of the nature of dark matter. Several Local

Group reviews have appeared in the past few years, including Mateo (1998), Grebel (1997, 1999, 2000a), and the very detailed recent reviews by van den Bergh (1999, 2000). Reviews dealing with dark matter in Local Group dwarf spheroidals include Mateo (1997) and Olszewski (1998).

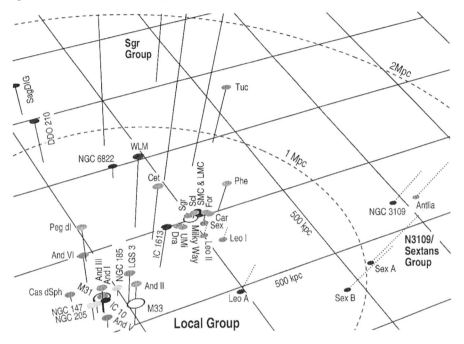

Figure 1. A scaled 3-D representation of the Local Group (LG). The dashed ellipsoid marks a radius of 1 Mpc around the LG barycenter (assumed to be at 462 kpc toward $l = 121.7$ and $b = -21.3$ following Courteau & van den Bergh 1999). Distances of galaxies from the the arbitrarily chosen plane through the Milky Way are indicated by solid lines (above the plane) and dotted lines (below). Morphological segregation is evident: The dEs and gas-deficient dSphs (light symbols) are closely concentrated around the large spirals (open symbols). DSph/dIrr transition types (e.g. Pegasus, LGS 3, Phoenix) tend to be somewhat more distant. Most dIrrs (dark symbols) are fairly isolated and located at larger distances. Also indicated are the locations of two nearby groups.

2. Local Group galaxy content and distribution

The most massive and most luminous Local Group galaxies are the two spirals, Milky Way and M31 ($\approx 10^{12} M_\odot$, $M_V \lesssim -21$ mag). The third, less luminous and less massive Local Group spiral M33 does not have any known companions and belongs to the M31 subsystem. About two thirds of the Local Group galaxies are found within 300 kpc of the two spirals. The majority of these close companions are dwarf spheroidal and dwarf elliptical galaxies. The ensemble of dwarf irregular galaxies, on the other hand, shows little concentration toward the two large spirals (although the two most massive Local Group irregulars, the Large and the Small Magellanic Cloud (LMC and SMC), are close neighbors of the Milky Way and interact with it as well as with each other). This correlation between morphological type and distance from massive galaxies is also known as morphological segregation and may be to some extent a consequence of evolutionary effects.

Whether a galaxy should be considered a dwarf galaxy is somewhat arbitrary, and different authors use different criteria. For the purpose of this review all galaxies with $M_V > -18$ mag will be considered dwarf galaxies, which results in 31 dwarfs, excluding only the three spirals and the LMC. We distinguish the following basic types of dwarf galaxies in the Local Group:

- Dwarf irregulars (dIrrs) with $M_V \gtrsim -18$ mag, $\mu_V \lesssim 23$ mag arcsec^{-2}, R $\lesssim 5$ kpc, $M_{HI} \lesssim 10^9 M_\odot$, and $M_{tot} \lesssim 10^{10} M_\odot$. DIrrs are irregular in their optical appearance, gas-rich, and show current or recent star formation. Several of the dIrrs contain globular or open clusters.

- Dwarf ellipticals (dEs) with $M_V \gtrsim -17$ mag, $\mu_V \lesssim 21$ mag arcsec^{-2}, R $\lesssim 4$ kpc, $M_{HI} \lesssim 10^8 M_\odot$, and $M_{tot} \lesssim 10^9 M_\odot$. DEs look globular-cluster-like in their visual appearance with a pronounced central concentration. All dEs are companions of M31. Two of the four dEs (M32, NGC 205) are nucleated. M32, a dE very close to M31, has a central black hole and follows the same scaling relations as large elliptical galaxies, whereas the other dEs resemble dSphs and are therefore called spheroidals by van den Bergh (1999, 2000). All dEs except for M32 contain globular clusters.

- Dwarf spheroidals (dSphs) with $M_V \gtrsim -14$ mag, $\mu_V \gtrsim 22$ mag arcsec^{-2}, R $\lesssim 3$ kpc, $M_{HI} \lesssim 10^5 M_\odot$, and $M_{tot} \sim 10^7 M_\odot$. These galaxies show very little central concentration and are dominated by old and intermediate-age stellar populations. Only three (Sgr, For, And I) contain globular clusters. With the exception of two isolated dSphs (Tuc and Cet) all known dSphs are close neighbors of M31 or the Milky Way. DSphs are gas-poor systems. Sensitive searches for H I in dSphs yielded only low upper limits, but recent studies detected extended H I clouds in the surroundings of some dSphs that may be associated with them judging from the similarity of their radial velocities (Carignan et al. 1998, Blitz & Robishaw 2000).

A few dwarf galaxies (Phe, LGS 3) are classified as "transition-type" objects and may be evolving from low-mass dIrrs to dSphs. These dIrr/dSph galaxies are found at distances of 250 kpc $< D_{\text{Spiral}} <$ 450 kpc. The Local Group does not contain blue compact dwarf galaxies, dwarf spirals, or massive ellipticals.

3. The Local Group spirals

The Local Group spirals have the most complex and varied star formation histories of all Local Group galaxies. Different subpopulations can be distinguished by their ages, metallicities, and kinematics. The oldest populations are found in the halos and thick disk components. Extremely metal-poor ([Fe/H] < −3 dex) halo stars are tracers of the earliest star formation events (Ryan et al. 1996), but it is difficult to derive ages for them.

The earliest significant star formation episodes in the Galactic thick disk appear to have occurred 13 Gyr ago, while the thin disk began to experience multiple bursts of star formation ∼ 9 Gyr ago (Rocha-Pinto et al. 2000). The metallicity in the thin disk depends more strongly on Galactocentric radius than on age and shows a large spread at any position and age (Edvardsson et al. 1993).

While halos may have largely formed through accretion of metal-poor Searle & Zinn (1978) fragments, bulges also host metal-rich old populations (mean metallicity of the Galactic bulge: −0.25 dex; Minniti et al. 1995), indicating that they experienced early and fast enrichment. M31 appears to have undergone rapid enrichment as a whole, whereas M33 shows a pronounced radial abundance gradient. The mean metallicity of M31's halo is −1 to −1.2 dex, more metal-rich than the halo of the Milky Way (∼ −1.4 dex) and of M33 (∼ −1.6 dex). While M31's bulge emits ∼ 30% of the visible light of this galaxy, M33 lacks a bulge.

M31's total number of globular clusters may be as high as ∼ 600. The Milky Way contains ∼ 160 globulars, and in the smaller M33 54 globulars are currently known (see Grebel 2000b for a review of star clusters in the Local Group). Main-sequence photometry of Galactic globular clusters suggests a range of ages spanning more than 3 Gyr. We lack such detailed information for M31's and M33's globulars, but the blue horizontal branch (HB) morphology observed in some of them may suggest similar ages as for the Milky Way globulars. On the other hand, the red HBs of M33's globulars may indicate that star formation was delayed by a few Gyr (Sarajedini et al. 1998).

The spiral arms in all three galaxies contain numerous OB associations and young star clusters. The UV line strengths of massive OB stars suggest that the young population of M31 is comparable to that of the Milky Way, whereas M33 resembles the Large Magellanic Clouds (Bianchi, Hutchings & Massey 1996). Present-day star-forming regions in the Milky Way range from very extended associations to compact starburst clusters such as the central cluster of NGC 3603 and the clusters Quintuplet and Arches near the Galactic center. M31's current star-forming activity is low. The increase in cluster formation in M33 over the past 10 − 100 Myr may be correlated with gas inflow into M33's center (Chandar, Bianchi & Ford 1999).

Warps in the stellar and H I disks of the Milky Way and M31 may have been caused by tidal interactions with the Magellanic Clouds and M32, respectively. The Milky Way disk may also have been significantly distorted by interacting with the currently merging Sagittarius dwarf galaxy (Ibata & Razoumov 1998). M33's stellar and H I disks are tilted with respect to each other, but no nearby companion is known that might be responsible.

4. Star formation histories of Local Group dwarf galaxies

The star formation histories of dwarf galaxies in the Local Group vary widely. No two galaxies are alike; not even within the same morphological type. The reasons for this diversity are not understood. It seems that both galaxy mass and environment play important roles in the evolution of these low-mass objects.

4.1. Methods and limitations

Star formation histories of resolved dwarf galaxies are commonly derived through photometric techniques. The most widely used method consists of sophisticated modelling of the observed color-magnitude diagrams (CMDs) through synthetic CMDs taking into account photometric errors, seeing, and crowding effects. For a recent review of procedures and techniques see Aparicio (1999). The methods are limited by the quality of the observations and by how closely theoretical evolutionary models reproduce observational features. For instance, Olsen (1999) notes that old red giant branches of evolutionary models may fit the observations poorly, which can lead to an underestimation of the contribution of the old population. Free parameters in modelling include the adopted initial mass function slope and the binary fraction.

Additional constraints can be imposed by using special types of stars as tracers of certain evolutionary phases. For instance, the presence of HB stars and RR Lyrae variables is a reliable indicator of an old population even when sufficiently deep main-sequence photometry is lacking. It is important to keep in mind that the age resolution that can be obtained is not linear and decreases strongly for older populations. Whereas young populations with short-lived, luminous massive stars can be accurately age-dated to within a few million years, the accuracy for the oldest, long-lived evolutionary phases is of the order of a few billion years. Relative ages of resolved old populations with high-quality, deep main-sequence photometry, on the other hand, can be established with a resolution of a Gyr or less through direct comparison with CMDs of Galactic globular clusters. In the following, "young" refers to populations with ages < 1 Gyr, "intermediate-age" denotes the age range from 1 Gyr to 10 Gyr, and "old" stands for ages > 10 Gyr.

Owing to the availability of 10-m class telescopes, spectroscopic measurements of stellar abundances are now feasible for individual supergiants and the brightest red giants in galaxies as distant as the M31 subgroup. Together with emission-line spectroscopy of H II regions, these data help to constrain the metallicity and metallicity spread in certain evolutionary phases. Still, accurate metallicity information as a function of time is lacking for almost all galaxies.

The increasing amount of data on internal kinematics and dwarf galaxy proper motions are beginning to constrain their dynamical history. Unfortunately accurate orbital data are not yet available for almost all of the Local Group galaxies, making it difficult to evaluate the suggested impact of environmental effects and interactions discussed later.

4.2. Old populations

A common property of all Local Group dwarfs studied in detail is the existence of an old population, whose presence can be inferred either from HB stars and/or from photometry reaching below the oldest main-sequence turnoff. Old populations may be difficult to detect in the central portions of galaxies with significant intermediate-age or young populations, as the location of these stars in a CMD may obscure an old HB. Also, coverage of only a small field of view may be insufficient to reliably detect a sparsely populated HB (compare the findings of Gallart et al. 1999 and Held et al. 2000 for Leo I). *Age dating* of the oldest populations is reliably possible only where high-quality photometry well below the oldest main-sequence turnoff exists; a challenge for present-day telescopes already for galaxies at the distance of M31. Definite statements about the existence of an old population are possible only where the photometry reaches at least the HB; feasible in principle with present-day telescopes out to distances ≈ 3 Mpc.

Deep main-sequence photometry based largely on *Hubble Space Telescope* data revealed that the ages of the oldest populations in the LMC (Holtzman et al. 1999), Sagittarius (Layden & Sarajedini 2000), Draco, Ursa Minor (Feltzing, Gilmore & Wyse 1999), Sculptor (Monkiewicz et al. 1999), Carina (Mighell 1997), Fornax (Buonanno et al. 1998), and Leo II (Mighell & Rich 1996) are as old as the oldest Galactic globular clusters and bulge populations. Thus all of these galaxies share a common epoch of early star formation. Similarly old ages were inferred from the existence of blue HBs in Sextans (Harbeck et al. 2000), Leo I (Held et al. 2000), Phoenix (Smith, Holtzman & Grillmair 2000), IC 1613 (Cole et al. 1999), Cetus (Tolstoy et al. 2000), And I (Da Costa et al. 1996), And II (Da Costa et al. 2000), NGC 185 (Geisler et al. 1999), NGC 147 (Han et al. 1997), Tucana (Lavery et al. 1996), M31 (Ajhar et al. 1996), potentially in M32 (Brown et al. 2000), and spectroscopically for one of NGC 6822's globular clusters (Cohen & Blakeslee 1998). Assuming that age is the second parameter determining HB morphology the apparent lack of a *blue* HB in M33 globular clusters (Sarajedini et al. 1998) and in the field populations of WLM (Dolphin 2000), Leo A (Tolstoy et al. 1998), DDO 210 (Tolstoy et al. 2000) and the Small Magellanic Cloud (SMC) may be interpreted as evidence for delayed formation of the majority of the old population in these galaxies. Furthermore, the oldest globular cluster in the SMC, NGC 121, is a few Gyr younger than the oldest Galactic globulars (Shara et al. 1998). A complete lack of an old population has so far not been established in any Local Group galaxy.

4.3. Spatial variations of stellar populations

Not surprisingly properties such as gas and stellar content, age structure, metallicity distribution, density, and scale height vary as a function of position within a galaxy. Spatial variations in the distribution of stellar populations of different ages are found in all types of galaxies, underlining the importance of large-area coverage when trying to determine the star formation history of a galaxy.

The oldest populations turn out to be spatially most extended. Spiral galaxies in the Local Group show pronounced population differences between disk, halo, and more intricate spatially and kinematically distinct subdivisions. In massive irregulars such as the LMC spatial variations are traced by, e.g. multiple distinct regions of concurrent star formation. These regions can remain active for several 100 Myr, are found throughout the main body of these galaxies, and can migrate.

In low-mass dIrrs and several dSphs the most recent star formation events are usually centrally concentrated. A radial age gradient may be accompanied by a radial metallicity gradient, indicating that not only gas but also metals were retained over an extended period of time. Occasionally evidence for shell-like propagation of star formation from the central to adjacent regions is found. DSphs that are predominantly old tend to exhibit radial gradients in their HB morphology such that the ratio of red to blue HB stars decreases towards the outer parts of the dwarfs. If such second-parameter variations are caused by age then this would indicate star formation persisted over a longer period of time in the centers of these ancient galaxies.

4.4. Differences in gas content

The H I in dIrrs is generally more extended than the oldest stellar populations and shows a clumpy distribution. Gas and stars in a number of low-mass dIrrs exhibit distinct spatial distributions and different kinematic properties. Shell-like structures, central H I holes, or off-centered gas may be driven by recent star formation episodes (Young & Lo 1996; 1997a,b). H I shells, however, do not always expand, which may argue against their formation through propagating star formation (Points et al. 1999, de Blok & Walter 2000).

Ongoing gas accretion appears to be feeding the starburst in the dIrr IC 10 (Wilcots & Miller 1998). An infalling or interacting H I complex is observed in the dIrr NGC 6822 (de Blok & Walter 2000).

DEs in the Local Group contain low amounts of gas (a few $10^5 M_\odot$; Sage, Welch & Mitchell 1998) or none (NGC 147). The apparent lack of gas in dSphs (e.g. Young 2000) continues to be hard to understand, in particular when considering that some dSphs show evidence for recent (Fornax: $\sim$ 200 Myr, Grebel & Stetson 1999) or pronounced intermediate-age star formation episodes (e.g. Carina: 3 Gyr; Hurley-Keller, Mateo & Nemec 1998; Leo I: 2 Gyr; Gallart et al. 1999). Gas concentrated in two extended lobes along the direction of motion of the Sculptor was detected beyond the tidal radius of this galaxy (Carignan et al. 1998). This gas may be moving inwards or away from Sculptor. Its amount is consistent with the expected mass loss from red giants, though that does not explain its location along the probable orbital direction of Sculptor. Blitz & Robishaw (2000) suggested the existence of similar gas concentrations with matching radial velocities in the surroundings of several other dSphs. Simulations by Mac Low & Ferrara (1999) suggest that *total* gas loss through star formation events can only occur in galaxies with masses of less than a few 10^{-6} $M_\odot$. Blitz & Robishaw discuss tidal effects as the most likely agent for the displacement of the gas. However, the absence of gas in Cetus and Tucana, two isolated dSphs in the Local Group, requires a different mechanism.

4.5. Star formation histories

Local Group dwarf galaxies vary widely in their star formation histories, chemical enrichment, and age distribution, even within the same morphological type. Despite their individual differences, however, they tend to follow common global relations between, e.g. mean metallicity, absolute magnitude, and central surface brightness. Galaxy mass as well as external effects such as tides appear to play major roles in their evolution.

Sufficiently massive irregulars and dIrrs exhibit continuous star formation at a variable rate. They can continue to form stars over a Hubble time and undergo gradual enrichment. Galaxies such as the LMC (Holtzman et al. 1999, Olsen 1999), SMC, and WLM (Dolphin 2000) have formed stars continuously and experienced considerable chemical enrichment spanning more than 1 dex in [Fe/H]. Their star formation *rate*, on the other hand, varied and shows long periods of low activity. Interestingly, in the LMC, cluster and field star formation activity show little correlation.

Low-mass dIrrs and dSphs often show continuous star formation rates with *decreasing* star formation rates. They typically show dominant old (or intermediate age) populations with little or no recent activity. A similar evolution appears to have occurred in dEs. DSph companions of the Milky Way tend to have increased fractions of intermediate-age populations with increasing Galactocentric distance, indicating that external effects such as tidal or ram pressure stripping may have affected their star formation history (e.g. van den Bergh 1994). The two closest dSphs to the Milky Way (other than the currently merging Sagittarius dSph) are Draco and Ursa Minor, which are dominated by ancient populations and are also the least massive dSphs known – possibly due to the early influence of Galactic tides, though present-day positions may not reflect early Galactocentric distances, and reliable orbital information is lacking.

The Local Group dwarf galaxy to show the most extreme case of *episodic* star formation with Gyr-long periods of quiescence and distinct, well-defined subgiant branches is Carina (Smecker-Hane et al. 1994, Hurley-Keller et al. 1998). It is unclear what caused the interruption and subsequent onset of star formation after the long gaps. Also, the apparent lack of chemical enrichment during these star formation episodes is surprising.

4.6. Potential evolutionary transitions

Fornax is the second most luminous dSph galaxy in the Local Group. The young age of its youngest measurable population ($\sim$ 200 Myr, Grebel & Stetson 1999) is astonishing considering its lack of gas. Just a few hundred Myr ago Fornax would have been classified as a dIrr. What caused Fornax to lose all of its gas after some 13 Gyr of continuous, decreasing star formation is not clear.

The presence of intermediate-age populations in some of the more distant Galactic dSphs, the possible detection of associated gas in the surroundings of several of them, indications of substantial mass loss discussed elsewhere in this paper, morphological segregation, common trends in relations between their integrated properties, and the apparent correlation between star formation histories and Galactocentric distance all seem to support the idea that low-mass dIrrs will eventually evolve into dSphs if their environment fosters this evolution. DSphs may be the natural final phase of low-mass dIrrs, and the type distinction may be artificial. The six dSph companions of M31 span a similar range in distances as the Milky Way dSphs (Grebel & Guhathakurta 1999). A study of whether their detailed star formation histories (not yet available) show a comparable correlation with distance from M31 would provide a valuable test of the suggested impact of environment.

The mass (traced by the luminosity) of a dwarf galaxy plays a major role in its evolution as indicated by the good correlation between luminosity and mean metallicity (e.g. Caldwell 1999). The observed lack of rotation in dSphs requires that its hypothesized low-mass dIrr progenitor must have gotten rid of its angular momentum, which may occur through substantial mass loss. However, this scenario does not account for the existence of isolated dSphs such as Tucana. Alternatively, the progenitor may have had very little rotation to begin with. Either way, the subsequent fading must have been low since otherwise dIrrs and dSphs would not follow such a fairly well-defined common relation. Several authors (e.g. Mateo 1998) suggested that the luminosity-metallicity relation is instead bimodal with separate loci for dIrrs and dSphs in the sense that at a given luminosity a dIrr tends to be more metal-poor than a dSph, excluding evolutionary transitions. Hunter, Hunsberger & Roye (2000) go a step further and suggest that a number of Local Group dIrrs might have formed as ancient tidal dwarfs that lack dark matter, are essentially non-rotating, and contribute to the increased scatter in the absolute magnitude–mean metallicity relationship for $M_B < -15$ mag.

5. Dark matter

Dark matter is a significant component of many Local Group galaxies. Spiral galaxies exhibit H I rotation curves that become approximately flat at large radii and that extend 2–3 times beyond the optically visible galaxy. Global mass-to-light ratios (M/L) inferred from rotation curves of spirals are typically ≤ 10 $M_\odot/L_\odot$ for the visible regions ($\sim 1 - 3$ $M_\odot/L_\odot$ in disks, $\sim 10 - 20$ $M_\odot/L_\odot$ in bulges), while the dark matter in halos seems to significantly exceed these values (Longair 1998). This motivates efforts to determine the nature of the dark matter through microlensing in the Galactic halo and toward the Galactic bulge as detailed elsewhere in this volume, and through pixel microlensing of stars in the disk of M31 by dark massive objects in M31's halo (Crotts 1992).

In gas-rich dwarfs the presence of dark matter is inferred as well from H I rotation curves. Some of the less massive dIrrs are rotationally supported only in their centers, while the majority of dSphs studied so far does not show evidence for rotation at all. Chaotic gas motions dominate in low-mass dIrrs, and the H I column density distribution is poorly correlated with the stellar distribution (Lo, Sargent & Young 1993). In the dE NGC 205, which is tidally interacting with M31, integrated light measurements revealed that the stellar component is essentially non-rotating though the H I shows significant angular momentum (Welch, Sage & Mitchell 1998). In gas-deficient dSphs kinematic information is based entirely on stars. Most dSphs show no rotation. Their velocity dispersions are typically ≥ 7 km s^{-1}. Assuming virial equilibrium velocity dispersions and rotation curves can be translated into virial masses. The derived total M/L ratios of Local Group dwarf galaxies present an inhomogeneous picture ranging from ~ 1 to ~ 80 (see compilation by Mateo 1998).

Compact high-velocity clouds (CHVCs) are a subset of high-velocity H I clouds with angular sizes of only about 1 degree on the sky. They show infall motion with respect to the barycenter of the Local Group. Preliminary estimates place them at distances of 0.5 to 1 Mpc in contrast to the extended nearby high-velocity-cloud complexes (Braun & Burton 1999). Their rotation curves imply high dark-to-H I ratios of 10–50 if distances of 0.7 Mpc are assumed, and masses of 10^7 $M_\odot$ (Braun & Burton 2000). CHVCs may be a significant source of dark matter and may represent pure H I/dark-matter halos prior to star formation. We are currently carrying out an optical wide-field survey to establish whether they also contain a low-luminosity, low-density stellar component, which would imply the discovery of a new, very dark type of galaxy, help to refine CHVC distances and allow detailed studies of their stellar populations.

5.1. Dwarf spheroidal galaxies and dark matter

Galactic dSphs are of particular interest in efforts to elucidate the nature of dark matter since they may be dark-matter-dominated and can be studied in great detail due to their proximity. From an analysis of the kinematic properties of Draco and Ursa Minor, Gerhard & Spergel (1992a) exclude fermionic light particles (neutrinos) as dark matter suspects because phase-space limits would then require unreasonably large core radii and masses for these two galaxies.

The initial measurements of velocity dispersions in dSphs were criticized for including luminous AGB stars and Carbon stars, whose radial velocities may reflect atmospheric motions, and for neglecting the impact of binaries (see Olszewski 1998 for details). Subsequent studies concentrated on somewhat fainter stars along the upper RGB, carried out extensive simulations to assess the impact of binaries (Hargreaves, Gilmore & Annan 1996; Olszewski, Pryor & Armandroff 1996), obtained multi-epoch observations (e.g. Olszewski, Aaronson & Hill 1995), and increased the number of red giants with measured radial velocities to more than 90 in some cases (Armandroff, Olszewski & Pryor 1995). These studies established that the large velocity dispersions in dSphs are not due to the previously mentioned observational biases. Kleyna et al. (1999) show that the currently available measurements for the two best-studied dSphs, Draco and Ursa Minor, are not yet sufficient to distinguish between models where mass follows light (constant M/L throughout the dSph) or extended dark halo models when interpreting the velocity dispersions as high M/L ratios due to large dark matter content. Mateo (1998) and Mateo et al. (1998) argue that the relation between total M/L and V-band luminosity for dSphs can be approximated well when adopting a stellar M/L of 1.5 (similar to globular clusters) and an extended dark halo with a mass of $2 \cdot 10^7$ $M_\odot$, suggesting fairly uniform properties for the dark halos of dSphs.

Luminosity functions (LFs) of old stellar systems can provide further constraints on the nature of dark matter. The main-sequence LFs of old field populations in the Galactic bulge (Holtzman et al. 1998), LMC and SMC (Holtzman et al. 1999), Draco (Grillmair et al. 1998), and Ursa Minor (Feltzing et al. 1999) are in excellent agreement with the solar neighborhood IMF and LFs of globular clusters that did not suffer mass segregation. Since globular clusters are not known to contain dark matter, one would expect to find differences in the LF of dark-matter-rich populations if low-mass objects down to 0.45 $M_\odot$ were important contributors to the baryonic dark matter content. Furthermore, these studies demonstrate that the LF in objects with a wide range of M/L ratios does not differ much. The possible contribution of white dwarfs (or lack thereof) is discussed elsewhere in these proceedings.

5.2. Tidal effects rather than dark matter?

Instead of a smooth surface density profile that one might expect from a relaxed population, Ursa Minor shows statistically significant stellar density variations (Kleyna et al. 1998). Fornax's four ancient globular clusters are located at distances larger than the galaxy's core radius. Dynamical friction should have led to orbital decay in only a few Gyr (much less time than the globular clusters' lifetimes) and have turned Fornax into a nucleated dSph. Simulations by Oh, Lin & Richer (2000) suggest that the best mechanism to have prevented this evolution is significant mass loss through Galactic tidal perturbation and the resulting decrease in the satellite galaxy's gravitational potential, which may have increased the clusters' orbital semimajor axes and efficiently counteracted the spiralling-in through dynamical friction. The detection of a possible extended population of extratidal stars around the dSph Carina might imply that this galaxy has now been reduced to a mere 1% of its initial mass (Majewski et al. 2000). If such significant tidal disruption is indeed real and widespread then the present-day stellar content of nearby dSphs cannot easily be used to derive evolutionary histories over a Hubble time. Furthermore, extended extratidal stars are not expected if the galaxy is dark-matter dominated (Moore 1996).

Additional indications in favor of the impact of galactic tides come from the structural parameters of dSphs (Irwin & Hatzidimitriou 1995), which seem to imply tidal disruption in several cases. Furthermore, substantial tidal disruption by the Milky Way is evidenced by the Magellanic Clouds and Magellanic stream, by the Sagittarius dSph, and by Galactic globular clusters (Gnedin & Ostriker 1997; Grillmair et al. 1995; Leon, Meylan & Combes 2000). Tracer features include gaseous and stellar tidal tails (Putman et al. 1998; Majewski et al. 1999; Odenkirchen et al. 2000). The conversion of velocity dispersions into M/L ratios and dark matter fractions assumes virial equilibrium, a condition that is violated in the case of severe tidal disruption.

Tidal heating due to resonant orbital coupling between the time-dependent Galactic gravitational field and the internal oscillation time scales of dSphs (Kuhn & Miller 1989; Kuhn 1993; Kuhn, Smith & Hawley 1996) may inflate the dSphs' velocity dispersions, but see Pryor (1996) for arguments against the efficiency of this mechanism.

As shown by Piatek & Pryor (1995) tides can, but need not, inflate the global M/L ratio to high values. Indeed, in a galaxy suffering tidal disruption the velocity dispersion can be sustained at its virial equilibrium value, and the central density is maintained even after substantial mass loss (Oh, Lin & Aarseth 1995). Pryor (1996) noted that a velocity gradient across a galaxy that is larger than the velocity dispersion is the clearest signature of tidal disruption, but such a gradient is not obvious in the Galactic dSphs.

Kroupa (1997) and Klessen & Kroupa (1998) proposed that stellar tidal tails may look like dSphs when seen along the line of sight, an orientation that follows naturally from their N-body simulations. The ordered motions in the tidal remnants would appear as increased velocity dispersion since they occur along the line of sight. These models can roughly reproduce the observed correlations between central surface brightness, absolute magnitude, and M/L. The predicted line-of-sight extension of the dSphs can be tested, in principle, through accurate measurements of the apparent width of their HBs. The predicted high orbital eccentricity, a consequence of the required radial orbits in the model, could be checked through accurate proper motion measurements with astrometric satellite missions such as *SIM* and *GAIA*. The tidal remnants may be leftovers from earlier mergers as suggested by the observation that the Galactic dSph galaxies appear to be located near at least two polar planes or great circles (the Magellanic Stream and the Fornax–Leo–Sculptor Stream; e.g. Kunkel & Demers 1996; Kunkel 1979; Lynden-Bell 1982; Majewski 1994). Such tidal remnants would not likely contain dark matter. The observed ages and abundances of galaxies potentially associated with "streams" constrain the time at which the break-up of a more massive parent could have occurred. This event must have happened very early on when the parent had not yet experienced significant enrichment. Siegel & Majewski (2000) suggest that galaxies potentially belonging to a stream may have originated from a common -2.3 dex progenitor and subsequently followed their own evolution.

The impact of Galactic tides remains a valid alternative to large amounts of dark matter in nearby dSphs. The determination of velocity dispersions of distant or even isolated dSphs, which are unlikely to be subject to tidal effects, is an important test of whether high M/L ratios in dSphs are largely caused by environmental effects (see, e.g. the discussion in Bellazzini, Fusi Pecci & Ferraro 1996). Stellar velocity dispersions indicative of high M/L were found in the most distant (~ 270 kpc) potential Milky Way dSph companion Leo I (Mateo et al. 1998) and in the outlying (~ 280 kpc) M31 transition-type satellite LGS 3 (Cook et al. 1999), but measurements of truly isolated Local Group dSphs such as Tucana and Cetus are still lacking.

5.3. Modified Newtonian dynamics

Modified Newtonian dynamics (MOND, Milgrom 1983a,b), which alters Newton's second law at low accelerations by introducing a multiplicative acceleration constant of $1.2 \cdot 10^{-8}$ cm s^{-1}, results in M/L ratios that do not require the presence of dark matter. While many attempts have been made to disprove MOND (e.g. Gerhard & Spergel 1992b), none of the presently existing measurements has been able to unambiguously refute MOND for either disk galaxies (van den Bosch & Dalcanton 2000) or dwarfs (e.g. Milgrom 1994; 1995; Côté et al. 1999). MOND remains a possible alternative to dark matter.

5.4. Implications for microlensing

Microlensing surveys are concentrating on the Galactic bulge, the Galactic halo through monitoring of sight lines toward the Magellanic Clouds, and on M31 through pixel lensing. All of these surveys concentrate on fields with high-density, luminous background populations. Results and constraints on dark matter from these surveys are discussed elsewhere in this volume.

The remaining Local Group dwarf galaxies are less well suited for classical microlensing studies. Advantages of using other dwarfs are that one can probe additional lines of sight and can take advantage of the large optical depth to microlensing since the sources are outside of the Milky Way halo. Also, in nearby dSphs crowding won't be much of a problem. However, the efficiency of such studies would be drastically reduced as compared to the ongoing studies since the targets are faint and stellar densities are low. This requires not only longer exposure times or larger telescopes but also implies much longer time scales before a significant number of events can be observed.

As discussed earlier the high velocity dispersions in low-mass dwarfs may arise from large amounts of dark matter. This increases the possibility that one may observe self-lensing when monitoring dwarfs rather than events in the Galactic halo, an effect that is negligible when turning to distant Galactic globular clusters instead (Gyuk & Holder 1998). As always, variable stars may act as contaminants. Future large survey telescopes such as the proposed *Dark Matter*

Telescope (Tyson, Wittman & Angel 2000) can provide routinely deep exposures of nearby Local Group dwarf galaxies once per night as a regular by-product of their search for cosmological weak lensing.

6. Summary

The Local Group, an ensemble of 35 galaxies most of which are dwarf companions of either M31 or the Milky Way, contains galaxies with a wide variety of masses, luminosities, star formation histories, and chemical and kinematic properties. No two galaxies in the Local Group experienced the same star formation history even within the same morphological type. Star formation episodes vary in length and times ranging from continuous star formation with variable star formation rates to gradually declining rates and episodic star formation, accompanied by either gradual chemical enrichment or almost no enrichment at all. Old populations are a common property of all Local Group galaxies studied in detail so far, though not all appear to share a common epoch of the earliest measurable star formation. Spatial variations in ages and abundances are observed in most Local Group galaxies ranging from widely scattered active regions in high-mass galaxies to centrally concentrated younger star formation episodes in low-mass dwarfs. Both galaxy mass and galaxy environment appear to have a major impact on galaxy evolution. Interactions such as ram pressure and tidal stripping seem to influence the evolution of less massive galaxies, contributing to the observed morphological segregation. Rotation curves and stellar velocity dispersions indicate the presence of dark matter in the majority of Local Group galaxies, although alternative explanations cannot be ruled out. The properties of the stellar populations in these galaxies as well as orbital and kinematic information can impose constraints on the nature and ubiquity of dark matter. Owing to faintness, low stellar density and hence low event probability as well as likeliness of self-lensing, Local Group dwarf galaxies other than the Magellanic Clouds are poorly suited for classical microlensing surveys but might become of interest for future large telescopes that routinely monitor a major fraction of the sky on a nightly basis.

Acknowledgments. This work was supported by NASA through grant HF-01108.01-98A from the Space Telescope Science Institute, which is operated by the Association of Universities for Research in Astronomy, Inc., under NASA contract NAS5-26555. I gratefully acknowledge support from the organizers who covered my local expenses. Last but not least I like to thank the editors for their patience while this manuscript was finished.

References

Ajhar, E.A., Grillmair, C.J., Lauer, T.R., Baum, W.A., Faber, S.M., Holtzman, J.A., Lynds, C.R. & O'Neil, E.J. 1996, AJ, 111, 1110

Armandroff, T.E., Olszewski, E.W. & Pryor, C. 1995, AJ, 110, 2131

Aparicio, A. 1999, in IAU Symp. 192, The Stellar Content of the Local Group, ed. P. Whitelock & R. Cannon (Provo: ASP), p. 304
Bellazzini, M., Fusi Pecci, F. & Ferraro, F.R. 1996, MNRAS, 278, 947
Bianchi, L., Hutchings, J.B. & Massey, P. 1996, AJ, 111, 2303
Blitz, L. & Robishaw, T. 2000, ApJ, submitted (astro-ph/0001142)
Braun, R. & Burton, W.B. 1999, A&A, 341, 437
Braun, R. & Burton, W.B. 2000, A&A, 354, 853
Brown, T.M., Bowers, C.W., Kimble, R.A. & Sweigart, A.V. 2000, ApJ, 532, 308
Buonanno, R., Corsi, C.E., Zinn, R., Fusi Pecci, F., Hardy, E., & Suntzeff, N.B. 1998, ApJ, 501, L33
Caldwell, N. 1999, AJ, 118, 1230
Carignan, C., Beaulieu, S., Côté, S., Demers, S. & Mateo, M. 1998, AJ, 116, 1690
Chandar, R., Bianchi, L. & Ford, H.C. 1999, ApJ, 517, 668
Cohen, J.G. & Blakeslee, J.P. 1998, AJ, 115, 2356
Cole, A.A., Tolstoy, E., Gallagher, J.S., Hoessel, J.G., Mould, J.R., Holtzman, J.A., Saha, A., Ballester, G.E., Burrows, C.J., Clarke, J.T., Crisp, D., Griffiths, R.E., Grillmair, C.J., Hester, J.J., Krist, J.E., Meadows, V., Scowen, P., Stapelfeldt, K.R., Trauger, J.T., Watson, A.M., & Westphal, J.R. 1999, AJ, 118, 1657
Cook, K.H., Mateo, M., Olszewski, E.W., Vogt, S.S., Stubbs, C.,& Diercks, A. 1999, PASP, 111, 306
Côté, P., Mateo, M., Olszewski, E.W. & Cook, K.H. 1999, ApJ, 526, 147
Courteau, S. & van den Bergh, S. 1999, AJ, 118, 337
Crotts, A.P.S. 1992, ApJ, 399, L43
Da Costa, G.S., Armandroff, T.E., Caldwell, N. & Seitzer, P. 1996, AJ, 112, 2576
Da Costa, G.S., Armandroff, T.E., Caldwell, N. & Seitzer, P. 2000, AJ, 119, 705
de Blok, W.J.G. & Walter, F. 2000, ApJ, 537, L95
Dolphin, A.E. 2000, ApJ, 531, 804
Edvardsson, B., Andersen, J., Gustafsson, B., Lambert, D.L., Nissen, P.E., & Tomkin, J. 1993, A&A, 275, 101
Feltzing, S., Gilmore, G. & Wyse, R.F.G. 1999, ApJ, 516, L17
Gallart, C., Freedman, W.L., Aparicio, A., Bertelli, G. & Chiosi, C. 1999, AJ, 118, 2245
Geisler, D., Armandroff, T.E., Da Costa, G., Lee, M.G. & Sarajedini, A. 1999, in IAU Symp. 192, The Stellar Content of the Local Group, ed. P. Whitelock & R. Cannon (Provo: ASP), p. 231
Gerhard, O.E. & Spergel, D.N. 1992a, ApJ, 389, L9
Gerhard, O.E. & Spergel, D.N. 1992b, ApJ, 397, 38
Gnedin, O.Y. & Ostriker, J.P. 1997, ApJ, 474, 223
Grebel, E.K. 1997, Rev. in Mod. Astron., 10, 29

Grebel, E.K. 1999, in IAU Symp. 192, The Stellar Content of the Local Group, ed. P. Whitelock & R. Cannon (Provo: ASP), p. 17

Grebel, E.K. 2000a, in 33rd ESLAB Symp., Star Formation from the Small to the Large Scale, ed. F.Favata, A.A. Kaas & C. Wilson (Noordwijk: ESA), ESA SP-445, in press (astro-ph/0005296)

Grebel, E.K. 2000b, in Observatoire de Strasbourg Workshop, Massive Stellar Clusters, ed. A. Lançon & C. Boily (Provo: ASP), in press (astro-ph/9912529)

Grebel, E.K. & Guhathakurta, P. 1999, ApJ, 511, L101

Grebel, E.K. & Stetson, P.B. 1999, in IAU Symp. 192, The Stellar Content of the Local Group, ed. P. Whitelock & R. Cannon (Provo: ASP), p. 165

Grillmair, C.J., Freeman, K.C., Irwin, M. & Quinn, P.J. 1995, AJ, 109, 2553

Grillmair, C.J., Mould, J.R., Holtzman, J.A., Worthey, G., Ballester, G.E., Clarke, J.T., Crisp, D., Evans, R.W., Gallagher, J.S. & Burrows, C.J. 1998, AJ, 115, 144

Gyuk, G. & Holder, G.P. 1998, MNRAS, 297, L44

Han, M., Hoessel, J.G., Gallagher, J.S., Holtzman, J. & Stetson, P.B. 1997, AJ, 113, 1001

Harbeck, D., Grebel, E.K., Holtzman, J., Geisler, D. & Sarajedini, A. 2000, AGM, 17, 45

Hargreaves, J.C., Gilmore, G. & Annan, J.D. 1996, MNRAS, 279, 108

Held, E.V., Saviane, I., Momany, Y. & Carraro, G. 2000, ApJ, 530, L85

Holtzman, J.A., Watson, A.M., Baum, W.A., Grillmair, C.J., Groth, E.J., Light, R.M., Lynds, R. & O'Neil, E.J. 1998, AJ, 115, 1946

Holtzman, J.A., Gallagher, J.S., Cole, A.A., Mould, J.R., Grillmair, C.J., Ballester, G.E., Burrows, C.J., Clarke, J.T., Crisp, D., Evans, R.W., Griffith, R.E., Hester, J.J., Hoessel, J.G., Scowen, P.A., Stapelfeldt, K.R., Trauger, J.T. & Watson, A.M. 1999, AJ, 118, 2262

Hunter, D.A., Hunsberger, S.D. & Roye, E.W. 2000, ApJ, in press (astro-ph/0005257)

Hurley-Keller, D., Mateo, M. & Nemec, J. 1998, AJ, 115, 1840

Ibata, R.A. & Razoumov, A.O. 1998, A&A, 336, 130

Irwin, M. & Hatzidimitriou, D. 1995, MNRAS, 277, 1354

Klessen, R.S. & Kroupa, P. 1998, ApJ, 498, 143

Kleyna, J.T., Geller, M.J., Kenyon, S.J., Kurtz, M.J. & Thorstensen, J.R. 1998, AJ, 115, 2359

Kleyna, J., Geller, M., Kenyon, S. & Kurtz, M. 1999, AJ, 117, 1275

Klypin, A., Kravtsov, A.V., Valenzuela, O. & Prada, F. 1999, ApJ, 522, 82

Kroupa, P. 1997, NewA, 2, 139

Kuhn, J.R. 1993, ApJ, 409, L13

Kuhn, J.R. & Miller, R.H. 1989, ApJ, 341, L41

Kuhn, J.R., Smith, H.A. & Hawley, S.L. 1996, ApJ, 469, L93

Kunkel, W.E. 1979, ApJ, 228, 718

Kunkel, W.E. & Demers, S. 1976, RGO Bull., 182, 241
Lavery, R.J., Seitzer, P., Walker, A.R., Suntzeff, N.B. & Da Costa, G.S. 1996, AAS, 188, 903
Layden, A.C. & Sarajedini, A. 2000, AJ, 119, 1760
Leon, S., Meylan, G. & Combes, F. 2000, A&A, 359, 907
Lo, K.Y., Sargent, W.L.W. & Young, K. 1993, AJ, 106, 507
Longair, M.S. 1998, Galaxy Formation (Heidelberg: Springer)
Lynden-Bell, D. 1982, Observatory, 102, 202
Mac Low, M.-M. & Ferrara, A., 1999, ApJ, 513, 142
Majewski, S.R. 1994, ApJ, 431, L17
Majewski, S.R., Siegel, M.H., Kunkel, W.E., Reid, I.N., Johnston, K.V., Thompson, I.B., Landolt, A.U. & Palma, C. 1999, AJ, 1709
Majewski, S.R., Ostheimer, J.C., Patterson, R.J., Kunkel, W.E., Johnston, K.V. & Geisler, D. 2000, AJ, 119, 760
Mateo, M. 1997, in Second Stromlo Symposium, The Nature of Elliptical Galaxies, ed. M. Arnaboldi, G.S. da Costa & P. Saha (San Francisco: ASP), p. 259
Mateo, M. 1998, ARA&A, 36, 435
Mateo, M., Olszewski, E.W., Vogt, S.S. & Keane, M.J. 1998, AJ, 116, 2315
Mighell, K.J. 1997, AJ, 114, 1458
Mighell, K.J. & Rich, R.M., 1996, AJ, 111, 777
Milgrom, M. 1983a, ApJ, 270, 365
Milgrom, M. 1983b, ApJ, 270, 384
Milgrom, M. 1994, ApJ, 429, 540
Milgrom, M. 1995, ApJ, 455, 439
Minniti, D., Olszewski, E.W., Liebert, J., White, S.D.M., Hill, J.M., & Irwin, M.J. 1995, MNRAS, 277, 1293
Monkiewicz, J., Mould, J.R., Gallagher, J.S., Clarke, J.T., Trauger, J.T., Grillmair, C., Ballester, G.E., Burrows, C.J., Crisp, D., Evans, R., Griffiths, R., Hester, J.J., Hoessel, J.G., Holtzman, J.A., Krist, J.E., Meadows, V., Scowen, P.A., Stapelfeldt, K.R., Sahai, R. & Watson, A. 1999, PASP, 111, 1392
Moore, B. 1996, ApJ, 461, L13
Odenkirchen, M., Grebel, E.K., Rockosi, C.M., Dehnen, W., Ibata, R., Rix, H.-W., Stolte, A., Wolf, C., Bahcall, N.A., Brinkmann, J., Csabai, I., Hennessy, G., Hindsley, R.B., Ivezić, Ž., Pier, J.R., York, D. 2000, ApJL, submitted
Oh, K.S., Lin, D.N.C. & Aarseth, S.J. 1995, ApJ, 442, 1420
Oh, K.S., Lin, D.N.C. & Richer, H.B. 2000, ApJ, 531, 727
Olsen, K.A.G. 1999, AJ, 117, 2244
Olszewski, E.W. 1998, in UC Santa Cruz Workshop, Galactic Halos, ed. D. Zaritsky (Provo: ASP), p. 70
Olszewski, E.W., Aaronson, E.W. & Hill, J.M. 1995, AJ, 110, 2120

Olszewski, E.W., Pryor, C. & Armandroff, T.E. 1996, AJ, 111, 750

Piatek, S. & Pryor, C. 1995, AJ, 109, 1071

Points, S.D., Chu, Y.-H., Kim, S., Smith, R.C., Snowden, S.L., Brandner, W., & Gruendl, R.A. 1999, ApJ, 518, 298

Pryor, C. 1996, in ASP Conf. Ser. 92, Formation of the Galactic Halo ... Inside and Out, ed. H. Morrison & A. Sarajedini (San Francisco: ASP), p. 424

Putman, M.E., Gibson, B.K., Staveley-Smith, L., Banks, G., Barnes, D.G., Bhatal. R., Disney, M.J., Ekers, R.D., Freeman, K.C., Haynes, R.F., Henning, P., Jerjen, H., Kilborn, V., Koribalski, B., Knezek, P., Malin, D.F., Mould, J.R., Oosterloo, T., Price, R.M., Ryder, S.M., Sadler, E.M., Stewart, I., Stootman, F., Vaile, R.A., Webster, R.L., & Wright, A.E. 1998, Nature, 394, 752

Rocha-Pinto, H.J., Scalo, J., Maciel, W.J. & Flynn, C. 2000, ApJ, 531, L115

Ryan, S.G., Norris, J.E. & Beers, T.C. 1996, ApJ, 471, 254

Sage, L.J., Welch, G.A. & Mitchell, G.F. 1998, ApJ, 507, 726

Sarajedini, A., Geisler, D., Harding, P. & Schommer, R. 1998, ApJ, 508, L37

Searle, L. & Zinn, R., 1978, ApJ, 225, 357

Shara, M.M., Fall, S.M., Rich, R.M. & Zurek, D. 1998, ApJ, 508, 570

Siegel, M.H. & Majewski, S.R. 2000, AJ, 120, 284

Smecker-Hane, T.A., Stetson, P.B., Hesser, J.E. & Lehnert, M.D. 1994, AJ, 108, 507

Smith, G.H., Holtzman, J.A. & Grillmair, C.J. 2000, in prep.

Tolstoy, E., Gallagher, J. S., Cole, A. A., Hoessel, J. G., Saha, A., Dohm-Palmer, R. C., Skillman, E. D., Mateo, M., & Hurley-Keller, D. 1998, AJ, 116, 1244

Tolstoy, E., Gallagher, J.S., Greggio, L., Tosi, M., de Marchi, G., Romaniello, M., Minniti, D. & Zijlstra, A. 2000, Messenger, 99, 16

Tyson, J.A., Wittman, D. & Angel, J.R.P. 2000, in Dark Matter 2000 (New York: Springer), in press (astro-ph/0005381)

Welch, G.A., Sage, L.J. & Mitchell, G.F. 1998, ApJ, 499, 209

Wilcots, E.M. & Miller, B.W. 1998, AJ, 116, 2363

Young, L.M. 2000, AJ, 119, 188

Young, L.M. & Lo, K.Y. 1996, ApJ, 464, L59

Young, L.M. & Lo, K.Y. 1997a, ApJ, 476, 127

Young, L.M. & Lo, K.Y. 1997b, ApJ, 490, 710

van den Bergh, S. 1994, ApJ, 428, 617

van den Bergh, S. 1999, A&ARv, 9, 273

van den Bergh, S. 2000, The Local Group (Cambridge: CUP)

van den Bosch, F.C. & Dalcanton, J.J. 2000, ApJ, 534, 146

Discussion

Evans: What are the prospects for measuring the radial velocities of And I, And III, And V, And VI, and And VII?

Grebel: In fact, Raja Guhathakurta and I just measured these radial velocities with Keck and hope to publish a paper on this soon.

Crotts: What is the lifetime of the dE and dSph satellites of M31 and the Milky Way and how many have been consumed in the halo?

Grebel: This depends on the direction of their orbits. It has been suggested that they are polar, but a more detailed model is that surviving satellites are on polar great circles, and others have been absorbed. In this case as many as 60 Carina-sized satellites may have already been absorbed and remain as tidal streams.

Zinnecker: What do we conclusively know about the dark matter content of the dwarf galaxies in the Local Group?

Grebel: HI rotation curves and/or stellar kinematics indicate high mass-to-light ratios and are commonly interpreted as implying large amounts of dark matter. Most Local Group galaxies show such evidence for large mass-to-light ratios. However, alternative explanations such as MOND have not been conclusively ruled out. Furthermore, tidal disruption by massive galaxies may lead to inflated velocity dispersions in dwarf spheroidal galaxies in close proximity of massive galaxies, although similarly high stellar velocity dispersions have also been measured in more isolated dwarfs.

Gould: Could you clarify why Fornax satellites have not been dragged into the center?

Grebel: Oh, Lin & Richer (2000) argue that globular cluster sedimentation in Fornax was probably prevented due to Fornax' undergoing tidal disruption and having lost a significant fraction of its original mass. In their model, Galactic tides increase the eccentricity of the clusters' orbits and isotropize their orbits such that the clusters stay in the outer parts of Fornax, where dynamical friction is low due to low stellar density. The authors also discuss alternative possibilities but judge Galactic tides to be the most likely one.

Cook: Would you comment on any evidence for differences in deep luminosity functions being related to the star-formation history in different Local Group galaxies.

Grebel: Deep luminosity functions were used for the Carina dSph in the beginning of the 1990s to derive its star-formation history (see papers by Mighell and collaborators). His results were later largely confirmed by detailed CMD studies. LFs can certainly give a first indication of the star-formation history of a Local Group galaxy. A recent study of the deep luminosity function of the Ursa Minor dSph (paper by Felting, Wyse and collaborators) showed no difference to deep luminosity functions of globular clusters. This luminosity function reaches far below the old main sequence turnoff and was the deepest ever obtained for a Local Group dSph. It confirmed U Mi as a predominantly old system (comparable to M92, I believe). In short, differences in star-formation history (and distinct episodes) show up in luminosity functions and can be used to derive said star-formation histories, though resolved CMDs will give more accurate and detailed results.

Bradbury: Is there any evidence that the dwarf galaxies, particularly those that are gas deficient, have an unusual Fe/C ratios?

Grebel: Not that I'm aware of.

The Haloes of the Milky Way and Andromeda Galaxies

N. Wyn Evans

Theoretical Physics, 1 Keble Rd, Oxford, OX1 3NP, UK

Mark I. Wilkinson

Institute of Astronomy, Madingley Rd, Cambridge, CB3 0HA, UK

Abstract. This article analyses the available data on the extents and masses of the haloes of the Milky Way and the Andromeda galaxies. As reckoned from the distant tracer populations, the mass of the Milky Way halo is $\sim 19^{+36}_{-17} \times 10^{11} M_\odot$, while the mass of the Andromeda halo is $\sim 12.3^{+18}_{-6} \times 10^{11} M_\odot$. The significance of this result for the pixel lensing experiments towards Andromeda is discussed.

1. Introduction

To ensure the future of microlensing, the most important project to complete over the next few years is the opening up of the new lines of sight towards the Andromeda galaxy (Crotts 1993; Baillon et al. 1993). The idea is to measure surface brightness variations caused by one of the many sources in a pixel being highly magnified by a foreground lens. Although high magnification events are comparatively rare, this is more than compensated by the fact that each pixel has many sources. This has motivated a number of groups to begin large-scale monitoring of fields in the Andromeda galaxy (for example, see the articles by the POINT-AGAPE and MEGA consortia in these proceedings). The lines of sight probe both the dark haloes of the Milky Way and the Andromeda galaxies. If the pixel lensing experiments are to fulfil their potential, it is important that good models of both haloes are available, with reasonably accurate estimates of their masses and their extents.

One worry is that inconsistent results for the mass of the Milky Way have been obtained when the sample of radial velocities of the satellites is analysed with and without the dwarf spheroidal Leo I. It is clearly disconcerting that the total mass falls by a factor of ~ 5 when just one datapoint is removed from the dataset (e.g. Little & Tremaine 1987). A second worry is that the mass of the Andromeda galaxy relative to the Milky Way has often been estimated by comparing their asymptotic rotation curves. Prior to 1985, the Andromeda galaxy was reckoned to be smaller than the Milky Way galaxy (e.g. Lynden-Bell & Lin 1977). Subsequent to 1985, the Andromeda Galaxy has been reckoned to be larger than the Milky Way Galaxy (e.g. Hodge 1992). This change occurred because the IAU reduced the recommended value of the local circular speed from 250 kms^{-1} to 220 kms^{-1} in 1985. However, the only trustworthy evidence comes *not* from the HI rotation curves (which cannot be traced further than ~ 30 kpc), but from the kinematics of the distant satellite galaxies. Here, the evidence that the Andromeda galaxy is the most massive member of the Local Group is surprisingly weak.

2. The Mass of the Milky Way Halo

Let us assume that the Milky Way halo can be described by a steady-state dynamical model with a mass density ρ that is isothermal out to an outer edge a, namely

$$\rho(r) = \frac{M}{4\pi} \frac{a^2}{r^2(a^2+r^2)^{3/2}}. \quad (1)$$

This model is attractive as it possesses a flat rotation curve in the inner parts ($r \ll a$) of amplitude $v_0 = \sqrt{GM/a}$.

The dataset consists of 17 distant ($r > 20$ kpc) globular clusters, two of which (Pal 3 and NGC 4147) have proper motions, and 10 distant dwarf satellites, five of which have proper motions (the LMC/SMC pair, Sculptor, Ursa Minor and Draco). The sample of dwarf galaxies is estimated to be almost complete, apart from perhaps a few galaxies hidden in the Zone of Avoidance. The number density of the satellites ρ_s is well described by a simple density law

$$\rho_s \propto \frac{1}{r^{3.4}}, \quad (2)$$

for $r > 20$ kpc. The distribution of velocities F depends only on the integrals of motion, namely the binding energy ε and the angular momentum l per unit mass. Let us assume that the tracer populations have a constant velocity anisotropy β given by:

$$\beta = 1 - \frac{\langle v_\theta^2 \rangle}{\langle v_r^2 \rangle}. \quad (3)$$

Then, the distribution of velocities F is given by (e.g. Wilkinson & Evans 1999, hereafter WE)

$$F(\varepsilon, l) = l^{-2\beta} f(\varepsilon), \quad (4)$$

where

$$f(\varepsilon) = \frac{2^{\beta-3/2}}{\pi^{3/2}\Gamma[m-1/2+\beta]\Gamma[1-\beta]} \frac{d}{d\varepsilon} \left[\int_o^\varepsilon d\psi \frac{d^m r^{2\beta} \rho_s}{d\psi^m}(\varepsilon-\psi)^{\beta-3/2+m} \right]. \quad (5)$$

Here, m is an integer whose value is chosen such that the integral converges.

Suppose for each of N satellites at positions r_i ($i = 1\ldots N$), we measure the radial velocity v_{ri}. Given a particular choice of model parameters (M, β), the probability of finding a satellite at radius r_i moving with radial velocity v_{ri} is just

$$P(r_i, v_{ri}|M, \beta) = \frac{1}{\rho_s} \int d^3v \, l^{-2\beta} f(\varepsilon) \delta(v_r - v_{ri}). \quad (6)$$

Haloes of the Milky Way and Andromeda Galaxies

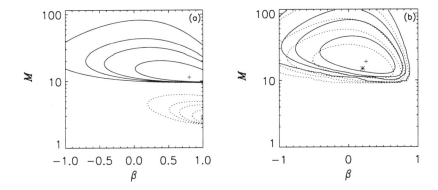

Figure 1. Likelihood contours for the total mass M (in units of $10^{11} M_\odot$) and the velocity anisotropy β obtained using Milky Way satellites and globular clusters with (a) radial velocities only and (b) radial velocities and proper motions. Results including Leo I (solid curves) and excluding Leo I (dotted curves) are shown. Contours are at heights of 0.32, 0.1, 0.045 and 0.01 of peak height.

If the proper motion of a satellite is available, then we can calculate its total velocity, v_i and hence its tangential velocity, $v_{ti}^2 = v_i^2 - v_{ri}^2$. In this case, the probability is simply

$$P(r_i, v_i | M, \beta) = \frac{f(\varepsilon) l^{-2\beta}}{\rho_s}, \qquad (7)$$

if $\varepsilon = \psi - (v_{ri}^2 + v_{ti}^2)/2 > 0$ and zero otherwise. In order to find the likelihood of a particular set of model parameters, we make use of Bayes' theorem. This gives us the fundamental formula of the algorithm, namely

$$P(M, \beta | r_i, v_{ri}, I) \propto P(M) P(\beta) \Pi_{i=1}^N P(r_i, v_{ri} | M, \beta). \qquad (8)$$

Here, $P(M, \beta | r_i, v_{ri}, I)$ is the probability of the model parameters taking the values M and β given the data (r_i, v_{ri}). I denotes the prior information, namely the prior probability distributions, $P(M)$ and $P(\beta)$. We usually choose these priors as

$$P(M) \propto 1/M^2, \qquad P(\beta) \propto 1/(3 - 2\beta)^2. \qquad (9)$$

Although these are our standard choices, we investigate the effects of changing the priors and find that this can change the inferred mass by $\sim 30\%$ (see WE).

Figure 1 shows likelihood contours in the plane of mass and orbital anisotropy. Panel (a) illustrates the Leo I problem (c.f. Little & Tremaine 1987). Only the data on the radial velocities is used, and the masses inferred with and without Leo I are grossly discrepant. The maximum of the probability surface is shown as a cross (asterisk) for the contours including (excluding) Leo I. Specifically, the most likely value of M is $11.4 \times 10^{11} M_\odot$ ($a \sim 100$ kpc) if Leo I is included and $2.7 \times 10^{11} M_\odot$ ($a \sim 25$ kpc) if Leo I is excluded. Even worse, the outermost contours (99 % confidence limits) do not overlap!

This impasse can be overcome with the addition of proper motion data. Although the radial velocities and positions are reasonably accurate, the proper motion errors are still large and we must allow for them. This is done by convolving with an error function to obtain the probability $P(r_i, v_{\text{obs},i}|M, \beta)$ of obtaining the observed full-space velocity $v_{\text{obs},i}$ given the values of the model parameters, namely:

$$P(r_i, v_{\text{obs},i}|M, \beta) = \int\int dv_\alpha dv_\delta E_{\text{L}}(v_\alpha) E_{\text{L}}(v_\delta) P(r_i, v_i|M, \beta), \quad (10)$$

where v_α and v_δ are the velocities tangential to the line of sight and v_i is the radial velocity. The observational errors are strongly non-Gaussian, so we assume the Lorentzian error convolution function E_{L} with broadening σ given by

$$E_{\text{L}}(v) = \frac{1}{\sqrt{2\pi}\sigma} \frac{2\sigma^2}{2\sigma^2 + (v - v_{\text{obs}})^2}. \quad (11)$$

Now, as Figure 1(b) illustrates, there is much better agreement between the contours including and excluding Leo I. The best estimate for the total mass of the halo is $\sim 1.9 \times 10^{12}$ $M_\odot$ with an extent of $a \sim 170$ kpc. The velocity anisotropy parameter has value $\beta \sim 0.25$ corresponding to mild radial anisotropy. The 32% contours in Figure 1(b) give a range of 1.5 - 4.8 $\times 10^{12} M_\odot$, though this is in fact an underestimate of the true uncertainty (WE).

It is interesting to ask how likely it is that a single satellite in a dataset of 30 objects with radial velocities drawn randomly from our halo model makes a substantial difference to the mass estimate. Monte Carlo simulations suggest that approximately $\lesssim 1\%$ of datasets contain such a satellite. Thus, Leo I is rather unusual and our prior expectation is not to find such an object. The addition of the proper motion data removes the paradox of Leo I, as the inner satellites have large tangential motions. The proper motion data therefore drives the mass of Milky Way halo upwards, which in turn allows Leo I to be bound despite its high radial velocity.

3. The Mass of the Andromeda Halo

The companion problem of estimating the mass of the Andromeda galaxy (M31) from its satellite subgroup has received scant attention. Hodge (1992) lists a number of determinations of the mass of M31, but almost all use either the optical and radio rotation curves or the inner globular clusters and so are really measurements of the mass within ~ 30 kpc. Recently, Evans & Wilkinson (2000) have examined this problem using all available data on the kinematics of objects outside ~ 20 kpc. There are 10 distant dwarf galaxy satellites with radial velocities and positions, but no proper motions. There are also 17 distant globular clusters and 9 planetary nebulae lying at projected radii $R \gtrsim 20$ kpc with published radial velocities.

Let us assume the same halo mass model as before [eq. (1)] and use a distribution of velocities F which depends on the binding energy ε and the angular momentum per unit mass l through the ansatz (e.g. Binney & Tremaine 1987)

$$F(\varepsilon, l) = f(Q), \qquad Q = \varepsilon - l^2/2r_{\text{a}}^2. \quad (12)$$

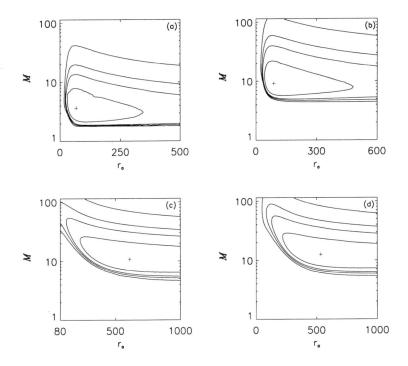

Figure 2. Likelihood contours for the total mass M (in units of $10^{11} M_\odot$) and the anisotropy radius r_a (in kpc). Contour levels as before. Data used: (a) Planetary nebulae only, (b) Globular clusters only, (c) Satellite galaxies only, (d) All data combined.

Here, r_a is the anisotropy length-scale; for $r \ll r_a$, the velocity distribution is isotropic, while for $r \gg r_a$, it tends towards radial anisotropy. The distribution of velocities F can be calculated explicitly by solving an Abel integral equation.

Figure 2(a) shows likelihood contours in the plane of the total mass M and the anisotropy radius r_a for the planetary nebulae. This dataset implies a low halo mass of $\sim 3.7 \times 10^{11} M_\odot$. Given that the planetary nebulae are all located at projected radii between 18 and 34 kpc, this mass estimate must be taken as a constraint only on the mass inside a three dimensional radius of ~ 31 kpc. A total mass of $3.7 \times 10^{11} M_\odot$ implies that the mass within ~ 31 kpc is $2.8 \times 10^{11} M_\odot$. Figure 2(b) shows the likelihood contours obtained from the globular cluster dataset. In this case, the mass estimate is $\sim 9.2 \times 10^{11} M_\odot$. The globular cluster data probe a range of projected radii between 19 and 34 kpc. They are more uniformly distributed than the planetary nebulae making this estimate more robust. A total halo mass of $9.2 \times 10^{11} M_\odot$ implies a mass inside 40 kpc of $4.7 \times 10^{11} M_\odot$. Figure 2(c) presents the results obtained using the data on the satellite galaxies. In this case, the best mass estimate is $\sim 10.8 \times 10^{11} M_\odot$. The large value of the anisotropy radius suggests that the velocity distribution is isotropic over the entire region probed by the satellites. Given that the satellites probe radii from ~ 10 kpc to ~ 500 kpc, this is an estimate of the total mass of the M31 halo. In Figure 2(d), all the data from our three populations are combined into a single dataset. The mass estimate is $\sim 12.3 \times 10^{11} M_\odot$ corresponding to a halo scalelength ~ 91 kpc. This is our preferred result, as it uses all the data. The value of the total mass implies masses interior to 31 kpc and 40 kpc of $\sim 4.0 \times 10^{11} M_\odot$ and $\sim 4.9 \times 10^{11} M_\odot$, respectively. It is reassuring to see that these values are in good agreement with the results obtained from the globular cluster data. The mass inside 31 kpc is greater than that implied by the planetary nebulae data – the non-uniform distribution of the planetary nebulae, which are clustered close to the optical disk, is the most likely cause of this discrepancy.

So, the kinematics of the tracers implies that the mass of the Andromeda halo is less than the mass of the Milky Way halo. There is one additional piece of evidence worth bringing forward. For an isotropic tracer population falling off like r^{-3} in a galaxy with a flat rotation curve of amplitude v_0, there is the simple result (e.g. Evans, Häfner & de Zeeuw 1997)

$$v_0^2 = 3 \langle v_r^2 \rangle \tag{13}$$

Here, $\langle v_r^2 \rangle$ is the mean square radial velocity of the sample (in the rest frame of the galaxy). Applying this formula to the dataset of the Milky Way satellites gives $v_0 \sim 220$ kms^{-1}. This is in excellent agreement with estimates of the local circular speed, from which we conclude that the Milky Way's halo is roughly isothermal out to the distances probed by the satellites. The dataset on the Andromeda satellites gives $v_0 \sim 140$ kms^{-1}, which is substantially less than amplitude of the HI curve (~ 240 kms^{-1}). This suggests that the isothermal region of the Andromeda halo is much less than the volume sampled by the satellite galaxies.

Our mass estimates are contrary to current opinion. Almost all recent authors have argued that M31 is the most massive member of the Local Group (e.g. Peebles 1989; Hodge 1992). These authors adduce a number of pieces of evidence to support their viewpoint. The asymptotic value of the rotation curve of M31 seems to be $\sim 10\%$ higher than that of the Milky Way. The number of globular clusters in M31 is more than double that in the Milky Way. The scalelength of the M31 disk exceeds that of the Milky Way. The face-on B-band absolute magnitude of M31 is $M_B = -21.1 \pm 0.4$, whereas that of the Milky Way is $M_B = -20.5 \pm 0.5$. On the other hand, the infrared luminosity of the Milky Way is much higher than that of M31. The mass in hydrogen gas in the Milky Way also exceeds that of M31. The difficulty in all this is judging whether any of these statements is properly comparing like with like. M31 has Hubble type Sb, whereas the Hubble type of the Milky Way is certainly later, perhaps Sbc or Sc (e.g. Gilmore, King & van der Kruit 1989). On such grounds alone, it is natural to expect that M31 has a larger bulge than the Milky Way. Late-type spirals are gas-rich and so it is also natural to expect the mass in gas in the Milky Way to exceed that of M31. The relative number of globular clusters is a poor argument, as it is only weakly correlated with total mass (e.g. Ashman & Zepf 1998). There is also no correlation between the satellite velocity dispersion in the outer halo of a galaxy and the rotation velocity of its disk (e.g. Zaritsky et al. 1997). Thus, a higher disk rotation velocity does not necessarily imply a more massive halo.

4. Microlensing towards Andromeda

The microlensing experiments towards the Large Magellanic Cloud have been bedevilled with uncertainties regarding the location of the lenses. The MACHO group interpret their results in terms of a halo partly composed ($\sim$ 20%) of compact objects. The EROS group interpret their results in terms of self-lensing by the Magellanic Clouds. This is despite the fact that both groups agree on similar values of the microlensing optical depth.

One hope of deducing the lens location is provided by microlensing exotica, such as the binary caustic crossing events. The stellar radius crossing time $t_\star$ enables the projected transverse velocity $v_p = R_s/t_\star$ at the source to be measured, provided the source radius R_s is known. This is an excellent diagnostic as to the lens location (e.g. Kerins & Evans 1999). Two such binary caustic crossing events have been detected. The first, MACHO LMC-9, occurred towards the Large Magellanic Cloud (LMC) and has $v_p \sim 20$ kms^{-1}. The second, 98-SMC-1, occurred towards the Small Magellanic Cloud and has $v_p \sim 80$ kms^{-1}. These low projected velocities imply that the lenses lie in the Clouds themselves. At first sight, this is strong evidence that the bulk of the lenses lie in the Clouds. Unfortunately, there is some uncertainty as to the MACHO LMC-9 event, as the first caustic crossing is poorly constrained by the data. It has been conjectured (Bennett et al. 1996) that both source and lens may be binary, in which case the projected velocity may be misleadingly low, although the likelihood of this seems quite small (Kerins & Evans 1999).

A second hope of deducing the lens location comes from the spatial geometry of events. If the lenses lie in the Milky Way halo, the events will trace the surface density of the LMC, whereas if the lenses lie in the Clouds, the events will be more concentrated towards the dense bar and central regions, scaling like the surface density squared. In propitious circumstances, Evans & Kerins (2000) showed that an experiment lifetime of $\lesssim 10$ years is sufficient to decide between the competing claims of Milky Way halos and LMC lenses. However, LMC disks can sometimes mimic the microlensing properties of Galactic haloes for many years and then decades of survey work are needed for discrimination. The difficult models to distinguish are Milky Way haloes in which the lens fraction is very low ($< 10\%$) and fattened LMC disks composed of lenses with a typical mass of low luminosity stars or greater.

In other words, it may be too difficult to solve the riddle of the location of the lenses with the present dataset. The value of studying microlensing towards the Andromeda galaxy is that it offers a reasonably unambiguous diagnostic of halo lenses. The near-far disk asymmetry is produced only by lenses lying in the Andromeda halo. This is because lines of sight to the far disk are longer and pass through more of the dense inner M31 halo than lines of sight to the near disk (Crotts 1993; Kerins et al. 2000). Although the pixel lensing experiments towards Andromeda are more difficult to carry out than conventional microlensing experiments, they offer the prospect of a reasonably unambiguous signature that will finally settle the controversies regarding the nature of the lenses!

5. Conclusions

The most important task for the immediate future is to open up the new microlensing windows on the Andromeda galaxy. This will teach us much about the structure of Andromeda and the present day mass function of its stars. If the near-far disk asymmetry is detected, it will firmly establish a lensing population residing in the Andromeda halo.

The magnitude of the near-far disk asymmetry depends on the relative sizes of the Milky Way and Andromeda haloes. We have argued that the mass of the Milky Way halo is $\sim 19^{+36}_{-17} \times 10^{11} M_\odot$ and its extent is ~ 170 kpc. By contrast, at least as reckoned from the kinematics of the satellite galaxies, the mass of the Andromeda halo is $\sim 12^{+18}_{-6} \times 10^{11} M_\odot$ and its extent is ~ 95 kpc. If these values are confirmed, it may be harder to detect the near-far disk asymmetry, as the contribution to the rate from lenses in the Milky Way halo may be larger than that from the M31 halo. For the field sizes monitored by POINT-AGAPE, typically ~ 6 events per year are expected from the M31 halo but ~ 9 from the Milky Way halo (scaling Kerins et al.'s (2000) results for a lens mass of $\sim 0.1 M_\odot$ and a baryon fraction of 20%). This may roughly double the time needed for a confident detection of the near-far disk asymmetry to $\gtrsim 5$ years.

References

Ashman K., Zepf S. 1998, Globular Cluster Systems, Cambridge University Press, Cambridge
Baillon P., Bouquet A., Giraud-Héraud Y., Kaplan J. 1993, A&A, 277, 1
Bennett D., et al. 1996 Nucl. Phys. B (Proc. Suppl.), 51B, 131
Binney J., Tremaine S. 1987, Galactic Dynamics, Princeton University Press, Princeton
Crotts A.P.S. 1992, ApJ, 399, L43
Evans N.W., Häfner R.M., de Zeeuw P.T. 1997, MNRAS, 286, 315
Evans N.W., Kerins E.J. 2000, ApJ, 529, 917
Evans N.W., Wilkinson M.I. 2000, MNRAS, in press (astro-ph/0004187)
Gilmore G., King I. R., van der Kruit P.C. 1989, The Milky Way as a Galaxy, University Science Books, Mill Valley, California
Hodge P. 1992, The Andromeda Galaxy, Kluwer Academic, Dordrecht
Kerins E.J., Evans N.W. 1999, ApJ, 517, 743
Kerins E.J., et al. 2000, MNRAS, submitted (astro-ph/0002256)
Little B., Tremaine S.D. 1987, ApJ, 320, 493
Lynden-Bell D., Lin D.N.C. 1977, MNRAS, 181, 37
Peebles P.J.E. 1989, ApJ, 344, L53
Wilkinson M.I., Evans N.W. 1999, MNRAS, 310, 645 (WE)
Zaritsky D., Smith R., Frenk C. S., White S. D. M. 1997, ApJ, 478, 39

Discussion

Popowski: How do the masses of M31 and the Milky Way depend on the assumed density profiles of the haloes?

Evans: The small size of the datasets forces us to use parametrised model fits rather than non-parametric techniques. Given this, the most attractive models to use are those with a flat rotation curve out to some unknown cut-off. It would be valuable to repeat our calculations with different models. Some confidence that this does not make major changes to the mass estimates can be gathered by comparing our results to Kochanek (ApJ, 457, 228) and to Kulessa & Lynden-Bell (MNRAS, 255, 105). These authors make different assumptions as to the density profiles for the Milky Way, and get broadly similar results.

Gould: The infall velocity of the Milky Way is about 70 kms^{-1}, which is more consistent with the Milky Way being lighter than M31. Also, lensing shows that early-type galaxies have large haloes. Since M31 has a higher rotation velocity, it should have a larger halo.

Evans: The calculation of the infall velocity depends on assuming that the ensemble of dwarf satellites has zero mean motion in the rest frame of the barycenter of the Local Group. It is an interesting calculation, but hardly conclusive. The statement regarding lensing of early-type galaxies is a rough generalisation. Given the importance of M31 in pixel lensing, we need to check it really does apply to M31. The only way to do this is through the analysis of the kinematics of the tracer populations. Finally, there is no correlation between the satellite velocity dispersion in the outer halo and the rotation velocity of the disk, as shown by Zaritsky et al. (ApJ, 478, 39). Thus, a higher disk velocity does not necessarily imply a more massive disk. A possible test is: Does the mass remain small when the missing five radial velocities of the M31 satellites are added to the dataset?

Glicenstein: Why don't you use a classical maximum likelihood method instead of Bayesian priors in your estimation of the masses of the Milky Way and the Andromeda masses. It seems to me that, since you have a multi-parameter model, you should not assume anything about the distribution of results.

Evans: Classical maximum likelihood *is* a special case of a Bayesian likelihood method and it too *does* assume something about the distribution of results. This is explained clearly on p. 495 of the paper by Little & Tremaine (ApJ, 320, 493) or in §19.2 of Kendall & Stuart's (1977) book *The Advanced Theory of Statistics*. Let me also note that there are only two parameters we estimate, not many parameters.

Graff: If the LMC has a constant velocity dispersion, then we expect the optical depth to be roughly constant across its face. Therefore, I disagree that one can ever rule out LMC self-lensing based on the spatial distribution of events.

Evans: I am not sure there is any real disagreement. Evans & Kerins (ApJ, 529, 917) reckoned that it could be difficult to distinguish between halo lensing and LMC self-lensing based on geometric effects like the spatial distribution of events.

Cook: I think the uncertainties on the proper motions of the Milky Way satellites are rather large. What uncertainties did you use and how did these affect the results?

Evans: This is a good point – all the proper motions come with large uncertainties. It is possible to allow for this by convolving the theoretical predictions with error distributions before comparing with the dataset. We examined both the case of Gaussian and Lorentzian error distributions with different broadenings. The main effect is to increase the error bars on our mass estimates. For example, our quoted errors on the mass of the Milky Way are larger than those for the mass of the Andromeda halo. This is a direct consequence of the large errors on the available proper motions for the Milky Way satellites.

Bennett: I want to make a couple of comments with regard to the implications of the binary lensing events towards the Magellanic Clouds. The LMC-9 event has a velocity that is so small that it has a low likelihood of being in the LMC. Also, I was particularly impressed by your argument that 98-SMC-1 was due to LMC self-lensing.

Evans: Your first statement is hardly fair! There is a famous example due to von Mises that illustrates the ludicrousness of such reasoning. Von Mises worked out the likelihood that the first bombing raid on Berlin in World War II would kill the elephant in Berlin Zoo. This has a negligibly small probability. Nonetheless, the first raid did kill the elephant. This shows how limited such arguments really are ... Your second statement is hardly fair! The Magellanic Clouds clearly have a complicated internal structure. Those of us worried about the size of the self-lensing contributions to the optical depth do so for both the SMC and the LMC. This seems more consistent than dismissing the contribution of self-lensing in the latter case based on simple models.

Pixel Lensing towards M31 in Principle and in Practice

Eamonn Kerins, for the POINT-AGAPE Collaboration
Theoretical Physics, University of Oxford,
1 Keble Road, Oxford OX1 3NP, UK

Abstract. The Andromeda galaxy (M31) provides a new line of sight for Galactic MACHO studies and also a signature, near-far asymmetry, which may establish the existence of MACHOs in the M31 halo. We outline the principles behind the so-called "pixel-lensing" experiments monitoring unresolved stars in M31. We present detailed simulations of the POINT-AGAPE survey now underway, which is using the INT wide-field camera to map the microlensing distribution over a substantial fraction of the M31 disc. We address the extent to which pixel-lensing observables, which differ from those of classical microlensing, can be used to determine the contribution and mass of M31 MACHOs. We also present some preliminary light-curves obtained from the first season (1999/2000) of INT data.

1. Andromeda: the new frontier

This meeting is in part a celebration of the success of the current pioneering generation of microlensing surveys, the discoveries of which are reported throughout these proceedings. However, as with all pioneering work, a number of puzzles remain to be solved including, crucially, whether MACHOs exist or whether current results towards the Large and Small Magellanic Clouds (LMC and SMC) can be explained by ordinary stellar populations. Whilst the MACHO Collaboration interprets the excess of events it finds as evidence of a $\sim 20\%$ halo contribution in objects of ~ 0.5 $M_\odot$ (Alcock et al. 2000) the EROS Collaboration has chosen to place only upper limits on the halo fraction from its sample of LMC/SMC events, even though the EROS dataset is formally consistent with that of MACHO (Lasserre et al. 2000). This dichotomy in approach reflects a community-wide uncertainty as to the contribution to the microlensing rate from the Magellanic Clouds themselves. As the current LMC/SMC surveys come to an end it is possible that the Magellanic satellite galaxies will continue to represent clouds of uncertainty for MACHOs.

Crotts (1992) and Baillon et al. (1993) proposed searching for MACHOs towards the Andromeda galaxy (M31). There are several aspects which make M31 very appealing for microlensing studies. First, it is a large external galaxy, with $\mathcal{O}(10^{10})$ sources, and has a halo of its own, providing an additional site for MACHOs. We should therefore expect many more events than towards the Magellanic Clouds. Small number statistics ought not to be an issue for M31 if the MACHO abundance is significant. Secondly, the additional line of sight

Figure 1. The concept of near-far asymmetry. Due to the tilt of the M31 disc the microlensing rate is larger towards the far disc than towards the near side if MACHOs are present in a spheroidal dark halo. The effect is less pronounced if the halo is flattened. The distributions of disc events and variable stars are symmetric.

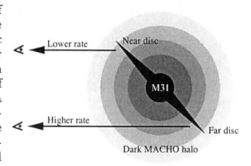

through our own halo that M31 provides may help to decide the viability of some of the stellar self-lensing scenarios, which have been proposed as non-MACHO alternatives to explain the LMC/SMC events. Alcock et al. (1995) have also noted that the ratio of microlensing rates between M31 and the LMC could be used in principle to probe the outer rotation curve of our Galaxy, though in practice the background from events in M31 makes this difficult. Lastly, our external viewpoint is advantageous in that we can map the MACHO distribution across the face of the M31 disc. The importance of this has been emphasized by Crotts (1992), who noted that the 77° inclination of the M31 disc should give rise to a noticeable gradient in the microlensing rate if MACHOs occupy a spheroidal distribution, as depicted in Figure 1. Such an asymmetry does not arise naturally for disc lensing or variable stars, and so would provide strong evidence for the existence of MACHOs if detected.

2. Pixel lensing

There are a number of difficulties in conducting a ground-based programme towards more distant targets like Andromeda. The major problem is that, in general, sources are resolved only whilst being lensed. Towards the M31 bulge the stellar surface densities may reach several thousand per arcsec^2, so after the event one often cannot identify the source to measure its baseline flux. In practice one must monitor pixel flux rather than individual source flux, hence the term *pixel lensing*. Because of the pixel flux contribution of the unlensed sources one typically requires the lensed source to be intrinsically bright or highly magnified, so the majority of microlensing events escape detection. In the high-magnification regime the light-curve suffers from a well-known near-degeneracy in which the excess flux due to microlensing is

$$\Delta F(t) \simeq \frac{A(t_0) F_s}{\sqrt{1 + \left[\frac{t - t_0}{t_e A(t_0)^{-1}}\right]^2}} \qquad (A \gtrsim 10), \qquad (1)$$

where t denotes observation epoch, t_0 is the epoch at which the magnification, A, attains its maximum, F_s is the baseline source flux and t_e the Einstein radius crossing time. As Woźniak & Paczyński (1997) have pointed out, equation (1) is invariant under transformations $F_s \to F_s/\alpha$, $A \to \alpha A$, and $t_e \to \alpha t_e$ for constants α preserving the high-magnification regime. Under such circumstances one is unable to unambiguously determine t_e, an important observable for conventional microlensing studies. All of this is made worse still by variations in seeing, sky background, detector position and point spread function which make it difficult to isolate the microlensing signal. Changes in seeing pose a severe problem because they induce localized flux variations on timescales comparable to microlensing.

These challenges, though difficult, have been met. Two pilot experiments able to detect true source flux variation on a routine basis have demonstrated the technical viability of pixel lensing (Crotts & Tomaney 1996; Ansari et al. 1997). The experiments employed different techniques, difference imaging and superpixel photometry, to compensate for changes in seeing. However, the limiting factor for both surveys has been the small field of view, yielding relatively modest statistics.

3. POINT-AGAPE

Building upon the pioneering French AGAPE (Andromeda Galaxy Amplified Pixels Experiment) programme, the Anglo-French POINT-AGAPE survey is employing the INT wide-field camera (WFC) to image a large fraction of the M31 disc (POINT is an acronym for Pixel-lensing Observations with INT). The survey began in August 1999 and is undertaking multi-colour observations in two fields, covering 0.6 deg^2 of the disc. The first season of data collection was completed in January 2000.

We have performed detailed simulations to model the sensitivity of the POINT-AGAPE survey, incorporating the effects of seeing and variable sky background, as well as our sampling (Kerins et al. 2000). The simulations include spherical near-isothermal models for both the Galaxy and M31 haloes, as well as a sech-squared disc and exponential bulge for the M31 stellar components. For the haloes we have simulated the expected signal for a wide range of MACHO masses, whilst for the stellar components we have assumed a Solar neighbourhood mass function.

Simulations of the expected spatial distribution of observed events after three seasons for 0.1, 1 and 10 M$_\odot$ MACHOs are provided in Figure 2. The figure assumes the Galaxy and M31 haloes are full of MACHOs. The POINT-AGAPE fields are shown by the dashed-line templates. Whilst the fields cover about one-fifth of the area of the disc their central location means that about half of all detectable pixel events across the disc should occur within them. The high concentration of events within the inner 5 arcmin is due mostly to bulge-bulge lensing, where as the MACHOs are dispersed over the entire disc. The near-far asymmetry in M31 MACHOs gives rise to a overall asymmetry in the spatial distribution. One can see from Figure 2 how the number of MACHOs increases and the near-far asymmetry becomes more prominent at small mass scales

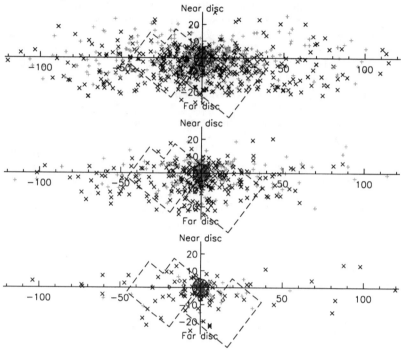

Figure 2. The simulated spatial distribution of pixel-lensing events for the POINT-AGAPE survey after three observing seasons. The axes are in arcmin. Galaxy MACHOs are shown as crosses, M31 MACHOs are depicted by an "x" and stellar events by diamonds. We assume here that MACHOs have a mass of 0.1 (*top*), 1 (*middle*) and 10 $M_\odot$ (*bottom*) and provide all the dark matter in the Galaxy and M31 haloes. The dashed lines indicate the POINT-AGAPE fields whilst the inner circle denotes the central region dominated by stellar self-lensing.

Figure 3. The combined M31 and Galaxy MACHO $t_{\rm FWHM}$ rate distributions for a range of MACHO masses, normalized to full MACHO haloes and averaged over the M31 disc. From the lightest to the darkest curve the Galaxy and M31 MACHO mass is 0.001, 0.003, 0.01, 0.03, 0.1, 0.3, 1, 3 and 10 $M_\odot$.

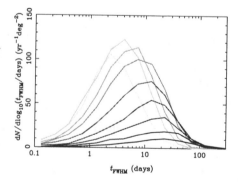

Pixel Lensing towards M31   313

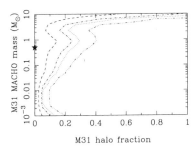 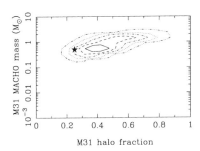

Figure 4. Simulated maximum-likelihood recovery of M31 MACHO parameters. The *left panel* shows the first-season sensitivity of POINT-AGAPE to an absence of near-far asymmetry, and hence an absence of M31 MACHOs (only stellar lensing events and variable stars have been generated in this simulation). In the *right panel* we assume a MACHO mass and contribution as indicated by the star. The contours here illustrate the survey sensitivity after three seasons. For both plots the solid, dashed, dot-dashed, dotted and triple-dot-dashed contours indicate 34%, 68%, 90%, 95% and 99% confidence levels, respectively.

Since the Einstein duration t_e is usually not measurable we have computed the expected event timescale distribution in terms of the full-width at half-maximum flux t_{FWHM}. Figure 3 displays the normalized combined Galaxy and M31 MACHO timescale distribution for a range of mass scales, assuming a maximal MACHO contribution and averaging the rate over the M31 disc. The distributions indicate that the average duration $\langle t_{FWHM} \rangle$ is sensitive to MACHO mass, though for our sampling it is much less sensitive than the underlying average Einstein duration $\langle t_e \rangle$, with $\langle t_{FWHM} \rangle \propto \langle t_e \rangle^{1/2}$. The rate peaks for $0.003 - 0.01$ M$_\odot$ MACHOs, giving ~ 100 events per season for M31 MACHOs and ~ 40 per season for Galaxy MACHOs within the two POINT-AGAPE fields if both haloes are full of MACHOs. If we instead use the most recent Galaxy MACHO mass estimate of 0.5 M$_\odot$, with a 20% halo contribution, then we should expect about 15 MACHO detections per season.

We have assessed the extent to which the MACHO mass and halo fraction can be recovered from the observables, specifically the event location and t_{FWHM}. Artificial datasets were constructed from our simulations and a maximum likelihood estimation of MACHO parameters computed using a Bayesian likelihood estimator. As well as M31 MACHOs the datasets include a fixed contribution from stellar lensing and may include Galaxy MACHOs with a different mass and density contribution to the M31 MACHOs. We also make allowances for dataset contamination due to variable stars passing our microlensing selection criteria. Our likelihood estimator is therefore computed over a five-dimensional likelihood space covering M31 MACHO mass and density; Galaxy MACHO mass and density; and a variable star contamination level. In Figure 4 we have summed the likelihood over three of the five dimensions to highlight the sensitivity of

Figure 5. Dual spatial and timescale asymmetry as a probe of velocity anisotropy. If MACHOs have strongly radial orbits then their motion will tend to be parallel to the near-disc sight-line but orthogonal to the far-disc sight-line. Hence near-side events should last longer and have a lower rate, enhancing the spatial asymmetry over that expected for isotropic haloes.

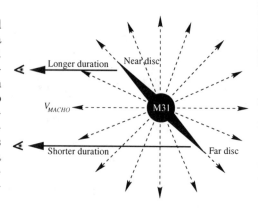

POINT-AGAPE to the M31 MACHO mass and halo fraction. The stars in each panel indicate the input parameters used to generate the artificial dataset whilst the contours indicate the parameter recovery from the dataset. In the left-hand panel our input MACHO fraction is zero, only stellar lensing events and variable stars have been generated. The contours indicate the POINT-AGAPE sensitivity after just one season. Powerful constraints are obtained over a wide range of mass scales due to the lack of any near-far asymmetry. The right-hand panel shows the sensitivity after three seasons for currently favoured MACHO parameters. The constraints on M31 parameters achievable by POINT-AGAPE are comparable to those being obtained for Galaxy MACHOs by the LMC/SMC surveys.

Pixel-lensing surveys towards M31 have the potential not just to probe the halo contribution and mass of MACHOs, but also their distribution function. The near-far asymmetry effect is sensitive to halo flattening, though for highly flattened haloes the asymmetry disappears, making it difficult to distinguish between M31 and Galaxy MACHOs, not to mention variable stars. In Figure 5 we illustrate how near-far asymmetry could also be used to probe the velocity anisotropy of M31 MACHOs. If the MACHOs have strongly radial orbits their trajectories should cut near-side and far-side lines of sight at different angles, giving rise to different characteristic event durations. Specifically, event durations should be longer for the near-side and the rate should be correspondingly reduced (since, for a given optical depth, the rate $\Gamma \propto \langle t_e \rangle^{-1}$) relative to isotropic halo models. In this case we would observe two signatures: an enhanced near-far spatial asymmetry and a near-far timescale asymmetry. This combination of signatures would be difficult to arrange by other means and would therefore represent a strong argument for MACHOs and for an anisotropic MACHO distribution function.

Analysis of the 1999/2000 INT dataset is yet to be finalized, but a preliminary reduction reveals many interesting light-curves, including a variety of variable stars and possible microlensing candidates. Some sample light-curves, spanning about a third of the 1999/2000 baseline, are shown in Figure 6. The light-curves have been processed using superpixel photometry, which is described

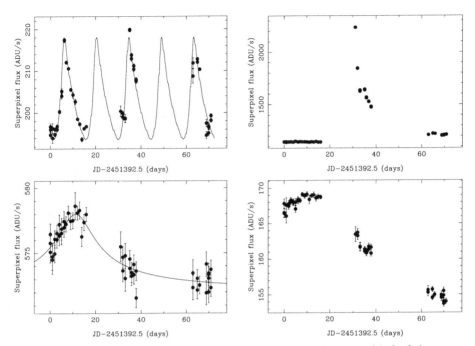

Figure 6. Preliminary r-band light-curves spanning one-third of the 1999/2000 POINT-AGAPE baseline. *Top left*: a possible Cepheid, together with a model fit. *Top right*: a probable nova. *Bottom left*: a light-curve consistent with microlensing, together with a model fit with $t_{FWHM} = 28$ days. *Bottom right*: another possible microlensing event. A longer baseline is required before either light-curve could be classified as a microlensing candidate.

in detail in Ansari et al. (1997). Briefly, the flux in each pixel is firstly represented by the summed flux over a pixel array, or superpixel, centred on the pixel. The size of the array is set by the size of the pixel relative to the seeing scale, and in the case of the INT WFC a 7×7 superpixel array was chosen. Though this binning dilutes any microlensing signal that may be present it helps to stabilize the effects of seeing variations to a point where a simple linear correction can be applied to the superpixel flux on each image in order to match their seeing characteristics. The method is simple and yet permits differential photometry practically down to the photon noise limit.

The top-left and top-right light-curves in Figure 6 are consistent with a Cepheid and a nova, respectively. They illustrate the high signal-to-noise ratio achievable with INT WFC superpixel photometry. The bottom panels show light-curves consistent with a microlensing interpretation. The light-curve at bottom-left peaks at about $r \simeq 22$ mag and has been fit with a theoretical degenerate microlensing curve with $t_{FWHM} = 28$ days for illustration. A longer baseline is required before either light-curve could be flagged as a microlensing candidate since long-period variable stars, such as Miras, pose a serious problem for pixel-lensing selection. To reject such variables one requires at least a three-year baseline. In the short term one can use the presence or absence of near-far asymmetry to statistically measure candidate contamination, since variable stars should have a symmetric spatial distribution.

There are of course limitations to pixel lensing. In common with Galactic microlensing, pixel-lensing constraints are sensitive to the assumed halo distribution function and, as already mentioned, the reduced near-far asymmetry for flattened haloes means that M31 MACHOs in such a configuration will be harder to detect or constrain. The pixel-lensing predictions are additionally sensitive to the shape of the bright end of the M31 stellar luminosity function, which is still to be properly determined. Near-far asymmetry nonetheless remains a very powerful discriminant between MACHOs and stellar self-lensing, an issue which currently hampers the interpretation of the LMC/SMC events. As well as searching for MACHOs, we also hope to be able to constrain the structure and mass function of the M31 bulge, where many self-lensing events are expected to occur. We are optimistic that, within about three seasons, the POINT-AGAPE survey will quantify and constrain both the stellar and MACHO mass function and distribution in M31, furthering our knowledge of the nature of dark matter, and providing a powerful probe of galactic structure.

References

Alcock, C., et al. 1995, ApJ, 449, 28
Alcock, C., et al. 2000, ApJ, submitted (astro-ph/0001272)
Ansari, R., et al. 1997, A&A, 324, 843
Baillon, P., Bouquet, A., Giraud-Héraud, Y., Kaplan, J., 1993, A&A, 277, 1
Crotts, A.P.S. 1992, ApJ, 399, L43
Crotts, A.P.S., Tomaney, A. 1996, ApJ, 473, L87
Kerins, E., et al. 2000, MNRAS, submitted (astro-ph/0002256)
Lasserre, T., et al. 2000, A&A, submitted (astro-ph/0002253)
Woźniak, P., Paczyński, B. 1997, ApJ, 487, 55

Discussion

Gould: The stellar halo of M31 is about 10 times denser than the Milky Way, so this will also produce an asymmetry.

Kerins: I take it that this spheroidal component varies with radius r something like r^{-3} or steeper, so I would be interested to know if the spheroid contribution remains significant at larger projected distances. It shouldn't be difficult to model this component in our simulations, but I agree this needs to be modelled if the halo fraction in M31 is small.

Hansen: If you are setting a high magnification cut, will your lens parameter estimates be much affected by binary events, which presumably contribute preferentially to high magnifications?

Sackett: With sufficient time resolution, you can tell the caustic crossings from high magnification point lens events.

Kerins: In the case of caustic crossing events you will have two peaks in the lightcurves, so I would expect the signature to be different in these cases. So I don't think they should confuse our lens parameter estimates. It remains an interesting question as to what extent we can characterize and study binary events in their own right with pixel lensing.

Gates: Given that a 20% MACHO halo is in conflict with many constraints, (from e.g. the implied baryon fraction in the progenitor population; deep galaxy counts; metal production; extragalactic background light, etc) the two currently indicated classes of models to explain MACHO LMC data are either a galactic component (extended protodisk, Gates & Gyuk 1999) or an LMC-related component (Kerins & Evans 2000 (self-lensing); Gould, Graff et al. 1999 (shroud)). Will POINT-AGAPE have the ability to distinguish between these two classes of models?

Kerins: I think there remains a question mark over pretty much all the currently proposed scenarios for explaining the LMC/SMC results. As for the signature of an extended protodisk, I haven't modelled it but I suspect the absence of a near-far asymmetric signal would make it difficult to identify without very good statistics.

Microlensing 2000: A New Era of Microlensing Astrophysics
ASP Conference Series, Vol. 239, 2000
John Menzies and Penny D. Sackett, eds.

Microlensing in M31 - The MEGA Survey's Prospects and Initial Results

Arlin Crotts, Robert Uglesich

Columbia University, Department of Astronomy, 550 W. 120th St., New York, NY 10027, U.S.A.

Andrew Gould

Ohio State University, Department of Astronomy, Columbus, OH 43210, U.S.A.

Geza Gyuk

University of California, San Diego, Department of Physics, 9500 Gilman Dr., La Jolla, CA 92093, U.S.A.

Penny Sackett, Konrad Kuijken

Kapteyn Astronomical Institute, 9700 AV Groningen, Netherlands

Will Sutherland

University of Oxford, Department of Physics, Oxford OX1 3RH, England, UK

Lawrence Widrow

Queen's University, Department of Physics, Kingston, ON K7L 3N6, Canada

Abstract. January 2000 completes the first season of intensive, wide-field observations of microlensing and stellar variability in M31 by MEGA ("**M**icrolensing **E**xploration of the **G**alaxy and **A**ndromeda") at the Isaac Newton 2.5-m telescope, the KPNO 4-m, and the 1.3-m and 2.5-m telescopes of MDM Observatory. In preliminary analysis, we detect ∼50000 variable objects, including some consistent with microlensing events. We present the level of sensitivity to be reached in our planned three-year program to test for the presence of a significant halo microlensing population in M31, as well as its spatial distribution and mass function. We also discuss our application of image subtraction to these wide fields and HST WFPC2 Snapshot follow-up observations to confirm candidates identified from previous years' surveys.

We present intermediate results from our smaller-field survey, on the MDM 1.3-m telescope and the 1.8-m VATT, from 1994 to 1998, wherein we have discovered eight additional probable microlensing events, over about one-half the time base of the project, in addition to confirming three of our original six microlensing candidates from 1995.

1. Introduction

Microlensing has become an interesting probe of dark matter in our Galaxy. Recent microlensing surveys have indicated that a large fraction of the matter in the Galactic halo may be dark objects with masses comparable to those of stars, but have not revealed what these objects might be. We have detailed how microlensing internal to M31 (Crotts 1992, Gyuk and Crotts 2000) might be used to test such results and show better how the microlensing matter is distributed in space and as a function of mass. A survey of small fields in M31 has revealed several such candidate events, at roughly the predicted rate (Crotts & Tomaney 1996). We discuss below what efforts have been required to further verify the microlensing nature of these events.

Many papers predict microlensing optical depths τ in M31, which should approach $\tau \approx 10^{-5}$, over ten times greater than towards the LMC, but previously none of these works studied variations in halo microlensing optical depth over the face of M31. Since M31 differs from the Galaxy in that many sightlines for microlensing are seen by an observer at Earth, the variation of τ depending on the spatial distribution of microlensing objects should be explored.

2. Testing Candidate Microlensing Events

Originally, Crotts & Tomaney (1996) identified six events from the 1995 observing season in M31 (using the Vatican Advanced Technology 1.8-m telescope on Mt. Graham, Arizona) over $\sim$ 60d covering a 125 arcmin^2 field. These events were characterized by full-width half-maximum timescales of 10d$< t_{fwhm} <$ 50d. The longer end of this range is troublesome, coinciding with pulse widths seen in Miras and other long term red variables. In fact, a Mira light curve, resembling a symmetric sawtooth in magnitude, with peak-to-valley amplitudes of about 5 magnitudes in the R band, appears similar to the light curve of a simple (point-mass, point-source) microlensing event during its maximum amplification. Furthermore, if the period of a Mira is about $\frac{2}{3}$ yr, one must monitor for two M31 seasons beyond (or preceding) the peak in the light curve in order to detect another peak, since proximate peaks occur when M31 is not easily observable. The peak of such a Mira has $t_{fwhm} \approx 40$d, which would corresponds to lensing masses $m \approx 0.5 M_\odot - 1 M_\odot$ for typical lensing geometries in our survey region. Hence Mira-like variables are troublesome contributors to a potential false event rate, and require multiple seasons of observations in order to be eliminated.

We have performed two tests of these six original candidates: (1) constructing well-sampled light curves over the M31 seasons of the three subsequent years (through 1998), and (2) obtaining HST WFPC2 snapshot observations of these sources in order to determine if their colors are consistent with Mira-like variables. (The latter test is impossible from the ground, since crowding does not allow one to resolve typical sources in average seeing conditions. Variable sources are made to appear isolated from one another by virtue of image subtraction (e.g. Tomaney & Crotts 1996)). The result of these two event filters is to eliminate three of the six events, with the remainder firmly inconsistent with Mira-like variables. For the remaining events, now that we have measured their baseline

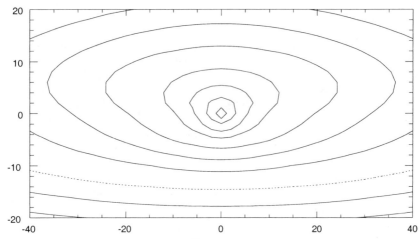

Figure 1. An 80×40 arcmin² plot of M31's center (major axis of M31 is horizontal, minor axis vertical) showing contours of the predicted event rate for bulge and halo microlensing in M31. The highest contour is for 50 events yr^{-1} arcmin^{-2} (near the center), with lower contours at 20, 10, 5, 2, 1, 0.5, 0.2 (dotted), 0.1 and 0.05 events y^{-1} arcmin^{-2}. This reasonable model, for an unflattened halo with a 5 kpc core radius, predicts over 100 detections during an M31 observing season.

magnitudes from WFPC2 images, we can calculate a more accurate peak amplification (assuming that the lensing mass rests as close as possible along the sightline to the core of M31). These persist with estimates for the lensing masses in the range $0.3M_\odot \lesssim m \lesssim 2M_\odot$, with two events possibly arising from stars in M31's bulge, but one almost certainly not a bulge lens, given its source position 2.5 kpc out into the disk.

3. Possible Results from a Larger Survey

Given the robust nature of some of the candidate microlensing events from the small area survey discussed above, it is worth considering possible results of a larger, wide-angle survey, especially since the advent of CCD imagers covering large fractions of a square degree. We present here representative results simulating a survey in which roughly 0.5 square degree is imaged for two hours every night on a 2-m telescope, in 1-arcsec seeing, requiring each event to be sampled at the 4σ level over at least 3 day timescales. The event rate predicted for such a survey is large. Figure 1 shows the predicted distribution of events over the field containing our survey area, for a typical model with 50% of the halo dark matter composed of 0.5 $M_\odot$ microlensing masses.

The halo fraction, halo flattening (q) and core radius (r_c) are allowed to vary between models, and then a maximum likelihood calculation is performed to yield resulting values for these parameters. These parameters can result in large changes in the distribution of microlensing optical depth across the face

MEGA: Microlensing in M31

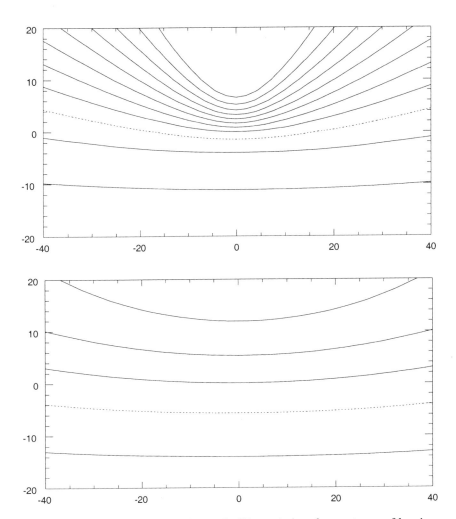

Figure 2. The same region as in Figure 1, but for contours of lensing optical depth (halo only) for two models of the spatial distribution of halo microlensing masses. The top panel corresponds to an unflattened model with 1.5 kpc core radius, and, on the bottom, a 3.3-to-1 flattening and a 10 kpc core radius. The dotted contours (just below center in each panel) correspond to 2×10^{-6}, increasing towards the top (far side of disk) to over 6×10^{-6} in the top panel and 3.5×10^{-6} in the bottom. In addition to the halo contribution shown, the central 10 arcmin diameter contains a $\tau \lesssim 5 \times 10^{-6}$ bulge signal (represented in Fig. 1); a uniform $\tau \lesssim 10^{-6}$ due to the Galaxy and a much smaller contribution from M31 disk-disk lensing (Gould 1994) are spread throughout.

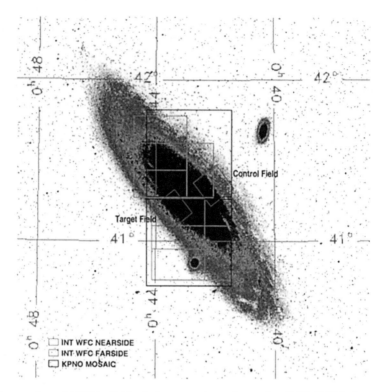

Figure 3. MEGA fields being monitored on the INT WFC (complex, light-shaded polygons) and KPNO 4-m/MOSAIC (large, dark-shaded squares), as well as the VATT/Columbia fields (smaller, diagonal squares: "Target" and "Control"). MDM fields are not shown.

of M31 (see Fig. 2), which significantly affects the distribution of microlensing event detections.

Our calculations show that this larger survey might easily observe ~100 such events per M31 observing season, which would allow the shape of a strong microlensing halo of M31 to be mapped. Since most masses reside near where the sightline passes the center of the galaxy, at a known source-lens distance, this survey would also allow a more exact determination of the masses doing the lensing. Selected fields in M31 might also serve as independent sightlines through the halo of our Galaxy. The preliminary epochs for a large survey in M31, over one-half square degree, have already been obtained, initiating the project MEGA: Microlensing Exploration of the Galaxy and Andromeda.

Our ability to measure r_c and q depend on the true value of r_c, with small values providing greater τ in the galaxy's center, where more sources exist. All models produce $\gtrsim 100$ events per season, with small r_c models producing more. After three seasons, r_c can be measured to within ~1.5 kpc (1σ), and q to ~0.1 (for $r_c < 5$ kpc), or ~2.5 kpc and ~0.2, respectively for $r_c > 10$ kpc. With q and r_c well-constrained, the data allow a superior estimate of lens mass distribution.

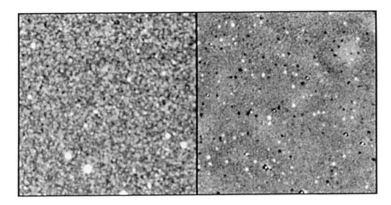

Figure 4. MDM 1.3-m CCD subimage in MEGA field, processed with our image subtraction technique. Left panel shows 110 arcsec square section from "raw" image taken 28 Oct 1998, a 3h exposure in ∼1 arcsec seeing. Right panel shows results of subtraction of average from the 1997 season of MDM data. A few saturated stars and ∼ 250 variables are seen.

This many events can result from a campaign using existing wide-field CCD arrays on 2-m+ class telescopes. Figure 3 shows the fields being covered by MEGA on some of the telescopes being used, compared to the fields for the earlier VATT/Columbia survey (described below). We have initiated this effort (MEGA) by establishing long baselines eliminating long-period variables, having obtained several epochs of such data in 1997 and 1998, and having begun more intensive observations in 1999 to detect microlensing events across much of M31 over the course of several seasons. At the time of the meeting, we have found approximately 40000 variables in the INT WFC data (some consistent with microlensing), which implies a sample of some 50000 when KPNO and MDM fields are included. (INT data are being collected in cooperation with AGAPE – see Kerins, this volume.)

Considerable improvements can be made in estimating halo shape parameters, as well as the microlens mass, if Einstein crossing times t_{ein} can be measured for events. Since the densest portion of the halo should sit over the bulge of M31, the lens-source distance is constrained by noting the impact parameter of the line of sight to the source relative to M31's center. Hence, t_{ein} can be converted, roughly, to a mass, at a much greater accuracy than in the LMC sightline situation. Furthermore, events that are due to halo lenses have preferentially longer t_{ein} than confusing bulge events, hence measurements of t_{ein} can lead to better determination of halo shape parameters beyond those accuracies quoted above.

The Einstein crossing time can be measured in two ways, the $t_{\sigma n}$ method of Baltz and Silk (2000), and by using HST imaging to determine baseline magnitudes by resolving the source star. As we showed at the meeting, neither of these is sufficiently reliable by itself to determine t_{ein} without a large fraction of outlying measurements, either due to misidentification of source stars in HST images, or poor S/N in the wings of microlensing light curves, where $t_{\sigma n}$ is

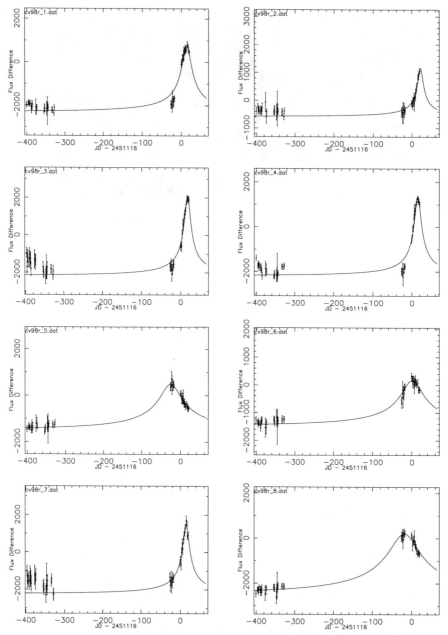

Figure 5. Eight new microlensing candidate light curves (ADUs vs. JD) from 1998 R-band MDM 1.3-m data. Data from other observatories (and MDM) show that these do not vary on extended baselines, and will fill in points during each event. Curves fit to the data are from Gould (1997) high-amplification fits, not full Paczyński (1986) fits.

determined. However, if a large field (~ 0.2 deg^2) is imaged with HST/ACS, requiring ~ 15 orbits, both methods will be available ($t_{\sigma n}$ from ground-based data), leading to t_{ein} for ~ 100 events, constraining the average microlensing mass to $\sim 0.1 M_\odot$ or better, and significantly improving halo shape parameter determinations.

4. Intermediate Results from the VATT/Columbia Survey

Uglesich, Crotts and Tomaney conducted a preliminary survey of two 125 arcmin^2 fields using the VATT 1.8-m and MDM 1.3-m. Data are reduced using the difference image photometry method of Tomaney and Crotts (1996), and example of which is shown in Figure 4 for typical MDM 1.3-m images. The S/N ratio in these difference images is limited by photon shot noise, not seeing fluctuations, as evidenced in several studies e.g. Uglesich et al. 1999. Quantities of data sufficient to find microlensing events for masses in the 0.1-1M$_\odot$ range where obtained in late 1995 through early 1999, with the best temporal coverage in the 1998-1999 season. Most of these data are now reduced, with the final season now having been searched for possible microlensing events. There are still $\sim$20 epochs to be added from additional observatories, but already 11 potential events have been found, with eight sampled over the peak of the best-fit microlensing curve, and points on either side. These eight candidate events are shown in Figure 5. (At the time of the meeting, we had only fit these with the high-amplification microlensing curve fit of Gould (1997), rather than the full parameter range implicit in general point-mass, point-lens microlensing curves of Paczyński (1986).) These sources have also been monitored in previous years' data, and have maintained a stable baseline, inconsistent with longterm variables e.g. Miras. We expect a large number of events to be found in this season's and other seasons' data. The VATT/Columbia survey has the potential for sensing the asymmetry in microlensing events across the face of M31, as seen in Figure 2, if a halo microlensing population actually exists as a significant fraction of the dark matter in spiral galaxy halos.

Acknowledgments. A.C. would like to express appreciation to the National Science Foundation for its support for this research under three separate grants.

References

Baltz, E.A. & Silk, J. 2000, ApJ, 530, 578
Crotts, A.P.S. 1992, ApJ, 399, L43
Crotts, A.P.S. & Tomaney, A.B. 1996, ApJ, 473, L87
Gould, A. 1994, ApJ, 435, 573
Gould, A. 1997, ApJ, 480, 188
Gyuk, G. & Crotts, A.P.S. 2000, ApJ, in press (astro-ph/9904313)
Paczyński, B. 1986, ApJ, 304, 1
Tomaney, A.B. & Crotts, A.P.S. 1996, AJ, 112, 2872
Uglesich, R., Mirabal, N., Sugerman, B. & Crotts, A. 1999, BAAS, 195, 132.06

Discussion

Kerins: Some of your candidates appear to be very sparsely sampled. What criteria are you using to select events?

Crotts: We examine, ultimately, any resolution element with two points in a light curve deviating by $> 4\sigma$, but here we have retained all curves, with total signal-to-noise ratio $> 20\sigma$, that rise and fall during the 1998 season and have a stable baseline during 1995-1996 (1997 is still a bit problematic for some epochs and some parts of the field).

Cook: 1) Were the three candidates which were elevated to event status, ones for which HST data identified a baseline flux?
2) Are you folding your candidate light curves to identify possible periods, if they are Miras, and thus plan future observing to rule out periodic variability?

Crotts: 1) All our original candidates, except one obvious, early Mira variable, were observed with the HST PC, including the surviving three candidates.
2) We *are* doing period analysis and period folding; I didn't go into that here.

Glicenstein: Is there any way of estimating the number of Miras and statistically subtracting them from the signal?

Crotts: We are cataloging all Miras. We hope to do a very good job of eliminating them from the signal.

Graff: Is there a statistically significant front-back asymmetry? Why do you observe for such a small fraction of the year?

Crotts: We will consider front-back asymmetry and the presence of a halo signal soon. We are still adding points to the 1998 season's light curves.

Microlensing 2000: A New Era of Microlensing Astrophysics
ASP Conference Series, Vol. 239, 2000
John Menzies and Penny D. Sackett, eds.

Things That Go Blip in the Night: Microlensing and the Stellar/Substellar Mass Function

I. Neill Reid

Dept. of Physics & Astronomy, University of Pennsylvania, 209 S. 33rd Street, Philadelphia, PA 19104

Abstract. The past five years have witnessed a veritable revolution in our understanding of the spatial frequency and evolutionary behaviour of low mass stars and brown dwarfs. Rather than debating their existence, we can now conduct statistical analyses. This talk reviews some highlights of those studies, notably estimates of the mass function, and considers how microlensing can further illuminate issues of interest.

1. Introduction

Microlensing experiments were designed with the goal of identifying compact, massive objects within the dark-matter halo of the Milky Way. However, observations towards the Galactic Bulge probe an extended range of distances within the Disk, and provide corresponding constraints on the stellar and substellar mass function for disk dwarfs. The aim of this review is to summarise our current knowledge of the mass function in the Galactic disk, as inferred from non-microlensing observations, and to consider the likely impact of future microlensing projects.

The mass function is defined as the number of stars per unit mass per unit volume, $\frac{dN}{dM}$. Masses, however, are seldom accessible to direct measurement, so one usually has to resort to combining measurement of the luminosity function with the luminosity/mass relation for the particular set of targets

$$\frac{dN}{dM} = \frac{dN}{dL} \cdot \frac{dL}{dM}$$

where we usually write $\frac{dN}{dM} = \Psi(M)$ and $\frac{dN}{dL} = \Phi(L)$ (the luminosity function, usually given per unit magnitude). The fact that the luminosity/mass relation for substellar-mass objects is not single-valued complicates the analysis, as discussed further below. The mass function is often represented as a power-law, index α,

$$\Psi(M) \propto M^{-\alpha}$$

where $\alpha = 2.35$ is the Salpeter mass function (Salpeter, 1956).

Low-mass M dwarfs and brown dwarfs have long been regarded as strong candidates for baryonic dark matter, and that prospect has stimulated most research in this area over the last three decades. However, the mass function is also of crucial importance for star formation theory and Galactic chemical

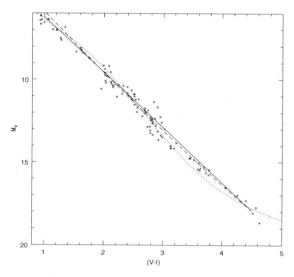

Figure 1. The (M_V, (V-I)) main sequence: crosses plot data for single stars within 8 parsecs of the Sun; the lines outline calibration relations adopted. Note that none of the latter takes full account of the feature at (V-I)$\sim$ 2.9.

evolution, since it provides a description of how a molecular cloud re-arranges itself to form fusion-powered and Kelvin-Helmholtz powered compact objects. That distribution determines how much mass is locked up forever, and how much is recycled through the interstellar medium. Finally, the stellar mass function together with the multiplicity function, the number of binary and multiple systems and the distribution of mass ratios and separations, is likely to influence the prevalence of planetary systems.

2. The stellar mass function in the Galactic disk

Low mass stars have low luminosities: the M7 dwarf Gl 644C (VB8) has $\frac{L}{L_\odot} \sim$ 10^{-3} (we shall refer to dwarfs with spectral types M7 and later as *ultracool dwarfs*). As a result, those sources are readily detected at relatively small distances from the Sun. Since the overwhelming majority of stars in the vicinity of the Sun are members of the Galactic disk, we are best informed on the mass function of that population. Three techniques have been used to investigate $\Psi(M)$ locally, with varying degrees of success.

2.1. Photometric analyses of field stars

Photometric parallax surveys for low mass stars came into vogue in the 1980s with the availability of deep, wide-field photographic Schmidt plates in the R and I passbands, and the development of automated scanning microdensitometers. The essence of the technique is using the colour of field stars as an absolute magnitude estimator; coupled with the measured apparent magnitude, distances can be estimated; stellar densities as a function of absolute magnitude ($\Phi(M_V)$) can be derived by setting a distance limit and including the appropriate statistical corrections (primarily Malmquist bias). Initial surveys were based on V- and I-band photometry (Reid & Gilmore, 1982); later studies employed (R-I) colours from both photographic plates (Hawkins & Bessell, 1988; Tinney, 1993) and, later, CCD imaging (Kirkpatrick et al. 1994).

The crucial step in this method is the colour-magnitude relation adopted for absolute magnitude calibration. All of the analyses derive broadly similar luminosity functions, with a strong peak at $M_V \sim 12$ and subsequent sharp decline in number density. The corresponding mass function, applying either theoretical mass-luminosity relations or empirical results from Henry & McCarthy's (1993) binary astrometry, has a maximum at 0.2 $M_\odot$ and decreasing number densities towards the hydrogen-burning limit. However, as has been discussed in detail elsewhere (Reid & Gizis, 1997), none of the colour-magnitude relations employed match the shape of the main sequence at $M_v \sim 12$ (Fig. 1). This leads to systematic errors in the inferred absolute magnitude and corresponding systematic bias in the inferred luminosity and mass functions. Given the large dispersion in absolute magnitude at these colours, photometric parallaxes cannot provide a reliable measure of the number density of mid-type M dwarfs. However, accurate photometric parallaxes can be derived for lower luminosity M dwarfs, where the partially degenerate structure leads to a shallower slope and a narrower main sequence. Analysis of data from the 2MASS JHK survey is currently underway, using this approach to identify ultracool M dwarfs.

2.2. A Solar Neighbourhood census

The conceptually simplest method of determining the luminosity function is to count the number of stars within a specific volume centred on the Sun. Nearby stars offer considerable advantages in detecting binary companions, since angular separation decreases linearly with distance; brighter apparent magnitudes allow high resolution spectroscopy; and accurate trigonometric parallaxes can be determine readily, allowing the definition of reliable distance limits. The main problem lies with completeness. The surface density of nearby stars is low – between 130 and 160 systems with absolute magnitudes in the range $-1 \lesssim M_v \lesssim 19$ lie within 8 parsecs of the Sun. Most are readily detected through their having substantial proper motions. However, some small proportion of nearby stars must have their motions primarily in the radial component. In general, identification is most difficult for systems with intrinsically low luminosity and for dwarfs near the Galactic Plane, where the number of background stars with similar apparent magnitude is highest and the available proper motion surveys least reliable.

The compilation of nearby star catalogues can generally be held as commencing with van de Kamp's (1953) list of 42 stars within 5 parsecs of the Sun. W. Gliese produced the first formal Nearby Star Catalogue ($r \leq 25$ parsecs) in the late 1950s, supplementing and revising the sample over the succeeding 30 years, latterly in collaboration with H. Jahreiss. The current catalogue includes data for ~ 3800 stars in ~ 3000 systems, but while the addition of data from modern sources, such as the Hipparcos catalogue (ESA, 1997), led to completeness at bright magnitudes ($M_V \lesssim 7$, $r \leq 25$ pc), the sample of late-type dwarfs

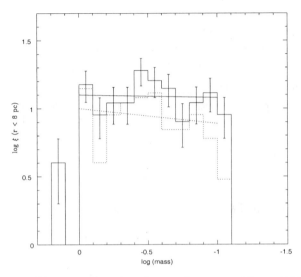

Figure 2. The mass function for nearby stars. The solid line plots data on a star-by-star basis the dotted line includes only single stars and primaries from multiple systems.

becomes incomplete at distances of less than 14 parsecs (Reid et al. 1995). Indeed, ultracool dwarfs may remain undetected even within 5 parsecs of the Sun.

Given these uncertainties, most nearby star analyses are based on samples within 5 to 8 parsecs from the Sun (Henry is currently compiling a catalogue including data or all known systems within 10 parsecs). Taking a distance cutoff of 8 parsecs, and limiting analysis to stars north of $\delta = -30°$ (the region covered most thoroughly by Luyten's Palomar proper motion and Henry's (1992) speckle imaging survey for companions), gives a sample of 150 stars (and 2 brown dwarfs - Gl 229B, Gl 670D) in 102 systems (Reid et al. 1999). One hundred and seventeen (78%) of the sample are M dwarfs, and the overall multiplicity fraction (including binaries at all separations) is only 31.5%, as compared with values of over 60% usually cited for solar-type stars. Thus, multiplicity clearly declines with spectral type if one includes binaries spanning the full range of separation (0.03 AU to >2000 AU).

The resultant mass function is shown in Figure 2. These data are calibrated using Baraffe et al.'s (1998) solar abundance models: the results are unchanged in the essentials if one adopts empirical calibrations. The best-fit linear relation to the star-by-star mass function at masses below $1.0 M_\odot$ is

$$log\xi(logM) = (-0.02 \pm 0.14)log(M) + 1.10$$

or $\alpha = 0.98 \pm 0.14$. There are few higher-mass stars in the immediate Solar Neighbourhood, but larger scale surveys indicate that the function steepens significantly, with $-2 < \alpha < -2.5$, close to the Salpeter value (Scalo, 1998). The systemic mass function for the Solar Neighbourhood (including only single stars and primaries from multiple systems) is consistent with an index of $\alpha =$

0.91 ± 0.13. One might argue that the most appropriate approach from the star formation perspective would be to combine masses in close binary systems, thus deriving a mass spectrum perhaps closer to the initial distribution of cloud cores. Given the extensive mass loss during formation, however, the utility of such an analysis remains unclear. Setting those considerations aside, Figure 2 provides the current best estimate of $\Psi(M)$ for disk dwarfs in the general field.

2.3. Open clusters and associations

Surveys of low-mass stars in the field include dwarfs spanning the full range of age and chemical abundance of the Galactic disk, and therefore provide a measure of the average mass function. Open cluster studies have the advantage of dealing with stars of known age and abundance, allowing the use of specific stellar models in the analysis. Moreover, young low mass stars and brown dwarfs are significantly more luminous, and therefore more readily detected. Complications rest with segregating low luminosity cluster members from foreground and background field stars; dealing with dynamical evolution – mass segregation and tidal stripping – in intermediate age clusters; obtaining adequate observations for sources in young (< 20 Myr), dusty star-forming regions, with patchy obscuration; and calculating accurate stellar models over the appropriate range of ages.

The two intermediate-age clusters which have received most observational attention are, unsurprisingly, the nearby Hyades and Pleiades. The former has an age of ∼ 625 Myrs and, at an average distance of 46 parsecs, covers an area of several hundred square degrees. It is clear that mass segregation is sufficiently pronounced to negate attempts to infer the initial mass function (Reid & Hawley, 1999). The Pleiades, with an age of ∼ 125 Myrs, is more amenable to study, with little evidence for mass segregation amongst lower main-sequence stars (Reid, 1998). Meusinger et al. (1996) have presented extensive analysis of the brighter cluster members, while Stauffer et al. (1996) and Hambly et al. (1993) probe the distribution of low mass stars. The resulting mass function is broadly similar to the field star distribution, well matched by power-laws with $\alpha = 2.5$ for $M > 1.3 M_\odot$, and $\alpha \sim 1$ at lower masses. The Tenerife group (Rebolo, Martín, Zapatero-Osorio and collaborators) have extended coverage to substellar masses, with the latest results suggesting $\alpha \sim 0.5$ below 0.1 $M_\odot$(Martín et al. 1998).

To date, most studies of young clusters derive $\Psi(M)$ by transforming the observed K-band luminosity function. That technique, however, can be biased if the dispersion in age in cluster members is significant compared with the mean age of the cluster: the mass-M_K relation is multi-valued. Reddening, both circumstellar and foreground, can also introduce significant uncertainties. A more reliable approach is to use spectroscopic observations to determine temperatures, placing stars on the HR diagram and deriving individual ages and masses. That approach is observationally intensive, and, so far, has been applied to only the Orion Nebula Cluster (Hillenbrand, 1997). The results for that cluster are significantly different from $\Psi(M)$ in the field and the Pleiades: the mass function is reminiscent of the Miller-Scalo log-normal form, peaking at ∼ $0.25M_\odot$, with decreasing number density towards lower masses. Hillenbrand & Carpenter (2000), Lucas & Roche (2000) and Luhman et al. (2000) have recently extended observations to objects fainter than K=18 mag, corresponding to masses of $\sim 0.01 M_\odot$. All three studies find that the ONC has a substantial brown dwarf population. Luhman et al. derive a mass function which is close in form to the field function over the full mass range, lacking any suggestion of a peak at $0.25 M_\odot$. They ascribe this difference to their use of more recent theoretical evolutionary tracks for mass calibration. Clearly, further more detailed observations of other young clusters are required to test whether a universal stellar mass function may indeed be appropriate for the disk.

3. L dwarfs, T dwarfs and the substellar mass function

Recent years have seen a spectacular advance in our understanding of the nature and number of substellar-mass brown dwarfs in the Solar Neighbourhood. While the confirmation of the existence of these very low-mass objects lay with the discovery of Gl 229B as a companion of a nearby M dwarf (Nakajima et al. 1995), the real breakthrough in understanding came with the first deep, wide-field near-infrared surveys, notably the Two Micron All-Sky Survey (2MASS) and, latterly, the Sloan Digital Sky Survey (SDSS). The energy distribution of ultracool dwarfs peaks between 1 and 2μm, so it is not surprising that the first surveys to extend beyond the third magnitude limit of Neugebauer & Leighton's Two-Micron Sky Survey have turned up interesting objects. The overwhelming majority have been identified using photometric criteria: extremely red colours at near-infrared wavelengths (J-K) or at optical/infrared wavelengths ((I-J), (i-z)). Recent observations have resulted in the characterisation of two new spectral classes: spectral class L, extending the temperature régime beyond type M9.5 (Kirkpatrick et al. 1999); and spectral class T, the even cooler methane dwarfs, typified by Gl 229B. This classification allows us to probe the field brown dwarf mass function.

3.1. Spectral evolution of ultracool dwarfs

Figure 3 shows how the emergent spectrum varies in the far-red as the spectral type changes from L0 to T. Following considerable theoretical and observational work, we now understand those changes as reflecting the changing atmospheric composition, in particular, the transition from gaseous to solid phase of the dominant molecular absorbers of spectral class M. Tsuji et al. (1996) originally pointed out that TiO is expected to combine with other metals to form particulate material (dust) at the temperatures present in late-type M dwarf atmospheres. Observations of the K I doublet suggest that dust first becomes important at spectral type M6.5 (Reid et al. 2000). Detailed chemical equilibrium calculations have been undertaken by Lodders (1999) and Burrows & Sharp (1999). At lower temperatures, VO solidifies. The removal of the two major opacity sources leads to more transparent atmospheres, with the photosphere located at lower physical depths. This, in turn, accounts for the strength of the alkali lines, Cs, Rb, Li (in lower-mass brown dwarfs) and, particularly, Na and K. The relatively high abundance of the latter two species leads to substantial pressure broadening of the resonance lines, with the Na D lines and the KI 7665/7699 Å doublet acquiring white dwarf-like equivalent widths of over 1000Å in the late-type L dwarfs and T dwarfs, sufficient to affect the broadband colours. The strongest molecular features at optical wavelengths in L dwarfs

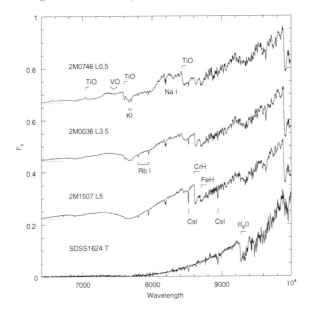

Figure 3. Optical spectra of L and T dwarfs

are the metal hydrides – CaH, MgH, FeH and CrH – and H_2O; only H_2O and, perhaps, FeH remain detectable in T dwarfs.

At near-infrared wavelengths, L dwarfs have spectral energy distributions which are similar to those of late-type M dwarfs (Fig. 4). Atomic features due to potassium, sodium and iron are present, mainly in the J passband, as are molecular features due to FeH. The strongest features are the steam bands, occupying the same wavelength intervals as the junction between the near-infrared JHK passbands, and the CO absorption feature at $2.03\mu m$. The crucial difference between L dwarfs and T dwarfs is the onset of strong CH_4 absorption in the H and K passbands, as the temperature drops below $\sim 1200K$ and carbon is bound preferentially in methane rather than carbon monoxide. This transition, originally highlighted by Tsuji (1964), leads to a correspondingly dramatic change in the (J-K) colours, from (J-K)=2.0 at spectral type L8 to (J-K)=0 for Gl 229B.

3.2. The mass function below the hydrogen burning limit

Figure 5 plots the near-infrared colour-magnitude diagram described by M, L and T dwarfs. Spectral type, like broadband colours, provides a measure of atmospheric temperature, and the M/L/T spectral sequence is a sequence of decreasing temperature. In the case of main-sequence stars, hydrogen fusion in the core leads to hydrostatic equilibrium, with internal pressure balancing self-gravity. Thus, the surface temperature correlates with the energy generated in the core, which in turn depends on the total mass. As a result, main sequence stars follow a mass-luminosity relation. Such is not the case for brown dwarfs. Indeed, substellar mass objects follow the evolutionary terminology originally

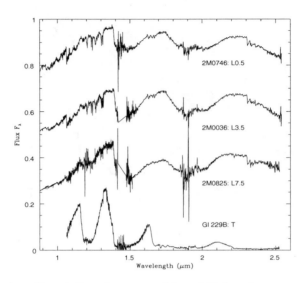

Figure 4.   Near-infrared spectra of L and T dwarfs

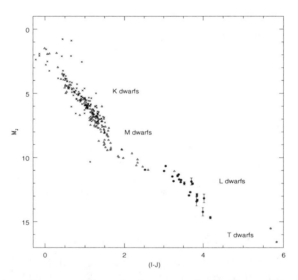

Figure 5. The near-infrared colour-magnitude diagram for M, L and T dwarfs. The majority of the lowest luminosity objects are either companions of known nearby stars (Gl 229B, Gl 670B, Gl 384C) or have parallax measurements by the US Naval Observatory. The much redder (I-J) colours of the two T dwarfs reflect strong KI absorption in the I-band.

associated (by Lockyer and others) with spectral type: young, *early-type* brown dwarfs (albeit, spectral type M, not O), evolve through types M and L to old, *late-type* T dwarfs. Since brown dwarf radii are set by degeneracy, rather than hydrostatic pressure, all brown dwarfs evolve along essentially the same tracks in the HR diagram (low-mass brown dwarfs are more luminous, since they have ~ 10% larger radii). With no calibrated gravity indicators at present, we lack the means to estimate masses for isolated brown dwarfs, even if their intrinsic luminosities and surface temperatures are known quantities.

Since we have no means of reliably estimating masses for individual brown dwarfs, we must rely on statistical techniques to infer the underlying mass distribution. Full details of our initial study are given in Reid et al. (1999). In brief, observational constraints are set by the observed surface densities of L dwarfs and T dwarfs with J < 16. Lesser constraints stem from the fraction of L dwarfs and ultracool M dwarfs with lithium absorption, since brown dwarfs with masses below 0.06 $M_\odot$ fail to achieve central temperatures sufficient to fuse primordial lithium through the PPI reaction $^1H + ^7Li \rightarrow 2.^4He$ (Rebolo et al. 1992).

We attempt to match these constraints by using Monte Carlo simulations to predict the local number/magnitude/temperature distribution of brown dwarfs, based on an assumed mass function, $\Psi(M)$; birthrate, B(t); bolometric correction/temperature relation; spectral type/temperature relation; and evolutionary models for low mass stars and brown dwarfs. Photometric and spectroscopic analyses indicate that the M9.5/L0 transition occurs at a temperature between 2200 and 2000K, and early-type L dwarfs are a mix of very low-mass stars and brown dwarfs. The location of the L/T transition is more problematic, but the fact that the coolest L dwarf currently known, Gl 384C, is only 0.4 magnitudes brighter in M_J than Gl 229B, suggests a temperature between 1200 and 1400K, and a relatively rapid transition. SDSS is well placed to address these issues, since (i-z) colours increase monotonically from L8 to T. Indeed, several L/T transition objects, with both CO and CH_4 absorption, have recently been identified (Leggett et al., in prep.).

Our preliminary analysis suggests that the observed surface densities – one L dwarf per 20 sq. deg. and one T dwarf per 200-400 sq. deg. to J=16 – are consistent with a power-law mass function with $\alpha \sim 1$. However, there are many caveats: apart from the obvious uncertainties in the evolutionary models, the temperature scale and bolometric corrections, it is important to recognise that bright L and T dwarfs are almost exclusively young, massive brown dwarfs. Low-mass brown dwarfs cool extremely rapidly, and therefore make little contribution to the total L dwarf population. Figure 6 shows the expected age distribution of the 2MASS-detected L dwarfs as a function of mass. These dwarfs represent less than 1% of the predicted brown dwarf population within 30 parsecs of the Sun: the overwhelming majority have temperatures between 300 and 500K and $M_J > 20$. Thus, we are using a small tail to wag a large dog, and are clearly vulnerable to significant variations in the recent star formation history. Nonetheless, all indications point to a local number density of brown dwarfs comparable with the number density of main sequence stars. The corresponding mass density is small - approximately 0.004 $M_\odot$ pc^{-3}, equivalent to the local

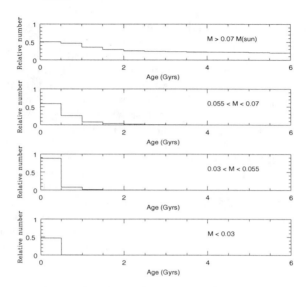

Figure 6. The predicted age distribution of L dwarfs as a function of mass. We assume temperature limits of 2000-1350K, and the relative numbers in each diagram have the same scaling. The majority of the M> $0.07 M_\odot$ L dwarfs are earlier than L4 in spectral type.

mass density of white dwarfs. Brown dwarfs are currently a poor bet as dark matter candidates.

4. Other stellar populations: the Bulge and the (stellar) Halo

We have only limited knowledge of the mass function in the non-disk populations of the Milky Way, and in both cases, studies are limited to hydrogen burning stars. The most detailed investigation of $\Psi(M)$ in the Bulge rests with Zoccali et al.'s (2000) analysis of deep HST NICMOS observations. Analysing those data using photometric parallaxes, the authors deduce a mass function consistent with a power law, $\alpha \sim 1.3$, somewhat steeper than the disk function, extending from $\sim 1.0 M_\odot$ to $0.15 M_\odot$. There is some evidence for a change in slope of $\Psi(M)$ at $M \sim 0.4 M_\odot$, but, given the previous experience in photometric analyses of disk dwarfs, it is possible that this feature might be introduced via either the colour-magnitude calibration or luminosity/mass conversion. Nonetheless, the derived bulge star mass function is not strikingly different from the disk results.

Observations of the stellar halo have concentrated in two areas: studies of nearby halo subdwarfs, usually based on proper motion surveys (Dahn et al. 1995; Gizis & Reid 1998); and, with the increasing availability of high spatial-resolution HST data, deep imaging of globular clusters (Piotto & Zoccali 1999). Gould et al. (1998) have also analysed high-latitude HST WFPC data, probing $\Psi(M)$ in the field halo at distances of several kiloparsecs above the Plane. With no empirical mass measurements of halo subdwarfs, the mass calibration rests entirely on theoretical models. In general, these analyses are in good agreement on the shape of the halo mass function, and are broadly consistent with a power-law function, $\alpha \sim 1.5$. There is some evidence for cluster to cluster variations, although residual effects of uncorrected dynamical evolution cannot be completely excluded as a possible alternative source. As with the Bulge, these results imply only relatively minor variations in $\Psi(M)$. Since the halo cluster population includes several clusters with abundances at or below $[m/H] = -2$, this suggests that the stellar mass function is remarkably insensitive to the chemical composition of the parent gas cloud.

5. Microlensing and the stellar mass function

How can microlensing observations complement mass function studies using other techniques? Conventional mass function analyses demand that we detect light from the target population. In contrast, in almost all cases the lens responsible for an amplification event is invisible. This is a disadvantage in interpreting individual lensing events; however, statistical analysis of large numbers of events offers the possibility of probing the low mass dwarf population in regions far from the Solar Neighbourhood. Clearly, lensing observations towards the Bulge are best suited to this type of study. The relatively small number of LMC/SMC sources, consistent with the expected low optical depth of the stellar halo, precludes application to this problem. However, a significant fraction of Bulge events are expected to arise from lenses in the foreground disk, with, presumably, well understood kinematics. Indeed, the scarcity of short timescale events has already been used to set limits on the frequency of brown dwarfs in the disk.

Lensing also provides a means of probing the binary fraction amongst low-mass stars at separations which are inaccessible to radial velocity studies. Again, the main limitation rests with the invisibility of the lens, and consequent ambiguities in the properties of individual sources. There is one régime, however, where both lens and background source are observable. High-resolution astrometric satellites, such as the Space Interferometry Mission, offer the prospect of micro-arcsecond accuracy, sufficient to measure the relative displacement of a lensed background source during an event. Targeted observations of nearby proper motion stars which are projected to undergo close encounters with background field stars will allow direct mass measurements. Since a significant fraction of high proper-motion stars are halo subdwarfs, these observations could result in the first empirical calibration of the mass-luminosity relation at subsolar abundances. High spatial resolution observations might provide the same information for globular clusters through the detection of self-lensing events.

6. Summary and conclusions

After a many decade drought, the new generation of deep, wide-field, near-infrared sky surveys is producing a glut of brown dwarfs in the immediate Solar Neighbourhood. However, the observed surface densities indicate that substellar-mass objects make no more than a $\sim 10\%$ contribution to the mass of the Galactic disk. The current best estimate of the form of the mass function in the disk is a two-component power law, with $\alpha \sim 2.3$ (the Salpeter value) at masses above $\sim 1.3 M_\odot$, and $\alpha \sim 1$ in the lower-mass régime. Observations of low luminosity stars in the Bulge and halo indicate slightly steeper mass function, with $\alpha < 1.5$. The implication is that brown dwarfs are not a candidate for baryonic dark matter and that the stellar mass function is relatively insensitive to chemical abundance for $[m/H] > -2$. Microlensing observations can reinforce those conclusions by probing the surface density of low-mass stars and brown dwarfs towards the Bulge, and, with the availability of high-precision astrometric satellites, provide accurate mass measurements for isolated stars and brown dwarfs, particularly low metallicity members of the halo population.

References

Baraffe, I., Chabrier, H., Allard, F., Hauschildt, P.H. 1998, A&A, 337, 403
Burrows, A, Sharp, C.M. 1999, ApJ, 512, 843
Dahn, C.C., Liebert, J.W., Harris, H., Guetter, H.C. 1995, in *The Bottom of the Main Sequence and Beyond*, ed. C.G. Tinney, Heidelberg: Springer, p. 239
ESA, 1997, The Hipparcos and Tycho Catalogues, ESA SP-1200 (Noordwijk: ESA)
Gizis, J.E., Reid, I.N. 1998, AJ, 117, 508
Gould, A., Flynn, C., Bahcall, J.N. 1998, ApJ, 503, 798
Hambly, N.C., Hawkins, M.R.S., Jameson, R.F. 1993, MNRAS, 263, 647
Hawkins, M.R.S., Bessell, M.S. 1988, MNRAS, 234, 177
Henry, T.J., 1992, Ph. D. thesis, University of Arizona, Tucson
Henry, T.J., McCarthy, D.W. 1993, AJ, 106, 773
Hillenbrand, L. 1997, AJ, 113, 1733
Hillenbrand, L., Carpenter, J. 2000, AJ, in press
Kirkpatrick, J.D., McGraw, J.T., Hess, T.R., Liebert, J., McCarthy, D.W. 1994, ApJS, 94, 749
Kirkpatrick, J.D., Reid, I.N., Liebert, J., Cutri, R., Nelson, B., Beichman, C.A., Dahn, C.C., Monet, D.G., Skrutskie, M.F., Gizis, J.E. 1999, ApJ, 519, 802
Lodders, K. 1999, ApJ, 519, 793
Lucas, P.W., Roche, P.F. 2000, MNRAS, in press
Luhman, K., Rieke, G.H., Young, E.T. et al. 2000, ApJ, in press
Martín, E.L., Basri, G., Gallegos, J.E., Rebolo, R., Zapatero-Osorio, M.R., Bejar, V.J.S. 1998, ApJ, 499, L61
Meusinger, H., Schilbach, E., Souchay, J. 1996, A&A, 312, 833
Nakajima, T., Oppenheimer, B.R., Kulkarni, S.R., Golimowski, D.A., Matthews, K., Durrance, S.T. 1995, Nature, 378, 463
Piotto, G., Zoccali, M. 1999, A&A, 345, 485
Rebolo, R., Martín, E.L., Maguzzu, A. 1992, ApJ, 389, L83
Reid, I.N. & Gilmore, G.F. 1982, MNRAS, 201, 73

Reid, I.N, Gizis, J.E. 1997, AJ, 113, 2246
Reid, I.N., Hawley, S.L., Gizis, J.E. 1995, AJ, 110, 1838
Reid, I.N. 1998, in *The 38th Herstmonceux Conference*, ed. G. Gilmore & D. Howell , A.S.P. Conf. Ser., 142, 121
Reid, I.N., Hawley, S.L. 1999, AJ, 117, 343
Reid, I.N., Kirkpatrick, J.D., Liebert, J., Burrows, A., Gizis, J.E., Burgasser, A., Dahn, C.C., Monet, D., Cutri, R., Beichman, C.A., Skrutskie, M. 1999, ApJ, 521, 613
Reid, I.N., Kirkpatrick, J.D., Gizis, J.E., Dahn, C.C., Monet, D.G., Williams, R.J., Liebert, J., Burgasser, A., 2000, AJ, 119, 369
Salpeter, E.E. 1956, ApJ, 121, 161
Scalo, J.M. 1998, in *The 38th Herstmonceux Conference*, ed. G. Gilmore & D. Howell , A.S.P. Conf. Ser., 142, 201
Stauffer, J.R., Klemola, A., Prosser, C., Probst, R. 1991, AJ, 101, 980
Tinney, C.G. 1993, ApJ, 414, 279
Tsuji, T. 1964, Ann. Tokyo Obs. ser. II, 9, 1
Tsuji, T., Ohnaka, K., Aoki, W., Nakajima, T. 1996, A&A, 308, L29
van de Kamp, P. 1953, PASP, 65, 17
Zoccali, M. et al. 2000, ApJ, 530, 418

Discussion

Evans: Do we expect the 8 pc sample to be representative of the disk, given the facts that the gravitational field is likely to be lumpy and the velocity distribution is dominated by clumps?

Reid: If the age distribution of the disk is such that a particular range of ages is represented only in clumps which don't intersect the Solar neighbourhood, then there would be a systematic bias in the inferred mass functions, below the hydrogen-burning limit. There should be much less of a problem for M dwarfs, whose age range matches the age of the disk. The L-dwarf analysis is very vulnerable to anything which affects the age distribution of local objects: changes in B (+) or sampling problems will have systematic repercussions.

Binney: We know that the solar neighbourhood is deficient in high-mass stars relative to a global average. Presumably standard IMF derivations are not using the correct star formation history. At the low-mass end star formation history is again important. How old are the objects you count at the sub-stellar end?

Reid: The oldest objects are the high-mass ($>0.06 M_\odot$) brown dwarfs, which are L dwarfs at age between ~ 1 and 5 Gyr. These objects make the largest contribution to both the L dwarf and T dwarf number counts.

Hauschildt: 1) Did you find any evidence for asymmetric line profiles (e.g. quasi-static broadening) in your spectra?
2) Is there any evidence for sub-solar metallicities in L and T dwarfs?

Reid: 1) We don't have significant evidence from the low resolution data.
2) All of the L dwarfs of a given spectral type have very similar spectra, so there's no clear evidence for any sub-solar metallicities yet.

Graff: What would be the number of brown dwarfs within 8 pc and in the Galaxy as a whole?

Reid: Approximately the same number as the number of main sequence stars in the disk, if our estimate of $\Psi(M) \propto M^{-1}$ is correct.

Gould: Gould et al. established a local correction to systemic MF to the IMF from Reid & collaborators) to be about 0.35, but you are showing 0.12, which would significantly change the apparent significance of the conflict between Zoccali et al. and Popowski's initial estimate of the MF slope from binary microlensing. Can you comment upon this discrepancy?

Reid: The 8-pc sample has changed slightly from the Reid/Gizis paper, which may make a slight difference. Another difference is that I had the data in log M rather than linear M converted to log M (see Zoccali et al., where the best fit to Reid/Gizis is $\alpha=0.8$ rather than $\alpha \sim 1.1$). Other than that, we need to compare results on a star by star basis.

Gould: Zoccali et al. took into account the wiggle in the color-magnitude relation, so even if we did not, it would produce a bump in the MF, not a change in slope.

Reid: It's entirely possible that the Zoccali et al. result is correct. However, having been bitten, I'd simply note that photometric analyses are liable to these systematic effects. The dip in $\Psi(M)$ from Zoccali et al. is coincident with a similar feature in Gould et al. (at $\log M \sim -0.4$), and the latter is based on standard photometric parallax analysis.

Microlensing 2000: A New Era of Microlensing Astrophysics
ASP Conference Series, Vol. 239, 2000
John Menzies and Penny D. Sackett, eds.

Discussion Session II:
Mass Functions/Budgets of Dark and Luminous Objects

edited by John Menzies[1] and Penny D. Sackett[2]

[1] SAAO, PO Box 9, Observatory 7935, South Africa

[2] Kapteyn Institute, 9700 AV Groningen, The Netherlands

1. Introduction

The second of two full-meeting discussions was held the morning of 25 February 2000, with Andy Gould presiding. Participants were asked to consider what was known and not known about the mass function and mass budget of compact objects in galaxies, and how microlensing had contributed already and could most effectively continue to do so to this field. The moderator was given the opportunity to open the session with a charge. Editorial remarks are in a slanting typeface.

2. Discussion

Gould: We have a couple of major issues to consider in this session. The first is the low-mass end of the Bulge mass function. It can be probed with microlensing, and maybe at least locally by other methods. These other methods have raised the spectre of a gap in the mass range of about 10 M_J to 100 M_J – there is a deficiency of objects there. This is a regime where perhaps microlensing can probe. Binary studies are now sensitive to orbital radii of a few AU, maybe eventually 10 AU but it is hard to believe any of us will be around when they get beyond that. It seems to be an area where microlensing is pushing on both sides – on the one hand with planets and and on the other with low-mass objects.

We have several initial observational results: we heard from Scott Gaudi today about the absence of planets, not quite to these radii, and then in Piotr Popowski's talk, which did not give an actual result yet, but points to what may soon be achieved with respect to the slope of the mass function and the cutoff. We seem to be converging on possibly interesting results. What we should be considering is where we are going with microlensing and how we get beyond these initial results in the future. Secondly there is the strong controversy over what exactly is being detected in the MACHO survey's 20%-halo objects – neutron stars, white dwarfs, LMC objects, or something like that. Then there is this new radio observation of microlensing in an external galaxy which is difficult to put into a common context with the optical microlensing results from our own Galaxy.

Carr: I would like to ask Andy why he sees a conflict between the radio and MACHO results. The mass quoted for the radio determination was 0.5 $M_\odot$, with rather large error bars, which on the face of it seems compatible with the mass estimate from the MACHO results.

Gould: It is totally incompatible, though it could probably be fixed up. If you bring the mass fraction down to 20% you drive up the mass by a factor of three to five, I think. Anyway, the result is not 0.5 $M_\odot$, it is actually 1 $M_\odot$ – you need everything in equal mass binaries to get 0.5 $M_\odot$. We would like to know how well MACHO and EROS are probing at several solar masses, which might be what we are actually seeing. They could be white dwarfs, black holes, or something else. It is not a very coherent picture at present.

Carr: You also have to tie it all up with the results for QSO 0957+561. Are you excluding MACHOs of 0.5 $M_\odot$ when the mass fraction is as low as 20%? It was not clear from the talk.

Wambsganss: I showed the exclusion diagram for 100% compact objects. We also did it for 50% and 25%. The limits got a little smaller but not significantly so – from 93%, say, at 0.1 $M_\odot$ it goes down to 80% or so for the exclusion probability.

Graff: It looks like you are not excluding 1 $M_\odot$ lenses. The results seem consistent with the radial velocity ones. I think 1 $M_\odot$ lenses are ruled out at the 100% level from the quasar microlensing work of Julianne Dalcanton. Does anyone know better than that?

Wambsganss: I think this is a different regime. Dalcanton looked at a cosmologically distributed population of compact objects. Here we are considering objects in the halo of a galaxy.

Carr: But the Dalcanton limit would not exclude the 20% limit for a halo, apart from the point just made.

Gould: Is it agreed then that 5 $M_\odot$ would avoid all constraints? *(Laughter.)*

Gates: I have just heard the radio result for the first time and have no immediate comments on it except that there is evidence for something doing some microlensing in the halo. But I would be cautious about the mass limit. If you have 1 $M_\odot$ primordial black holes in the halo you still have all the problems with pollution from producing them and the progenitors. I think we can reject $\geq$ 1 $M_\odot$ primordial black holes. There could be something in the LMC population itself. There could be a Galactic component, but not a full halo – which brings the mass down; something at 6 kpc with a concentration representative of a 20% halo, but more complicated. It could have some sort of stellar remnants.

Gould: Leon Koopmans says his galaxy has a rotational speed of $\sim$280 km/s, which makes it a sort of super-Milky Way, and 6 kpc is quite close in. Are there any more comments on candidates?

Popowski: I think there are some more bizarre options to consider – things like 'mirror' stars and quark stars. We would not be limited to primordial black holes if the mass turns out to be 3 $M_\odot$.

Carr: I would like to remind you all again of the red halo problem. This is partly in defence of Evalyn Gates's model. While there is no evidence for large red halos beyond 10 kpc, there is a problem with what is happening inside. One of the features of extended protodisk models is that there is a concentration inside a small radius. There is still the possibility of some red halo contribution at a smaller radius. It could not be white dwarfs – this would give the wrong colour features. It would be an important clue if it has anything to do with the dark matter.

Reid: To switch to the other topic: it is clear that microlensing towards the Bulge has already accumulated a large number of events. This is probing a different mass regime from that of the solar neighbourhood stellar mass function, and a comparison would be interesting. There is no way of detecting room-temperature brown dwarfs at present. A proposal for a satellite experiment to do this at around 5 μm was not accepted and SIRTF won't cover a large enough area to be useful for this. For finding these very small mass objects microlensing is unrivalled.

Zinnecker: You mean a free-floating population, not companions in a binary?

Reid: Yes ... When I said yes, I meant no! It is very difficult to find nearby brown dwarfs. The most recent one, just found with HST, has a mass ratio of 2:1. These nearby objects have a semi-major axis of a few AU or more. You can go to much smaller separations with microlensing, but they are too faint to observe by the radial velocity technique. There is this small separation regime where microlensing could make a significant contribution.

Graff: You could change your strategy to one more like EROS I, i.e. do fewer fields but get more exposure per night to increase your sensitivity to these short timescale events.

Popowski: When you do this you encounter problems with your photometry not being accurate enough. It is necessary to follow microlensing objects for a long time to be sure you have an actual microlens. At short timescales it is even more important so that you don't have false detections.

Sackett: The way to do this might be to go to a strategy of surveying and doing follow-up observations at the same time by using a wide-angle instrument on one field. The VST telescope to go on Paranal is expected to have first light in 2002. With its large field of 1 degree, in the direction of the Bulge you can calculate that there will be 14 microlensing events ongoing at any one time. You could concentrate on a few fields, accumulate images rapidly, and analyse them later.

Gould: But isn't the plan to waste the time on that telescope doing non-microlensing science?

Sackett: Perhaps Roberto can say something about that.

Gilmozzi: The fraction of time dedicated to Bulge and LMC microlensing observations with this telescope is planned to be 20%.

Gould: So it will be only 80% wasted! *(Laughter.)*

Zinnecker: It is not easy to find brown dwarf stars in the solar neighbourhood. For microlensing they need to be halfway to the Galactic Centre to be observable. Is it not the case that you can't confirm the local brown dwarfs by microlensing?

Gould: That is right. You couldn't.

Reid: I was not really talking about local L dwarfs in the context of microlensing. Unless you have a very large satellite and put it in front of a local brown dwarf you won't get microlensing. I was concerned with the low-mass binary fraction amongst the short-timescale events that have been detected in the Bulge.

Gould: There are two sets of microlensing results about to be announced, from MACHO and from EROS, that are more or less independent. From these we should get a fairly accurate picture of where the cutoff in the mass function occurs.

Bradbury: In connection with M31 pixel lensing and the brown dwarf problem: where does occultation astronomy fit in? You are observing fields where you can't separate stars; what if some dark thing goes in front of them?

Crotts: We see objects that behave like eclipsing binaries, though we can't resolve them.

Bradbury: What I mean is, if you have a brown dwarf in front of M31, what would it look like?

Gould: It would look like a microlensing event. There is a characteristic distance of the order of 1 light year below which you get an occultation; beyond that you get microlensing.

Popowski: Another point to add to the two mentioned at the beginning is the contribution microlensing can make to Galactic structure studies. It would be profitable to combine microlensing data with that from other techniques as I indicated in my talk. Perhaps this will help provide a more coherent picture of the Galaxy.

Crotts: Microlensing can be used to look for intra-cluster or intra-group MACHOs. It is difficult to do, though there is a trial experiment in Cycle 9 to look for intra-cluster MACHOs in M87. If you had things along the way to M31 you would look for long timescale events. The optical depth would not be large, but even so they would have such a strange signature that it would be obvious if you found one.

Sackett: Could you use radio microlensing to search for intra-cluster MACHOs?

Koopmans: You would need a high density, say near the critical surface density, to see anything. Unless all the dark matter in the universe was in the form of compact objects there would only be a very small chance. In the radio you need to monitor large parts of the sky. You would go for very faint radio objects, a few mas, and monitor them for many years. If they show super-luminal motion, in which case the optical depth would be dominated by the area the source sweeps across the sky rather than by the lensing cross section (determined by the Einstein radius), then radio observations would be more efficient than optical ones. So, for example, the work by Hawkins would be more efficiently done in the radio because you sweep out a larger area in the same amount of time.

Gould: For local objects the radio source would be too big for the Einstein ring.

Koopmans: No, the Einstein radius locally is a few mas, which helps a lot since the radio source will be much smaller than that. But the optical depth would be small and you would need about a million sources spread all over the sky. It would be difficult to do locally.

Gould: I was thinking of Virgo-type radio sources, but you are obviously thinking of extragalactic ones.

Wambsganss: Optical quasar microlensing can also test for an intra-cluster population of compact objects, if any. For example, in the double quasar, the lens is a central cluster galaxy, part of a relatively massive cluster at z = 0.36 which does most of the lensing. Compact objects in front of the quasar could be in two populations, one in the halo, one intra-cluster. If there is no microlensing then you can exclude both at once, or at least show they cannot contribute much to the total mass doing the lensing.

Kerins: An ongoing experiment by Tadros, Warren and Hewett, who are looking for MACHOs in the direction of Virgo covers a mass range $\sim 10^{-3}$ $M_\odot$, which is perhaps not particularly interesting since it is well-constrained by observations in our Galaxy. However it is sensitive to small clumps of gas, which could look like point lenses over those kinds of scales and could give complementary constraints on the kinds of baryonic scenarios in our Galaxy that microlensing can't constrain.

Carr: These constraints from Virgo are already available, and there is another season of observations to come that will improve them. The gas cloud scenario has been considered extensively from a theoretical point of view. It has many signatures that could be exploited. While it wouldn't produce microlensing towards the LMC, it would in the context of quasars.

Zinnecker: I am interested in gas clouds in the context of star formation. It is conceivable that a facility like the Atacama mm array, with a high resolution, might be able to probe the structure of the interstellar medium and molecular clouds which are very clumpy. Could such clouds, with masses ~ 1 $M_\odot$, though they are quite fluffy, perhaps be seen against a background radio source and produce amplification?

Carr: I can't comment on the Atacama array, but these clouds have a number of other observational signatures. One is the gamma ray signature induced by cosmic rays, which can be calculated. This could provide a constraint.

Hansen: Walker and Wardle proposed that molecular hydrogen clouds were a source of extreme scattering events in quasars, but they were considering objects with masses of the order of Jupiter's. If there is a significant mass in this form in Virgo, the clouds would undergo shock heating and appear as X-ray sources, which would be resolved by Chandra, so this would also constrain models of star formation.

Zinnecker But they are relatively fluffy. Would they microlens?

Gould: If they are larger than a couple of AU there would not be any microlensing. It would be possible to evade local constraints if you found them externally. You could say they were gas clouds.

Gould: Perhaps we should vote on whether the more robust black hole of the two we heard about yesterday will actually turn out to be one when SIM observes it. This is a testable point, so we should vote on it.

The result of the vote was: Yes 18, No 6, Abstain 6.

Han: You, Andy and Penny, amongst others, voted against a black hole. What else can it be?

Sackett: Maybe the interpretation would change if different assumptions were made for the structure and kinematics of the Milky Way.

Kerins: Are there any comments on the possibility of explaining the excess of events towards the Bulge and of refuting the MACHO scenario towards the LMC by perhaps contamination by some other variable phenomena? For me it is an unlikely possibility, because the source stars evenly populate the HR diagram, and the distribution of impact parameters is as would be expected, but I would be interested to hear any other opinions.

Graff: Looking at the LMC, it looks to me as though six of the MACHO events are just below the 'bumper' cutoff. Perhaps half of those could be variables?

Discussion Session II: Mass Functions and Budgets

Sackett: For the Bulge, I am concerned about the long duration events, which hold a large fraction of the optical depth and there are three in the same field. There may be something odd going on there. Perhaps we need to take a hard look at these events to check whether there may perhaps be some kinematic sub-structure influencing things.

Wambsganss: I should like to hear some discussion of the evidence for one, two or no planets having been found by microlensing. For example, we have this recent paper by Rhie et al. This is important since it affects our outreach efforts. Many of you will have had some contact with the press on this subject – the public are well aware of these matters. I have no answer, but I am still sceptical. There might be extra information or unpublished results that someone here has? Perhaps a vote amongst those present would provide some useful guidance?

Gould: I have an answer. There is no clear evidence for detection of any planet. But don't just trust my word, let us have a vote. We should exclude the 'free-floating' planet. Something was detected but we have no real evidence as to what it was.

The result of the vote was:

No. planets	No. votes
2	0
1	2
0	most of those present
abstain	non-negligible number

The meeting being in a voting frame of mind, it was felt useful to consider the nature of the events detected towards the LMC. On the question of 'What is the dominant source of MACHOS?' the results were as tabulated here:

Source of MACHO events	No. votes
LMC self-lensing	13
Galactic objects	22
Other	3

Asked to specify the nature of the objects in each of the above categories, and following some discussion to narrow down the possible candidates, the meeting voted as follows:

Nature of object	No. votes
<u>LMC</u>	
Polar ring (coherent structure; unvirialized)	6
Virialized untraced halo (stars)	3
<u>Galactic</u>	
White dwarfs	10
Primordial black holes	6
Astro-engineered objects	1
Other (astrophysical black holes,'mirror' matter)	3
<u>Other</u>	
(Known populations of stars + mostly variable stars)	3

Voting was conducted in a friendly atmosphere, and the delegates were in a jovial mood when they adjourned for lunch.

Part 5: And Beyond ...

Cosmological Microlensing, Large Telescopes

Cosmological Microlensing

Joachim Wambsganss

Universität Potsdam, Institut für Physik, Am Neuen Palais 10, 14067 Potsdam, Germany
and
Max-Planck-Institut für Gravitationsphysik, "Albert-Einstein-Institut", Am Mühlenberg 1, 14476 Golm, Germany

Abstract. Variability in gravitationally lensed quasars can be due to intrinsic fluctuations of the quasar or due to "microlensing" by compact objects along the line of sight. If disentangled from each other, microlens-induced variability can be used to study two cosmological issues of great interest, the size and brightness profile of quasars on one hand, and the distribution of compact (dark) matter along the line of sight. In particular, multi-waveband observations are useful for this goal.

In this review recent theoretical progress as well as observational evidence for quasar microlensing over the last few years will be summarized. Comparison with numerical simulations will show "where we stand". Particular emphasis will be given to the questions microlensing can address regarding the search for dark matter, both in the halos of lensing galaxies and in a cosmologically distributed form. A discussion of desired observations and required theoretical studies will be given as a conclusion/outlook.

1. What is cosmological microlensing?

1.1. Mass, length and time scales

The lensing effects of cosmologically distant compact objects in the mass range

$$10^{-6} \leq m/M_\odot \leq 10^6$$

on background objects is usually called "cosmological microlensing". The "source" is typically a background quasar, but in principle other distant sources can be microlensed as well, e.g. distant supernovae or gamma-ray bursters. The only "condition" is that the source size is comparable to or smaller than the Einstein radius of the respective lenses.

The microlenses can be ordinary stars, brown dwarfs, planets, black holes, molecular clouds, or other compact mass concentrations (as long as their physical size is smaller than the Einstein radius). In most practical cases, the micro-lenses are part of a galaxy which acts as the main (macro-) lens. However, microlenses could also be located in, say, clusters of galaxies or they could even be imagined "free floating" and filling intergalactic space.

The relevant length scale for microlensing is the Einstein radius of the lens:

$$r_E = \sqrt{\frac{4GM}{c^2}\frac{D_S D_{LS}}{D_L}} \approx 4 \times 10^{16}\sqrt{M/M_\odot}\,\text{cm},$$

where "typical" lens and source redshifts of $z_L \approx 0.5$ $z_S \approx 2.0$ were assumed for the expression on the right hand side (G and c are the gravitational constant and the velocity of light, respectively; M is the mass of the lens, D_L, D_S, and D_{LS} are the angular diameter distances between observer – lens, observer – source, and lens – source, respectively).

This length scale translates into an angular scale of

$$\theta_E = r_E/D_S \approx 10^{-6}\sqrt{M/M_\odot}\ \text{arcsec}.$$

It is obvious that the image splittings on these angular scales cannot be observed directly. What makes microlensing observable anyway is the fact that observer, lens(es) and source move relative to each other. Due to this relative motion, the micro-image configuration changes with time, and so does the total magnification, i.e. the sum of the magnifications of all the micro-images. And this change in magnification over time can be measured: microlensing is a "dynamical" phenomenon.

There are two time scales involved: the standard lensing time scale t_E is the time it takes the source to cross the Einstein radius of the lens, i.e.

$$t_E = r_E/v_\perp \approx 15\sqrt{M/M_\odot}\,v_{600}^{-1}\ \text{years},$$

where the same typical assumptions are made as above, and the relative transverse velocity v_{600} is parametrized in units of 600 km/sec. This time scale results in discouragingly large values for stellar mass objects. However, we can expect fluctuations over much shorter time intervals. Due to the fact that the magnification distribution is highly non-linear, the sharp caustic lines separate regions of low and high magnification. So if a source crosses such a caustic line, we will observe a large change in magnification during the time it takes the source to cross its own diameter:

$$t_{cross} = R_{source}/v_\perp \approx 4 R_{15} v_{600}^{-1}\ \text{months}.$$

Here the quasar size R_{15} is parametrized in units of 10^{15}cm. This time scale, t_{cross}, can be significantly shorter than t_E.

1.2. Geometry

The typical geometry of a cosmological microlensing situation is displayed in Figure 1. The main lensing agent is a galaxy, which consists partly of stars (and other compact objects). The "macro"-lensing situation produces two (or more) images of the background quasar, separated by an angle of order one arcsecond. However, because of the graininess of the main lens, each of these macro-images consists of many micro-images, which are separated by angles of order microarcseconds, and hence are unresolvable. However, due to the relative motion between source, lens and observer, the micro-image configuration changes, and so does the total magnification, which is observable.

Cosmological Microlensing

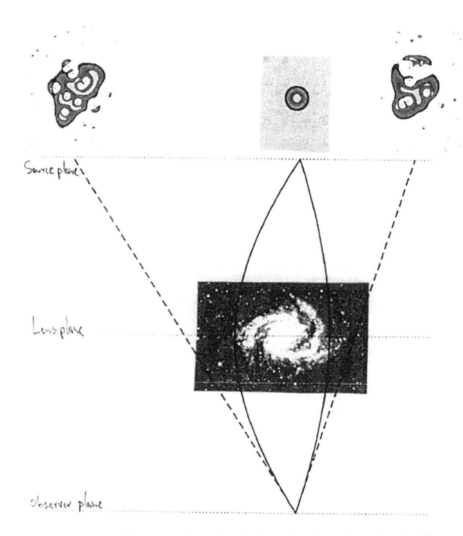

Figure 1. Geometry of a microlensing situation: the galaxy in the lens plane acts as a "macro-lens", producing two separate images of the background quasar. Due to the graininess of the matter distribution in the galaxy, each macro-image is split into many micro-images. Only the total magnification of all the microimages is measurable.

1.3. History

Only a few months after the detection of the first multiply imaged quasar was published by Walsh, Carswell and Weymann (1979), Kyongae Chang and Sjur Refsdal suggested in their paper "Flux variations of QSO 0957+561 A, B and image splitting by stars near the light path", (Chang & Refsdal 1979):

> "If the double quasar QSO 0957+561 A, B is the result of gravitational lens actions by a massive galaxy, stars in its outer parts and close to the line of paths may cause significant flux changes in one year."

Only two years later, J. Richard Gott III asked the question: "Are heavy halos made of low mass stars? A gravitational lens test", Gott (1981). He suggests that a heavy halo made of low mass stars in the range $4 \times 10^{-4} M_\odot$ to $0.1 M_\odot$

> " ... should produce fluctuations of order unity in the intensities of the QSO images on time scales of 1-14 years."

He went on to propose:

> "Observations of QSO 0957+561 A, B and other quasars over time can establish whether the majority of mass in the heavy halo is in the form of low mass stars."

In a number of further papers, the lensing effect of individual stars on the background quasar was explored, e.g. Young (1981) did some numerical simulations and applied them to the double quasar. In the year 1989, the first observational evidence for quasar microlensing was presented: Irwin et al. (1989) showed that fluctuations in image A of the quadruple quasar Q2237+0305 could not be due to intrinsic variability of the quasar. Such fluctuations could be explained by the lensing action of low-mass main sequence stars and allowed conclusions on the quasar size to be of order a few times 10^{14} cm (Wambsganss, Paczyński & Schneider 1990).

1.4. Early Promises

Fluctuations in the brightness of a quasar can have two causes: they can be intrinsic to the quasar, or they can be microlens-induced. For a single quasar image, the difference is hard to tell. However, once there are two or more gravitationally lensed (macro-)images of a quasar, we have a relatively good handle to distinguish the two possible causes of variability: any fluctuations due to intrinsic variability of the quasar have to show up in all the quasar images, after a certain time delay. In fact, time delays of quasars are only measurable *because* quasars are variable intrinsically. (This argument could even be turned around: the measured time delays in multiple quasars are the ultimate proof of the intrinsic variability of quasars.) So once a time delay is measured in a multiply-imaged quasar system, the incoherent fluctuations can be attributed to microlensing.

There is another possibility of distinguishing the two causes of fluctuations: even without measuring the time delay, it is possible to tell whether measured fluctuations are intrinsic or not. In some quadruple lens systems, the image arrangement is so symmetrical around the lens that any possible lens model predicts very short time delays (of order days or shorter), so that fluctuations in individual images that are longer than the (theoretical) time delay and not followed by corresponding fluctuations in the other images, can be safely attributed to microlensing. This is in fact the case of the quadruple system Q2237+0305 (see below).

Early on, the papers exploring microlensing made four predictions concerning the possible scientific successes. With microlensing we should be able to

1. determine the effects of compact objects between the observer and the source,

2. determine the size of quasars,

3. determine the two-dimensional brightness profile of quasars,

4. determine the masses of lensing objects.

In Section 3 the observational results to date will be discussed in some detail. It can be stated here that (1) has been achieved. Some limits on the size of quasars have been obtained, so (2) is partly fulfilled. We are still (far) away from solving promise (3), and concerning point (4) it is fair to say that it was shown that the observational results are consistent with masses of the lensing objects corresponding to low-mass stars.

1.5. "Local Group" Microlensing versus Extragalactic Microlensing

This contribution deals mainly with quasar microlensing, where in most cases the surface mass density (or optical depth) is of order the critical one. In contrast to that, most other papers at these proceedings are concerned with the "local group" or low optical microlensing. Since there are a number of similarities, but as well quite some differences between these two regimes, in Table 1 the various quantities are compared to each other.

2. Theoretical Work on Cosmological Microlensing

In the situation of a multiply imaged quasar, the surface mass density (or "optical depth") at the position of an image is of order unity. If this matter is made of compact objects in the range described above, due to the relative motion of source, lens(es) and observer, microlensing is expected to be going on basically "all the time". In addition, this means that the lens action is due to a coherent effect of many microlenses, because the action of two or more point lenses whose projected separations are of order their Einstein radii adds in a very non-linear way.

The lens action of more than two point lenses cannot be easily treated analytically any more. Hence numerical techniques were developed in order to simulate the gravitational lens effect of many compact objects. Paczyński (1986) had used a method to look for the extrema in the time delay surface. Kayser, Refsdal & Stabell (1986), Schneider & Weiss (1987) and Wambsganss (1990) all developed and applied an inverse ray-shooting technique that produced a two-dimensional magnification distribution in the source plane. An alternative

Table 1. The important lensing properties for the two regimes of microlensing – local group vs. cosmological – are compared to each other. On the left the various properties of interest are named; the middle and right-hand columns list either whether these properties are known or not, or the values for the Milky Way and the lensing galaxy, respectively.

Lensing galaxy:	Milky Way	Lens in Q0957+561
distance to Macho?	no	yes
velocity of Macho?	no	(no)
mass?	???	???
optical depth?	$\approx 10^{-6}$	≈ 1
Einstein angle (1 $M_\odot$)?	≈ 1 milliarcsec	≈ 1 microarcsec
time scale?	hours to years	weeks to decades
event?	individual/simple	coherent/complicated
default light curve?	smooth	sharp caustic crossing
when/who proposed?	Paczyński 1986	Gott 1981
first detection?	EROS/MACHO/OGLE 1993	Irwin et al. 1989

technique was proposed by Witt (1993) and by Lewis et al. (1993). They solved the lens equation along a linear source track. All the recent theoretical work on microlensing is based on either of these techniques.

Fluke & Webster (1999) explored analytically caustic crossing events for a quasar. Lewis & Belle (1998) showed that spectroscopic monitoring of multiple quasars can be used to probe the broad line regions. Wyithe et al. (2000a, b) explored and found limits on the quasar size and on the mass function in Q2237+0305.

Agol & Krolik (1999) and Mineshige & Yonehara (1999) developed techniques to recover the one-dimensional brightness profile of a quasar, based on the earlier work by Grieger et al. (1988, 1991). Agol & Krolik showed that with frequent monitoring of a caustic crossing event in many wave bands (they used of order 40 data points in eleven filters over the whole electromagnetic range), one can recover a map of the frequency-dependent brightness distribution of a quasar! Yonehara (1999) in a similar approach explored the effect of microlensing on two different accretion disk models. In another paper, Yonehara et al. (1998) showed that monitoring a microlensing event in the X-rays can reveal structure of the quasar accretion disk as small as AU-size.

With numerical simulations and limits obtained from three years of Apache Point monitoring data of Q0957+561, and based on the Schmidt & Wambsganss (1998) analysis, we extend the limits on the masses of "Machos" in the (halo of the) lensing galaxy: the small "difference" between the time-shifted and magnitude-corrected light curves of images A and B excludes a halo of the lensing galaxy made of compact objects with masses of $\leq 10^{-2} M_\odot$ (Wambsganss et al. 2000).

3. Observational Evidence for Cosmological Microlensing

3.1. The Einstein Cross: Quadruple Quasar Q2237+0305

In 1989 the first evidence for cosmological microlensing was found by Irwin et al. (1989) in the quadruple quasar Q2237+0305: one of the components showed fluctuations, whereas the others stayed constant. In the meantime, Q2237+0305 has been monitored by many groups (Irwin et al. 1989; Corrigan et al. 1991; Østensen et al. 1996; Lewis et al. 1998; Woźniak et al. 2000).

The most recent and most exciting results (Woźniak et al. 2000) show that all four images vary dramatically (but incoherently!), going up and down like a rollercoaster in the last three years:

- $\Delta m_A \approx 0.6$ mag,

- $\Delta m_B \approx 0.4$ mag,

- $\Delta m_C \approx 1.3$ mag (and rising?),

- $\Delta m_D \approx 0.6$ mag.

This is very encouraging news, and it calls for continuing and expanding monitoring programs for lensed quasars.

3.2. The Double Quasar Q0957+561

The microlensing results for the double quasar Q0957+561 are at the moment not as exciting as those for Q2237+0305. In the first few years, there appears to be an almost linear change in the (time-shifted) brightness ratio between the two images ($\Delta m_{AB} \approx 0.25$ mag over 5 years). But since about 1991, this ratio stayed more or less constant to within about 0.05 mag, so not much microlensing has been going on in this system recently (Schild 1996; Pelt et al. 1998; Schmidt & Wambsganss 1998; Wambsganss et al. 2000). At his moment, the possibility for some small amplitude rapid microlensing cannot be excluded; however, one needs a very well determined time delay and very accurate photometry, in order to establish that (Colley & Schild 1999).

3.3. Other multiple quasars

A number of other multiple quasar systems are being monitored more or less regularly. For some of them microlensing has been suggested (e.g. H1413+117, Østensen et al. 1997; or B0218+357, Jackson et al. 2000) In particular the possibility for "radio"-microlensing appears very interesting (B1600+434, Koopmans & de Bruyn 2000 and Koopmans, these proceedings), because this was not expected in the first place, due to the presumably larger source size of the radio emission region. This novel aspect of microlensing is definitely worth pursuing in more detail.

4. Unconventional Microlensing

4.1. Microlensing in individual quasars?

There were a number of papers interpreting the variability of individual quasars as microlensing (Hawkins 1996, 1998; Hawkins & Taylor 1997). Although this is an exciting possibility and it could help us detect a population of cosmologically distributed lenses, it is not entirely clear at this point whether the observed fluctuations can be fully or partly attributed to microlensing. After all, quasars *are* intrinsically variable (otherwise we could not measure time delays), and the expected amount of microlensing in single quasars must be smaller than in multiply imaged ones, due to the lower surface mass density. Only more observations can help solve this question.

4.2. "Astrometric Microlensing" Centroid shifts

An interesting aspect of microlensing was explored by Lewis & Ibata (1998). They looked at centroid shifts of quasar images due to microlensing. At each caustic crossing, a new very bright image pair emerges or disappears, giving rise to sudden changes in the "center of light" positions. The amplitude could be of order 100 microarcseconds or larger, which should be observable with the next generation of astrometric satellites, like SIM.

5. Microlensing: Now and Forever?

Monitoring observations of various multiple quasar systems in the last decade have clearly established qualitatively that the phenomenon of "cosmological" microlensing exists. Uncorrelated variations in multiple quasar systems with amplitudes of more than a magnitude have been observed, on time scales of weeks to months to years. However, in order to get close to a really quantitative understanding, much better monitoring programs need to be performed.

On the theoretical side, there are two important questions: what do the light curves tell us about the lensing objects, and what can we learn from them about the size and structure of the quasar. In response to the first question, the numerical simulations are able to give a qualitative understanding of the measured light curves (detections and non-detections), in general consistent with "conservative" assumptions about the object masses and velocities. But due to the large number of parameters (quasar size, masses of lensing objects, transverse velocity) and due to the large variety of light curve shapes possible even for a fixed set of parameters, no "unique" quantitative explanation or even predictions could be achieved. The prospects of getting much better light curves of multiple quasars, as shown by the OGLE collaboration, should be motivation enough to explore this regime more quantitatively in the future.

The question of the quasar structure deserves much more attention. Here gravitational lensing is in the unique situation to be able to explore an astrophysical field that is unattainable by any other means. Hence much more effort should be put into attacking this problem. This involves much more ambitious observing programs, with the goal to monitor caustic crossing events in many filters over the whole electromagnetic spectrum, and to further develop numerical techniques to obtain useful values for the quasar size and (one-dimensional) profile from unevenly sampled data in (not enough) different filters.

In summary it can be said that cosmological microlensing – though still a young discipline – has already achieved some of its original goals in attacking the questions of compact (dark) matter and quasar size and structure. But there is still a lot of very interesting astrophysics out there, "in reach". The field is definitely worth pursuing with more effort in the future – both theoretically and observationally.

Acknowledgments. It is a pleasure to thank the organisers, Penny Sackett and John Menzies and their colleagues for their excellent macro- and microplanning and running of "Microlensing 2000: A New Era of Microlensing Astrophysics".

References

Agol, E., Krolik, J. 1999, ApJ, 524, 49
Chang, K., Refsdal, S. 1979, Nature, 282, 561
Colley, W., Schild, R. 1999, A&AS, 195, 4801
Corrigan, R.T., Irwin, M.J., Arnaud, J., Fahlman, G.G., Fletcher, J.M. et al. 1991, AJ, 102, 34
Fluke, C.J., Webster, R.L. 1999, MNRAS, 302, 68
Gott, J.R. 1981, *Astrophys. J.* 243, 140,
Grieger, B., Kayser, R., Refsdal, S. 1988, A&A, 194, 54
Grieger, B., Kayser, R., Schramm, T. 1991, A&A, 252, 508
Hawkins, M.R.S. 1996, MNRAS, 278, 787
Hawkins, M.R.S., Taylor, A.N. 1997, ApJ, 482, L5
Hawkins, M.R.S. 1998, A&A, 340, L23
Irwin, M.J., Webster, R.L., Hewett, P.C., Corrigan, R.T., Jedrzejewski, R.I. 1989, AJ, 98, 1989
Jackson, N., Xanthopoulos, E., Browne, I.W.A. 2000, MNRAS, 311, 389
Kayser, R., Refsdal, S., Stabell, R. 1986, A&A, 166, 36
Koopmans, L.V.E., de Bruyn, A.G. 2000, A&A, submitted (astro-ph/0004112)
Lewis, G.F., Miralda-Escudé, J., Richardson, D.C., Wambsganss, J. 1993, MNRAS, 261, 647

Lewis, G.F., Belle, K.E. 1998, MNRAS, 297, 69
Lewis, G.F., Ibata, R.A.. 1998, ApJ, 501, 478
Lewis, G.F., Irwin, M.J., Hewett, P.C., Foltz, C.B. 1998, MNRAS, 295, 573
Mineshige, S., Yonehara, A. 1999, PASJ, 51, 497
Østensen, R., Refsdal, S., Stabell, R., Teuber, J, Emanuelsen, P.I. et al. 1996, A&A, 309, 59
Østensen, R., Remy, M., Lindblad, P.O., Refsdal, S., Stabell, R. et al. 1997, A&AS, 126, 393
Paczyński, B. 1986, ApJ, 301, 503
Pelt, J., Schild, R., Refsdal, S., Stabell, R. 1998, A&A, 336, 829
Schild, R. 1996, ApJ, 464, 125
Schmidt R., Wambsganss, J. 1998, A&A, 335, 379
Schneider, P., Weiss, A. 1987, A&A, 171, 49
Walsh, D., Carswell, R.F., Weyman, R.J. 1979, Nature, 279, 381
Wambsganss, J. 1990, PhD thesis (Munich University), preprint MPA 550
Wambsganss, J., Brunner, H., Schindler, S., Falco, E. 1999, A&A, 346, L5
Wambsganss, J., Schmidt R., Kundić, T., Turner, E. L., Colley, W. 2000, in preparation
Witt, H.J. 1993, ApJ, 403, 530
Woźniak, P. R. , Alard, C., Udalski, A. , Szymański, M., Kubiak, M., Pietrzyński, G., Żebruń, K. 2000, ApJ, 529, 88
Wyithe, J.S.B., Webster, R.L., Turner, E.L. 2000a, MNRAS, 315, 51
Wyithe, J.S.B., Webster, R.L., Turner, E.L., Mortlock, D.J. 2000b, MNRAS, 315, 62
Yonehara, A., Mineshige, S., Manmoto, T., Fukue, J., Umemura, M., Turner, E. L. 1998, ApJ, 501, L41
Yonehara, A. 1999, ApJ, 519, L31
Young, P. 1981, ApJ, 244, 756

Discussion

Graff: Has there been any improvement on the Dalcanton et al. limit on cosmological MACHOs?

Wambsganss: There is certainly much more and much better data available by now, and it is very worthwhile to repeat this work. I don't know whether anyone is working on a similar study at this moment.

Carr: Can you comment on the lens mass required for Q2237+0305? The original estimate was very low (~ 0.01 $M_\odot$); more careful calculations then gave a larger mass (~ 1 $M_\odot$), but the recent estimate you quoted is low again (0.01-0.1$M_\odot$). Is the mass compatible with the MACHO value of 0.5 $M_\odot$ for our own Galaxy?

Wambsganss: In the original detection of microlensing in Q2237+0305, Irwin et al (1989) interpreted it as produced by a very small mass object ($\sim 10^{-5} M_\odot$). This was based on the interpretation as an individual lens, which is not quite justified in this system with optical depth of order 0.5. In Wambsganss et al (1990) we showed that the fluctuations are consistent with a lens population in the mass range [0.1 $M_\odot$, 1.0 $M_\odot$], with a Salpeter mass function. The new OGLE data are interpreted by Wyithe et al. (2000a) as being produced by objects of masses from $\sim 0.03\ M_\odot$ to $\sim 0.5\ M_\odot$. Due to the relatively short time scales of the events a lens population of only 1 $M_\odot$ objects or larger is probably excluded. With only 0.5 $M_\odot$ objects it may be marginally possible to produce the current data.

Crotts: With QSO lens photometric monitoring programs so far, do people use standard filter bands, or do they choose filters to isolate continuum or emission lines?

Wambsganss: This would be a very good idea, but up to now people have used one or two standard bands based on what they have available.

Zinnecker: You mentioned X-ray monitoring of multiply lensed quasars. Can you say something about the expected amplifications and time scales?

Wambsganss: The OGLE team has seen fluctuations of 1 mag within a few weeks. If our understanding of accretion disks in the centers of quasars is correct, the x-ray emitting regions must be (much) smaller than the optical continuum region. That means the time scale for a caustic crossing event in x-rays can be significantly smaller and the amplitude significantly larger, than what is observed in the optical. I wouldn't be surprised if we would see x-ray high magnification events in Q2237+0305 with a time scale of hours and magnifications of many tens. Chandra can resolve the images individually, whereas XMM could still discover such events in the combined light curve of the four images which opens up really promising opportunities for the near future.

Binney: It looks as if you have powerful evidence for non-baryonic DM: the lines of sight through 0957 pass through the halo and don't show microlensing, while those through 2237 pass through the visible galaxy and do. Is this summary too simplistic?

Wambsganss: This implication is certainly in agreement with the current (still very limited) data: the four images in Q2237+0305 pass 0.7 arcsec from the center of a z = 0.04 galaxy, and the assumption of an old stellar population in the mass range [0.1 $M_\odot$, 1.0 $M_\odot$] can explain the observed microlensing observations. In Q0957+561 the inner image passes about 1 arcsec from a z = 0.36 galaxy; this one should feel at least some microlensing effects of ordinary stars. Maybe the $\Delta m = 0.25$ mag change in the 1980s reflects this. The other image at about 5 arcsec distance from the center of the galaxy passes through the halo and does not show much variation, which you would expect for continuously distributed non-baryonic dark matter. However, in Q0957+561 the typical time scale is (due to the geometrical factors in the definition of the Einstein radius) seven times larger than in Q2237+0305, which means that the fact that we haven't seen much microlensing in Q0957+561 could still be a statistical fluke.

Gould: Popowski and I looked at the Schild data covering about 5 years and found one definite event with a duration of 30 days and an amplitude of about 2%. Can we learn anything positive from this?

Wambsganss: In this optical depth regime ($\sim$0.5) it is not possible to directly infer the lens mass or the source size from an individual event. However, the duration and amplitude of the event you found definitely point towards a small lens mass (my offhand guess is $\lesssim$0.1 $M_\odot$) and a relatively small quasar source size ($\lesssim$ few x 10^{14} cm). More quantitative statements require knowledge of the frequency and shape of such events, and comparisons with numerical simulations.

Carr: Would you comment on the credibility of Schild's claim to have found planetary mass lenses in Q0957+561?

Wambsganss: Rudy Schild is the pioneer of monitoring the two images of the double quasar Q0957+561. He had established the correct time delay of around 417 days long before it was accepted in the community. He continues to monitor this system and has concluded that there are very short time scale microlensing fluctuations which could be interpreted as produced by planetary systems along the line of sight. Well, in order to detect microlensing events of order a few days, the time delay has to be known much better than that; I am not sure whether we have reached this accuracy yet.

… Microlensing 2000: A New Era of Microlensing Astrophysics
ASP Conference Series, Vol. 239, 2000
John Menzies and Penny D. Sackett, eds.

A Radio-microlensing Caustic Crossing in B1600+434?

L.V.E. Koopmans[1,4], A.G. de Bruyn[2,1], J. Wambsganss[3,5] and C.D. Fassnacht[6]

Abstract. First, we review the current status of the detection of strong "external" variability in the CLASS gravitational lens B1600+434, focusing on the 1998 VLA 8.5-GHz and 1998/9 WSRT multi-frequency observations. We show that these data can best be explained in terms of *radio-microlensing*. We then proceed to show some preliminary results from our new multi-frequency VLA monitoring program, in particular the detection of a strong feature ($\sim$30%) in the light curve of the lensed image which passes predominantly through the dark-matter halo of the lens galaxy. We tentatively interpret this event, which lasted for several weeks, as a *radio-microlensing caustic crossing*, i.e. the superluminal motion of a μas-scale jet-component in the lensed source over a single caustic in the magnification pattern, that has been created by massive compact objects along the line-of-sight to the lensed image.

1. Introduction

Optical microlensing has unambiguously been detected both in our own Galaxy (e.g. the EROS, MACHO and OGLE collaborations; see these proceedings), as well as in external galaxies (e.g. Q2237+0305; Irwin et al. 1989). However, no convincing detections of microlensing in other wavebands have been claimed thus far. In our Galaxy this is mostly due to the very low surface number density of bright sources, other than stars, that are compact enough (i.e. $\lesssim$few mas) to be significantly microlensed. For cosmological microlensing there are similar arguments. Because the angular Einstein radius of a point-mass scales approximately as $\propto D^{-1/2}$, where D is the distance to the lens, the lensed source must be $\lesssim$fewμas in angular size to by microlensed by compact stellar-mass lenses at cosmological distances (few Gpc). It was thought until recently that only the optical to γ-ray emitting regions of quasars and AGNs had these extremely small angular scales and would still be observable over cosmological distances.

In this proceeding we will show, however, that 'radio-microlensing' has most likely been detected in the CLASS gravitational lens B1600+434 (Jackson et al. 1995) and that it promises to be a new and exciting technique for the study of massive compact objects in galaxies at intermediate redshifts.

---

[1]Kapteyn Astronomical Institute, P.O.Box 800, NL-9700 AV Groningen, The Netherlands; [2]NFRA-ASTRON, P.O.Box 2, NL-7990 AA Dwingeloo, The Netherlands; [3]University of Potsdam, Institute for Physics, Am Neuen Palais 10, 14469 Potsdam, Germany; [4]Jodrell Bank Observ., Lower Withington, Macclesfield, Cheshire SK11 9DL, UK; [5]Max-Planck-Institut für Gravitationsphysik, "Albert-Einstein-Institut", Am Mühlenberg 1, 14476 Golm, Germany; [6]NRAO, P.O.Box 0, Socorro, NM 87801, USA

2. Macro & microlensing of flat-spectrum radio sources

The JVAS/CLASS gravitational-lens survey (e.g. Browne et al. 1997) has discovered at least 17 radio-bright gravitational lens systems. All systems were selected to have a flat spectrum (α<0.5 with $S_\nu \propto \nu^{-\alpha}$) between 5 and 1.4 GHz. This ensures that most of these sources are dominated by a compact radio-bright core. These sources are often variable on short time scales and in many cases show superluminal motion (e.g. Vermeulen & Cohen 1994).

Strong variability seems to imply that these radio sources are very compact. Based on a simple light-travel-time argument, one expects a source at a cosmological distance (few Gpc) to have an angular scale of only a few μas if it varies significantly on a time scale of say one month. However, superluminal motion with a high Doppler-boosting factor ($\mathcal{D}$) complicates this argument, because the intrinsic variability time scale is reduced by a factor $\mathcal{D}^3$, whereas the intrinsic angular scale of the source is reduced by only $\mathcal{D}^{1/2}$ (e.g. Lähteenmäki, Valtaoja & Wiik 1999). Based on variability arguments, the angular size of the source could therefore be severely underestimated if $\mathcal{D} \gg 1$. Furthermore, based on the Rayleigh–Jeans law and a typical surface brightness temperature of 10^{11-12} K (e.g. Lähteenmäki et al. 1999), one expects an angular size as much as several tens of μas for those sources that have been observed in the JVAS/CLASS survey. This is much larger than the typical Einstein radius of a solar-mass object. Overall, it appears that microlensing of these sources is at most marginal, especially for the brighter sources (tens of mJy).

However, matters are not as bad as they seem! First of all, lensed sources are often magnified by the lensing potential. Their intrinsic flux-density is therefore less than the observed flux-density and, because surface brightness (temperature) is conserved in lensing, also their intrinsic angular source size is smaller. Second, flat-spectrum sources at high redshifts often exhibit jet-structures with superluminal motion (e.g. Vermeulen & Cohen 1994). If these jets contain many distinct "bullets" or shock-fronts, their angular sizes can be very small, especially if they are significantly Doppler-boosted ($\mathcal{D} \gg 1$). We expect this to be the case near the core of these flat-spectrum sources, although components further along the jet might grow in size and have lower velocities. Hence, even though the integrated flux-density of the source might be large, suggesting a large angular size, in reality the likely presence of compact substructure with high Doppler-boostings makes this argument significantly weaker. The expected angular size of these flat-spectrum subcomponents is

$$\Delta\theta \sim \sqrt{\frac{S_{\mathrm{mJy}}(1+z_\mathrm{s})}{\mu \mathcal{D} T_{12}}} \left(\frac{\lambda}{1\,\mathrm{cm}}\right) \mu\mathrm{as},$$

where T_{12} is the surface brightness temperature of the component in units of 10^{12} K, μ is the magnification by the lens and z_s is the source redshift. For subcomponent flux-densities (S_{mJy}) of a few mJy, an angular size of a few μas can be expected. This is small enough to be appreciably microlensed!

Below we will illustrate this by the CLASS gravitational lens system B1600+434, in which believe to have detected radio-microlensing of precisely this type of μas-scale jet component (Koopmans & de Bruyn 2000; KdB00 hereafter).

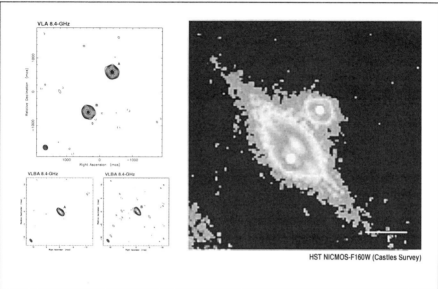

Figure 1. **Left:** upper – VLA A-array 8.5-GHz image of B1600+434, showing the two compact flat-spectrum components, separated by 1.4 arcsec. lower – VLBA 8.5-GHz images of the two components, showing that they are compact down to the milli-arcsec level. **Right:** HST NICMOS H–band image of B1600+434 from the CASTLES survey, showing the edge-on spiral lens galaxy and the two compact optical components (A and B), corresponding to the two radio images.

3. CLASS B1600+434

The CLASS gravitational lens B1600+434 consists of two compact radio components separated by 1.4 arcsec (Fig. 1). The lens galaxy is an edge-on spiral at a redshift of 0.41, whereas the source has a redshift of 1.59 (Koopmans, de Bruyn & Jackson 1998). Image A passes predominantly through the dark-matter halo of the lens galaxy (~ 6 kpc above the plane of the galaxy), whereas image B passes through its stellar component (i.e. disk/bulge). The extended optical emission around image A is thought to be associated with its host galaxy at $z=1.59$. Both radio components are compact ($\lesssim 1$ mas) and have a flat radio spectrum. The source is highly variable at frequencies between 1.4 and 8.5 GHz (Koopmans et al. 1998).

To determine the time delay between the lensed images, we monitored B1600+434 with the VLA in A- and B-arrays at 8.5 GHz, during a period of about 8 months in 1998 (Koopmans et al. 2000). From these light curves we determined a time delay of 47^{+5}_{-6} days (68%), using the minimum-dispersion

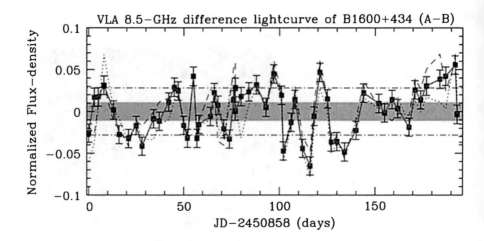

Figure 2. The normalized difference light curve between the two lensed images, corrected for both the time-delay and flux density ratio (Sect.2.1). The shaded region indicates the expected 1–σ (1.1%) region if all variability were due to measurement errors. The dash–dotted lines indicate the observed modulation-index of 2.8%. The dotted and dashed curves indicate the normalized difference curves for a time delay of 41 and 52 days (68% confidence region), respectively.

method from Pelt et al. (1996). The PRH method (Press, Rybicki & Hewitt 1992) gives the same time delay within the error range. Shifting the light curve of image B back in time over this time-delay and subsequently subtracting it from the light curve of image A, taking the proper flux-density ratio in to account, results in the difference light curve shown in Fig.2. The rms variability of the difference light curve (2.8%) is significantly larger than the expected rms variability due to noise only (1.1%). The difference light curve is consistent with the presence of non-intrinsic variability (i.e. "external" variability) at the 14.6–σ confidence level (KdB00). Because the time delay was determined using a minimum-dispersion method, time delays other than 47 days would yield an even larger rms variability in the difference light curve.

4. 'External' variability in B1600+434

What can be the origin of the presence of "external" variability in the image light curves? There are several plausible causes:

- Scintillation by the Galactic ionized interstellar medium.

- Microlensing of the background source by massive compact objects in the lens galaxy.

Before proceeding, however, let us first summarize the results on B1600+434 from our VLA & WSRT monitoring data (until late 1999), that might be relevant in uncovering the true nature of the observed external variability (see KdB00):

1. The short-term rms variability (i.e. the first order modulation index) of the lensed images A and B are 2.8% and 1.6%, respectively.

2. The difference light curve has an rms variability of 2.8%.

3. The VLA 8.5-GHz image light curves, as well as the difference light curve, seem to show long-term ($\gg$1 d) variability, next to some faster variability. The precise time scales are hard to determine, because of the average light-curve sampling rate of once per 3.3 days.

4. Several 5-GHz WSRT 12-h integrated flux-density lightcurves of B1600+434 (June-July 1999), with a sampling rate of once every 5 min, show no evidence for variability $\gtrsim$2% on time scales less than 12 hours.

5. Multi-frequency monitoring with the WSRT shows a decrease in the short-term rms variability from 5 to 1.4 GHz by a factor of $\sim$3.

Can this considerable body of multi-frequency data be explained in terms of Galactic scintillation (Section 4.1) or radio-microlensing (Section 4.2)?

4.1. Galactic Scintillation

From "standard" scintillation theory (e.g. Narayan 1992), we expect a strong anti-correlation between the time-scale of variability and its amplitude (i.e. rms variability). According to the Taylor & Cordes (1993) model for the Galactic ionized medium, B1600+434 should be in the weak-scattering regime and have a variability time scale of at most $\sim$1 day for an observed rms variability of 1.6–2.8%. However, the WSRT data show no evidence for this short-term variability (point 4). In fact a number of flux-density variations appear to have relatively long time scales (several weeks; see point 3 and also Section 5). These time scales would normally correspond to an rms variability that is well below the noise level ($\ll$1%). Hence, the longer variability time scales are *not* compatible with the observed amplitude of variability in the lensed images (KdB00).

Beside this, the images are separated by only 1.4 arcsec but show a factor $\sim$1.75 difference in their rms variabilities (point 1). This either results from a difference of a factor $\sim$3.1 in their Galactic scattering measures, or a 90-μas scatter-broadening of image B in the lens galaxy at $8.5\times(1+z_{\rm s})\approx$12 GHz (KdB00). This requires a scattering measure $\sim$1 kpc m$^{-20/3}$ in the lens galaxy, which is very large compared to typical lines-of-sight through our Galaxy. Scatter-broadening at a lower level can not be excluded, however.

Finally, in the case of scintillation the observed rms variability should increase towards longer wavelenghts if the source is extended (i.e. larger than the scattering disk). In the case of B1600+434, however, we find that the rms variability decreases by almost a factor $\sim$3 between 5 and 1.4 GHz (KdB00; point 5). One might argue that the source contains a compact component that scintillates *and* a more extended component that does not vary. The compact component would show less variability at 1.4 Ghz than at 5 GHz, because it would be in the strong-scattering regime. However, the time-scale of variability at 5 GHz for this component would remain only a few hours, whereas the 12-h WSRT light curves show no evidence for this (point 4).

Overall, we conclude that the observational evidence does not seem to favor scintillation as the explanation of the observed external variability in B1600+434 (KdB00). To save this hypothesis a considerable rethinking of the properties of the Galactic ionised medium is required, at least in the direction of B1600+434.

4.2. Radio-microlensing

How about radio-microlensing? There are different "ingredients" that play a role:

- **Source Structure:** We assume that the source consist of a stationary core with a jet structure. This jet contains (several) condensations, either physical "bullets" or shock-fronts. These jet components move with near- or super-luminal motion and are compact (μas scale), especially near the core. Components further along the jet might grow in size.

- **Compact Objects in the Lens Galaxy:** The lens galaxy consists (partly) of massive compact objects. These objects create a magnification pattern on the source-plane, if they constitute a considerable fraction of the lens-galaxy mass inside a radius of 0.5–1 arcsec.

When these two "ingredients" are combined, one will have jet-components that move over the magnfication pattern, being continuously magnified and/or demagnified, resulting in variability in the integrated flux density of the lensed images. The time-scale for these jet components to cross the Einstein diameter of a point-mass is

$$\Delta t \sim 4\,(1 + z_\mathrm{s})\,\beta_\mathrm{app}^{-1} \cdot \sqrt{(M/M_\odot)}\sqrt{(D_\mathrm{ds}D_\mathrm{s}/D_\mathrm{d}\ \mathrm{Gpc})} \quad \mathrm{weeks},$$

where z_s is the source redshift, β_app is the apparent jet-component velocity in units of c, M is the lens mass, and D_ds, D_s and D_d are the angular diameter distances between lens-source, observer-source and observer lens, respectively. More complex microlensing simulations show similar time-scales of the order of weeks to months for typical jet velocities of a few time c and compact-object masses of $\sim 1\,M_\odot$. The longer observed variability time-scales for B1600+434 are therefore consistent with the radio-microlensing hypothesis (KdB00).

The difference in rms variability between the lensed images can be explained either by a difference in the compact-object mass function in their respective lines-of-sight and/or by moderate scatter-broadening ($\gtrsim$few μas) of the image passing through the disk/bulge. If for example the average mass of compact objects in the disk/bulge is much lower than that for the halo line-of-sight, the resulting magnification pattern will show a more dense caustic network. This reduces the expected rms variability for a given angular source size (e.g. KdB00).

The strongest argument in favor of microlensing, however, is the frequency dependence of the observed rms variability. Using the constraints on the jet-structure that we derived from the 8.5-GHz VLA observations in 1998, we predict a decrease in modulation index from 5 to 1.4 GHz by a factor $\sim$3, if we assume

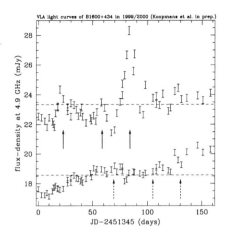

Figure 3. Preliminary results (at 5 GHz) from the 1999/2000 VLA monitoring campaign of B1600+434. The upper light curve (image A) passes through the dark-matter halo of the edge-on spiral lens galaxy (Fig.1). Note several strong (up to 30%) events (solid arrows) in the upper lightcurve and the complete absence of these events in the lower light curve (image B) after the time delay of ∼47 days (dashed arrows).

that the jet-components are synchrotron self-absorbed (i.e. grow linearly with wavelength). This decrease has indeed been observed in the independent WSRT observations from 1998/1999 (Section 4; KdB00).

5. A caustic crossing in the radio?

During the 1999/2000 VLA A- and B-array configurations we have been monitoring B1600+434 at 8.5, 5 and 1.4 GHz (Koopmans et al. in prep). Some very preliminary results at 5 GHz are shown in Fig. 3. These data show several strong "events" (10–30%) at both 5 and 8.5 GHz, whereas they are absent in the 1.4-GHz light curve. What is their origin? All distinct events in image A do not re-occur in image B (after the time-delay of ∼47 days) and are therefore *not* intrinsic source variability.

Can the strongest feature (around day 80) be an "extreme scattering event" (ESE; Fiedler et al. 1987)? Plasma clouds in our Galaxy move with typically ∼30 km/s. Event durations of several weeks then imply an angular size for these clouds of ∼1 mas, if they are located at ∼0.5 kpc distance from us, thereby covering the entire source. An event of ∼30% at 5 GHz would then be detectable at 1.4 GHz as well, whereas no evidence for this is seen in the VLA 1.4-GHz light curves. Moreover, these very rare ESEs are not expected to occur several times in the same light curve of image A within a time span of only ∼150 days. Similarly, also scintillation does not show this extreme behavior in amplitude over time scales of several weeks (see Section 4.1).

Rather, we think that this event is caused by a μas-scale jet component in the lensed source, which has recently been ejected from the core. This component subsequently moves over a single caustic in the magnification pattern that is created by compact objects in the line-of-sight towards lensed image A, causing observable variability in the integrated flux-density of the lensed image.

6. Conclusions

The CLASS gravitational lens B1600+434 shows strong "external" variability in the VLA 5 and 8.5-GHz light curves of lensed image A, which passes predominantly through the halo of the lens galaxy. Neither Galactic scintillation nor Extreme Scattering Events (ESE) can satisfactorily explain all the observations. Only "radio microlensing" can explain the observed variability time-scales and amplitudes, as well as its frequency dependence, without invoking extreme assumptions. It also offers a natural explanation for the strong ($\sim$30%) event that we more recently observed, as being a "radio-microlensing caustic crossing".

Based on the 1998 VLA and 1998/9 WSRT data, we already attempted to place constraints on the mass-function of compact objects in the lens-galaxy halo, finding a lower limit of $\gtrsim 0.5\,M_\odot$ (KdB00). However, these constraints are still weak and depend strongly on assumptions about scatter-broadening in image B. The way foreward is to use *only* the lightcurve of image A to put constraints on the mass function in its line-of-sight. With the detection of distinct radio-microlensing events, this task has become a realistic possibility. We have started to work on this (Koopmans et al. in prep) and hopefully we will soon be able to place significantly better constraints on the mass function of compact objects in the lens-galaxy halo, bringing cosmological radio-microlensing up to Galactic standards!

Acknowledgments. We thank Roger Blandford, Konrad Kuijken, Jean-Pierre Marquart, Penny Sackett and Jane Dennett-Thorpe for critical discussions during the course of this work. We also thank the participants of the Microlensing 2000 conference for positive feedback. LVEK and AGdeB acknowledge the support from an NWO program subsidy (grant number 781-76-101). This research was supported in part by the European Commission, TMR Program, Research Network Contract ERBFMRXCT96-0034 'CERES'.

References

Browne, et al., 1997, in Observational Cosmology with the New Radio Surveys, eds. M. Bremer and Jackson, N. J., p305

Fiedler, R. L., Dennison, B., Johnston, K. J. & Hewish, A. 1987, Nature, 326, 675

Irwin, M. J., Hewett, P. C., Corrigan, R. T., Jedrzejewski, R. I. & Webster, R. L. 1989, AJ, 98, 1989

Jackson, N., et al. 1995, MNRAS, 274, L25

Koopmans, L. V. E., de Bruyn, A. G. & Jackson, N. 1998, MNRAS, 295, 534

Koopmans, L. V. E., de Bruyn, A.G., Xanthopoulos, E., Fassnacht, C.D. 2000, A&A, 356, 391

Koopmans, L. V. E. & de Bruyn, A. G. 2000, A&A, in press (astro-ph/0004112) [KdB00]

Lähteenmäki, A., Valtaoja, E. & Wiik, K. 1999, ApJ, 511, 112

Narayan, R. 1992, Phil. Trans. Roy. Soc., 341, 151

Pelt, J., Kayser, R., Refsdal, S. & Schramm, T. 1996, A&A, 305, 97

Press, W. H., Rybicki, G. B. & Hewitt, J. N. 1992, ApJ, 385, 416
Taylor, J. H. & Cordes, J. M. 1993, ApJ, 411, 674
Vermeulen, R. C. & Cohen, M. H. 1994, ApJ, 430, 467

Discussion

Hansen: Based on optical images of the lensing galaxy, can you cast your lensing population in terms of a mass-to-light ratio? Given the inferred ~ 1 $M_\odot$ it would help to distinguish stars from remnants.

Koopmans: The surface-density ratio derived from our microlensing model is of order 4-5. Although I have not determined the M/L ratios at the positions of the lens images, the surface brightness falls off exponentially, where we assume the mass model falls off isothermally. The M/L ratio thus increases as a function of height above the disk. In any case the M/L ratio is significantly higher near the image passing through the halo and the high average mass of compact objects there would imply a non-stellar nature.

Binney: Can you clarify your picture of why image B, which passes through the galaxy is so much more constant than image A? Is the dominant effect averaging over many caustics or scattering in the interstellar medium of the lensing galaxy?

Koopmans: If one assumes that all external variability is dominated by microlensing, then the "low" observed rms variability of image B (passing through the disk or bulge) implies a lower limit on the source component, given the higher density of caustics near that image. This directly places a limit on the separation of caustics near the other image, in order to explain *its* observed rms variability. This basically places a lower limit on the average mass of compact objects in the halo line of sight. If scatter broadening *is* important the image passing through the disk/bulge is larger, thereby weakening the lower limit on the average mass of compact objects, possibly bringing it down. We assume no scatter broadening, but have done a recent global VLBI experiment that will place an upper limit on the scattering disk of the bulge/halo image 10 times more stringent than available now.

Graff: What mass function was used for the B image? What is the allowed fraction of the halo in MACHOs? How firm is the $M > 1$ $M_\odot$ estimate? Is 0.8 ruled out?

Koopmans: We assumed single-mass (delta-function) mass functions. The difference with power-law mass functions is negligible at the level of our current models and observations. The allowed fraction can be lowered from 100% to probably $\sim$20-30%, but that might imply a somewhat higher lower-mass limit on the objects in the halo line-of-sight. If it is still lower, the caustic network might break up, making it much more difficult to explain the high rate of microlensing events. The lower limit is at present not very firm and a more careful statistical analysis will shed more light on this, after we have taken all uncertainties into account.

Telescopes of the Future

R. Gilmozzi

Paranal Observatory, ESO, Alonso de Cordova 3107, Vitacura, Casilla 19001, Santiago

Abstract. The next generation of ground based telescopes will break the 20th century paradigm of the "factor of two" diameter increase. Taking adantage of the enormous advances in technology that the present generation of 8-10m telescopes has fostered, they will be fully adaptive, fully stearable behemoths of up to 100m diameter performing at the diffraction limit in the optical and near infrared.

At ten times the collecting area of every telescope ever built put together, they will have limiting magnitudes of 37-38, angular resolutions of 1-2 milliarcseconds, and a price tag that does **not** follow the historical $D^{2.6}$ cost law.

I will discuss some of the possible science cases (ranging from the determination of H [not H_0] unencumbered by local effects, to the study of every SN ever exploded at any $z < 10$, and including spectroscopy of extrasolar planets, studies of ultrahigh frequency phenomena, detection of brown dwarfs in external galaxies, etc) and show the design status of some of the proposed projects, in particular ESO's 100m concept (called OWL, for its sharp night vision and for OverWhelmingly Large telescope).

Microlensing observations with the 4-m International Liquid Mirror Telescope

Jean-François Claeskens[†], Christophe Jean and Jean Surdej[‡]
Institut d'Astrophysique et de Géophysique, Université de Liège, Avenue de Cointe 5, B-4000 Liège, Belgium.
[†] *Chargé de recherches du F.N.R.S. (Belgium)*
[‡] *Directeur de recherches du F.N.R.S. (Belgium)*

Abstract. The 4-m International Liquid Mirror Telescope (ILMT) is a zenithal telescope dedicated to a direct imaging survey in two broad spectral bands. It will be located in the Atacama desert and will become operational two years from now. At such geographical latitudes and due to the rotation of the Earth, the field of view of the ILMT will scan the Galaxy from the Southern Pole to the bulge and the central regions. Very precise photometric and astrometric data of millions of stars will be obtained in the drift scan mode night after night, so that microlensing events will inevitably be detected towards the bulge of the Galaxy.

1. Introduction

It is well known that the surface of a liquid spinning in a constant gravitational field takes the shape of a perfect paraboloid, which can be used as the primary mirror of a telescope. However, such a telescope cannot be pointed in the sky, except to the zenith, and can not even track the stars. Nevertheless, Borra et al. (1993) showed that the image quality required for astronomy could be achieved with such mirrors. On the other hand, technology can circumvent the tracking problem by moving the charges on the CCD at the apparent speed of the stars in the field (this is known as time delay integration (TDI) or drift scan mode). This technique was first demonstrated by Hickson et al. (1994) with a 2.7-m diameter liquid mirror telescope.

Despite the restricted access of the sky to a narrow strip and the fixed exposure time, a liquid mirror telescope remains very attractive! Indeed, first it is much cheaper than a conventional telescope: there is no need for a rotating dome nor for a precise mount, as a simple tripod suffices to support the CCD camera and the lens corrector at the prime focus (see Fig. 1). Secondly, it also allows specific scientific goals to be achieved, thanks to precise and regular photometry and astrometry measurements of each object present in the strip. Detection of microlensing events represents such a potential application.

After briefly presenting the 4-m International Liquid Mirror Telescope project, we discuss the rate of microlensing events expected from such observations.

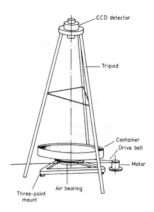

Figure 1.   A liquid mirror telescope.

2. The 4-m International Liquid Mirror Telescope (ILMT)

The 4-m International Liquid Mirror Telescope (ILMT) project is conducted by an international consortium of European, Argentinian, Canadian and Chilean astronomers. Detailed information on the project, the participating institutions and the numerous scientific drivers may be found at the Internet address: http://vela.astro.ulg.ac.be/lmt. Here, we describe some important ILMT specifications in the context of microlensing observations.

ILMT specifications related to microlensing:

- Located in a high quality astronomical site, in the Atacama desert (Chile), at latitude $\varphi \simeq -22.5°$ (El Toco) or $\varphi \simeq -29°$ (La Silla, European Southern Observatory).

- Field of view: $30' \times 30'$. For $\varphi = -29°$, this corresponds to a strip of sky of $\simeq$140 sq. degrees, with $\simeq$ 100 sq. degrees at high galactic latitudes and one crossing of the Galactic plane is very close to the Galactic center.

- Detector: mosaic of thinned high quantum efficiency 2048×2048-pixel CCDs, equivalent to a 4096×4096 chip; scale: $0.4''$/pixel.

- Survey: at least two bands (equivalent to B and R), down to a limiting magnitude of about 23 per scan.

- Effective integration time: the time delayed integration (TDI) is equivalent to $\simeq 140$ sec for one channel or $\simeq 70$ sec for two channels. A semi-classical, on-axis corrector must be installed to remove the TDI distortions.

- On line detection of photometric variations and possible follow up with a small robotic telescope.

- First light at the end of 2001 and operation of the telescope for at least 5 years.

3. Microlensing with the ILMT

3.1. Modellling

Qualitatively, microlensing events are expected within the ILMT observations because the field of view is going to cross the bulge very close to the Galactic center and because each detected star will be monitored every night, for about 120 consecutive nights. This will lead to well-sampled light curves for lensing masses in the range $10^{-3} - 4\,M_\odot$. This mass range is known to be responsible for the majority of the observed microlensing events (e.g. Alcock et al. 1997, 2000).

To obtain more quantitative results, we proceed in two steps:

1. The *star counts* are estimated along each galactic direction (l, b) scanned by the ILMT ($\varphi = -29°$ or $\varphi = -22.5°$), as a function of their distance and magnitude using the Besançon model (Robin et al. 1995), which takes into account the contributions from the halo, the thick and the thin disks; mean galactic extinction is used. A conservative limiting magnitude is computed by requiring a $S/N \geq 10$, as this should allow a significant ($\geq 3\sigma$) detection with the image subtraction technique (Alard & Lupton 1998; Alard 2000) of all lensing events with impact parameters smaller than the Einstein radius. First empirical estimates indicate that the limiting magnitude will be 0.5 mag brighter in the crowded fields.

2. *Modelling the Galaxy:* the Galaxy is assumed to be entirely composed of dark compact objects. The optical depth for microlensing along the direction (l, b) is computed from the Galactic matter density distribution and the latter is estimated by the addition of three components: the *halo* (Griest 1991), the *disk* (Peale 1998; Bahcall & Soneira 1980) and the *bulge* (Peale 1998; Zhao 1996).

3.2. Predicted results

Simulations have been performed for a transverse velocity $v_t = 200$ km/s and a lensing mass $M = 1\,M_\odot$. Observations are assumed to be done in the B band (where stars counts are the lowest) with the 4-m ILMT located either at latitude $\varphi = -29°$ or at $\varphi = -22.5°$ and with a S/N=10, a FWHM=0.8″ and an exposure time $t = 70$s or $t = 140$s. The resulting total number of stars is estimated to be between 15 and 19 million for the different locations and exposure times ($B_{\text{lim}} \simeq 22.5$).

As expected, both the star counts and the optical depth have a maximum towards the bulge and the Galactic center. Consequently, the rate of microlensing events also has a maximum in that direction, as can be seen from its distribution as a function of Right Ascension (or Sidereal Time of the observations) in Figure 2. On the basis of an observing time of 8 hours per night, the efficiency of the ILMT is simply approximated to 33% for every star brighter than B_{lim}.

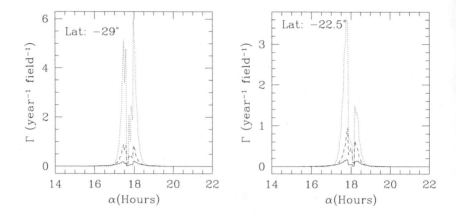

Figure 2. Number of events per year and per field as a function of Right Ascension for La Silla ($\varphi = -29°$; left) and El Toco ($\varphi = -22.5°$; right); $t = 70$ s. Individual contributions are represented for the halo (thick lines), the disk (dashed lines) and the bulge (dotted lines).

The total number of observed microlensing events per year amounts to approximately 30, with 75% due to the bulge, 20% to the disk and only 5% to the halo.

4. Conclusions

Conservative estimates predict the detection of approximately 30 microlensing events per year due to 1 $M_\odot$ lenses among the ILMT observations. By their nature, the latter will scan all southern galactic latitudes b, so that the number of microlensing events as a function of b will probably help in constraining the Galactic structure. The bulk of events is indeed expected towards the bulge.

Finally, let us also note that the ILMT observations might constitute a precious contribution to the detection of microlensing events in the near future when the EROS and MACHO experiments are over.

References

Alard, C. 2000, A&AS, 144, 363
Alard, C., Lupton, R.H. 1998, ApJ, 503, 325
Alcock, C., Allsman, R.A., Alves D. et al. 1997, ApJ, 479, 119
Alcock, C., Allsman, R.A., Alves D. et al. 2000, ApJ, submitted (astro-ph/0001272)
Bahcall, J.N., Soneira, R.M. 1980, ApJS, 44, 73
Borra, E.F., Content, R., Girard, L. 1993, ApJ, 418, 943
Griest, K. 1991, ApJ, 366, 412

Hickson, P., Borra, E.F., Cabanac, R. et al. 1994, ApJ, 436, L201
Peale, S.J. 1998, ApJ, 509, 177
Robin, A., Haywood, M., Gazelle, F. et al. 1995,
   http://WWW.obs-besancon.fr/www/modele/modele_ang.html
Zhao, H.S. 1996, MNRAS, 283, 149

Discussion

Graff: Since your telescope is so cheap, it seems you could easily build more, or larger ones, to cover a wider area and go dimmer.

Claeskens: Yes, of course!! This technique may be applied by individual groups or institutions, with dedicated scientific projects. However, considering area and depth the quantity of data is already 10 Gbytes per night and co-addition of frames will allow us to reach 26th magnitude for extragalactic research. On the other hand, a better corrector lens will soon provide a corrected field of view of a few tens of degrees.... So, a single LMT is already a very good source of data.

Vermaak: Is there any fundamental limit to the size of the "mirror"?

Claeskens: Not to my knowledge, at least for the 6- to 8-m class. A 4-m mirror is the largest for which technology already exists (container, air-bearing), so that no R&D is needed.

Zinnecker: Did you talk to ESO to get permission to put your LMT on La Silla (or El Toco)?

Claeskens: Yes, contacts with ESO have been established to get permission to install the LMT at La Silla. We should receive a definite answer in the very near future.

Monitoring Light Variations from Space with the OMC

Alvaro Giménez

LAEFF, INTA-CSIC, Apartado 50.727, 28080 Madrid, Spain,
and
Instituto de Astrofísica de Andalucia, CSIC, Apartado 3.004, 18080 Granada, Spain

Abstract. The OMC is a small optical instrument to fly onboard the European high-energy mission INTEGRAL in less than 2 years' time. A refractive optical system with an aperture of only 50 mm, focused on a large format CCD working in frame transfer mode, will allow the study of variations in the V band of high-energy objects simultaneously with the hard X-ray and gamma-ray domains. In addition, a large number of serendipitous sources will be observed allowing the study of objects emitting only in the optical range with short and long periodicities. The main scientific performance of the OMC and the importance of massive photometric surveys to add candidates to its input catalogue are briefly reviewed.

1. Introduction

The purpose of this presentation is to point out, once more, the importance of some time neglected scientific by-products to other areas of Astrophysics. The Optical Monitoring Camera, OMC, is certainly not intended for microlensing studies. It is an optical monitor for the European high-energy mission INTEGRAL which will observe in the energy range between 4 KeV and 20 MeV. Nevertheless, the OMC is a good example of serendipitous science since it will provide a tool for astronomers studying variable stars. In addition to monitoring the optical light of high-energy sources, the OMC will follow the light variations of many other optical sources simultaneously with an X-ray monitor which has the same field of view. The two main instruments, an imager and a spectrometer, together with the X-ray and optical monitors are coaligned and thus observing the same sources in different wavelength ranges and sensitivities.

Unfortunately, only previously selected targets can be monitored and therefore the importance of building up an input catalogue. The OMC will not in principle find new objects. It can only download around 1% of each image during normal operations. It will therefore use predefined positions to download windows of 11 x 11 pixels around some selected 100 objects (or 7 x 7 pixels in 200 objects). For such a purpose, the input catalogue of the OMC contains a list of 130,000 high-energy objects and 105,000 known or suspected optical variables as well as 150,000 reference stars for photometric calibration.

Microlensing surveys, on the other hand, have an important role to play in the discovery of new and interesting variable objects that should be later studied in detail either from space or with ground-based telescopes. Targets discovered in crowded fields will not be useful for the OMC but for ground-based telescopes. The OMC nevertheless will benefit from ongoing massive surveys using wide field cameras searching for planetary transits, optical transients and optical counterparts of GRBs.

2. The Optical Monitor on board INTEGRAL

The OMC is designed to monitor, for extended periods of time, the optical emission of high-energy targets within its field of view, simultaneously with the gamma-ray instruments. This will allow the collection of optical light curves for comparison of variability patterns with the hard X-ray and gamma-ray measurements as well as detailed studies of the multiwavelength behaviour of the sources. It will also provide simultaneous calibrated V photometry for the comparison of levels of activity of the targets with previous or future ground-based optical observations. The OMC will also analyse and locate the optical counterparts of high energy transients. In addition, it will monitor many other optically variable sources within the field of view that may require long uninterrupted observations for their physical understanding, thanks to the large field of view covered.

The design of the OMC is adapted to its objectives as a monitor for high-energy instruments, with very large fields of view and relatively poor angular resolution. A wide FOV, of 5 x 5 square degrees, ensures continuous monitoring and a large format CCD (2048 x 1024 pixels), working in frame transfer mode and passively cooled to -90 C, provides good sensitivity despite the small aperture of the optical system (5 cm). The photometric performances of the OMC will cover a magnitude range from around magnitude 7, with mmag precision, to around 18, with precisions of the order of tenths of a magnitude. Standard integration time for each image will be 1 minute but it can be changed to any value between 1 and 100 seconds. Larger integrations will be obtained through addition of individual measurements. With standard integrations, the OMC will reach the same photometric precision as the Hipparcos astrometric satellite but at a level 2-3 magnitudes fainter. The optical design focuses stellar light within one pixel to minimize source confusion. Each pixel is 13 microns in size corresponding to 17.6 arcsec. The limiting magnitude for a 3σ detection in 10 integrations of 100 sec is V = 18.5. Close to the Galactic plane nevertheless the limiting magnitude will be set by source confusion to approximately 17.5 magnitude.

On-board calibration of the CCD will be performed by means of LEDs installed in the optical cavity, while absolute photometric calibration will be done through the continuous measurement of standard reference stars within the field of view.

3. Microlensing and Eclipsing Binaries

As mentioned, the OMC can monitor a variety of optical sources with no high-energy counterpart. In the case of the now very popular field of extrasolar planets, for example, stars can be monitored to mmag precision looking for transits during intervals longer than the orbital period in most known cases with giant planets. Moreover, solar-type stars can be monitored for activity cycles, if any, spots or even flares.

But my favourite area of research is binary stars for the accurate determination of absolute dimensions. For this purpose, of course, we need double-lined eclipsing binaries. The OMC will monitor many of these stars and provide light curves, better periods, and in particular information about variations due to activity in the component stars.

The number of known well-detached eclipsing binaries is nevertheless still very small, providing reasonable statistics only within the main sequence for stars between 1 and 15 $M_\odot$. Outside this range the cases studied are almost nonexistent. There are two specific problems that need further discoveries, for which microlensing surveys can help: looking for eclipsing binaries in the Magellanic Clouds and M31, or within our Galaxy, searching for evolved eclipsing binaries and massive stars or late type dwarfs.

Binaries in other galaxies provide measurements of distance independent of calibrations and lead to a better knowledge of the important distance scale in the neighborhood of our Galaxy. Moreover, accurate absolute dimensions allow the study of stellar structure models in different chemical environments and specially for low metal abundances (SMC). If we look for binary systems in our own Galaxy, we badly need, as mentioned, new candidates in the most evolved parts of the HR diagram, to check in particular how relevant may be overshooting in convective stellar cores. For the calibration of the scale of temperatures, we still lack a good number of well-behaved eclipsing binaries in the hottest part of the main sequence, beyond 15-20 $M_\odot$, as well as in the lower mass end, for stars lighter than the Sun, specially below 0.5 $M_\odot$.

Getting accurate absolute dimensions requires a lot of observing time in order to obtain complete high-precision multicolour light curves as well as radial velocity curves. Therefore, we want to observe the good candidates, not just a randomly selected sample. But, how do we get more candidates, either for the OMC input catalogue or for extended ground-based observations? For this purpose, astronomers interested in detached eclipsing binaries follow the work of other people doing photometric surveys with different aims.

In the old times, photographic surveys searching for Cepheids provided many eclipsing binaries, but the quality of the light curves was not good enough and most follow-up detailed studies based on these surveys failed. In the study of eclipsing binaries in the Magellanic Clouds it was found some years ago that, after long observing campaigns, many of the selected objects were actually semidetached systems and thus not good for comparison with evolutionary models. More recently, some wide-field photometric surveys with CCD cameras have been started to search for transits of planets in front of their host star or looking for the optical counterparts of transient events, like GRBs. Few results on binary stars are yet available from these cameras. On the other hand, the Hipparcos satellite has surveyed the photometric behaviour of a large sample of bright stars, with a limit around V = 10. In fact it was proven how poor our previous knowledge was since 30% of all the eclipsing binaries, and 50% of the variable stars, were new discoveries even for such bright stars. The OMC will follow up all these sources but can go significantly fainter, and the sample is still not large enough to find good candidates in the missing parts of the HR diagram mentioned above.

Microlensing surveys have now discovered thousands of new eclipsing binaries providing a perfect data sample to be harvested if enough details of the available measurements are published. In fact, microlensing surveys produce an unbiased selection of eclipsing binaries with light curves of a quality which is adequate enough for the rejection of uninteresting systems. Obviously, crowded regions have been selected in the Magellanic Clouds, the Galactic bulge, and even M31. This implies a difficulty for follow-up observations with wide-field cameras like the OMC but not for detailed ground-based studies. Telescopes used for the detection of optical transients, like GRBs, are better suited for the OMC purposes, also because of the link with gamma-ray sources. But this is just an example of how different research activities can help each other in their scientific objectives. In summary, variable stars observers behave like predators, looking for unused material thrown away by others. Before microlensing studies, the total number of known eclipsing binaries was not larger than 3,000 objects, 200 of them outside our Galaxy. These numbers have been completely changed with results from microlensing studies.

Telescope Design, Instrumentation and Status of SALT

R. S. Stobie and D. O'Donoghue

South African Astronomical Observatory, PO Box 9, Observatory 7935, South Africa

D. A. H. Buckley and K. Meiring

SALT Project, PO Box 9, Observatory 7935, South Africa

Abstract. SALT is a 10-m class telescope for optical/infrared astronomy to be sited at Sutherland, the observing station of the South African Astronomical Observatory. This telescope will be based on the principle of the Hobby-Eberly Telescope (HET) at McDonald Observatory, Texas. This cost-effective design is a tilted-Arecibo concept with a segmented spherical primary mirror of diameter 11 metres. The telescope has a fixed gravity vector but with full 360 degrees rotation in azimuth. A spherical aberration corrector mounted on a tracker beam at the prime focus enables a celestial object to be followed for 12 degrees across the sky. The SALT design enables over 70% of the sky to be accessed for about 20% of the cost of a conventional telescope of similar aperture. The telescope will be used primarily for spectroscopic studies of celestial objects with a low-dispersion imaging spectrograph mounted at the prime focus and other higher-dispersion instruments fibre-fed and mounted in an environmentally controlled basement. The concept design for SALT is presented with emphasis on the design changes between HET and SALT and the status of the SALT project is described.

1. Introduction

South Africa, together with international partners, plans to construct the Southern African Large Telescope (SALT), a 10-m class telescope for optical/infrared astronomy. SALT will be based on the design of the Hobby-Eberly Telescope (HET), recently completed at McDonald Observatory, Texas. The HET is a 10-m class telescope that represents a radical departure in design of optical/infrared telescope. The design is a tilted-Arecibo concept with a segmented spherical primary of diameter 11m and spherical aberration corrector mounted on a tracker beam at the prime focus. The southern hemisphere equivalent of HET, to be sited at the Sutherland outstation of the South African Astronomical Observatory, will give South Africa and partners access to the (single) most powerful optical/infrared telescope in the southern hemipshere.

The SALT telescope has been designed primarily as a spectroscopic survey instrument but will have enhanced imaging capability compared to HET. It has a scientific scope that will extend from searching for planets around neighbouring stars to the study of the most distant objects in the Universe. Because of a conscious decision not to build a general purpose telescope but rather a telescope with specialised spectroscopic capabilities, the whole design is extremely cost-effective. The estimated cost of constructing the SALT telescope/facility (excluding instrumentation) is US$ 16.5 million, about one-fifth the cost of a general-purpose telescope of similar aperture.

The SALT partnership includes South Africa, the HET Board, Poland, Rutgers University, Göttingen University, Wisconsin University, Carnegie Mellon University and University of Canterbury, New Zealand. A major milestone in the SALT project occurred on 1 June 1998 when the South African Minister of Arts, Culture, Science and Technology announced in Parliament that the South African Cabinet had approved the construction of SALT on condition that sufficient international funding was forthcoming. The initial proposal was that the project be funded 50:50 by South Africa : International Partners. By 25 November 1999 sufficient international funding had been committed that the Minister was able to announce the "green light" for SALT, thus enabling the construction of SALT to proceed. As the project is not quite fully funded a number of other potential partners are in the process of actively fund raising.

2. Design of SALT

2.1. General Description

SALT will be the southern hemisphere equivalent of the HET at McDonald Observatory, Texas. The HET has been described in detail in a number of publications (Ramsey, Sebring & Sneden 1994, Sebring et al. 1994, Sebring and Ramsey 1997, Ramsey et al. 1998). Changes to the HET design, however, will be incorporated as a result of scientific requirements, advances in technology, or modifications as a result of the HET experience.

Figure 1 gives an overall view of the proposed telescope facility. The mounting is based on a tilted-Arecibo concept, with the telescope at a fixed angle to the zenith and full 360° rotation in azimuth. The rotation is achieved through an air bearing system with a precision encoder on the central azimuth pintle bearing. During an observation the telescope remains stationary and a tracker beam at the top end enables an object to be followed for 12° across the sky. Thus an object can only be observed in an annulus 12° wide, centred on the tilt angle of the telescope. As a result of the fixed tilt, the telescope and primary mirror system have a constant gravity vector, thus considerably reducing the demands on the support structure.

The angle of tilt of the HET is 35° to the zenith. At the latitude of Sutherland (32°22'46" S) and with the ±6° travel of the tracker beam, this means that the southerly limit of the telescope would be 73°22' S. This will accommodate the Large Magellanic Cloud and 47 Tucanae, but unfortunately passes nearly through the middle of the Small Magellanic Cloud (SMC). With this tilt the southern half of the SMC would be inaccessible. As the SMC is likely to be one of the prime targets for SALT, the angle of tilt in the baseline design has been changed from 35° to 37° in order to accommodate most of the SMC. A preliminary study has shown that only a relatively small modification to the structure

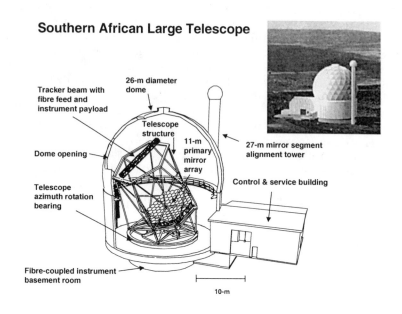

Figure 1. The planned SALT telescope facility, illustrating (1) the telescope structure, (2) the spherical primary mirror consisting of 91 hexagonal segments, (3) the tracker beam that contains the spherical aberration corrector and prime focus instrument platform, and (4) the alignment tower that houses the primary mirror alignment system located at the centre of curvature of the primary mirror array. The inset is a rendering of how SALT may look at Sutherland.

of the telescope is needed to increase the tilt to 37°. This change in tilt enables celestial objects to be accessed in the declination range $-75°22' < \delta < +10°37'$. The HET at McDonald Observatory can access objects in the declination range $-10°20' < \delta < +71°40'$. Thus there is a declination zone of 20° centred on the equator wherein both SALT and HET can observe objects.

The 12° wide annulus of accessibility means that in the worst case on the celestial equator an object can be followed for up to 48 minutes. This observation time, however, increases as one moves away from the equator and, scaling as sec δ, can exceed 90 minutes for declinations south of $-58°$. In addition towards the southern and northern limits it is possible for increased observation time beyond this by allowing for repeated moves in azimuth and re-acquiring the object.

2.2. Primary Mirror and Alignment System

The primary mirror, of spherical shape, has a maximum diameter of 11m and consists of 91 hexagonal segments each of 1m inscribed diameter and radius of curvature 26.165m. In the HET design the maximum diameter of primary

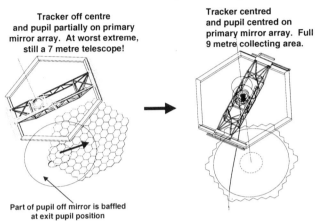

Figure 2. The tracker beam assembly and image pupil on the primary mirror. During an observation the pupil traverses the primary mirror. With the SALT design of spherical aberration corrector it is possible to increase this pupil size to include the whole primary mirror array when the tracker is centred.

that can be imaged is 9.2m. During an observation the image pupil traverses the primary mirror and in the worst case (6° off axis) the primary area imaged corresponds to a 7-m telescope (Fig. 2). On SALT with the improved design of spherical aberration corrector (O'Donoghue 2000) it is possible for the whole primary mirror array to be imaged, at least during the central part of the tracker motion. The total reflecting area of the primary mirror is 77.84m^2, equivalent to a monolithic primary of diameter 9.96m. With this larger pupil size the design of the baffle (to exclude stray light off the mirror) is no more complex; however, the requirements on the design of spherical aberration corrector are more severe and careful consideration will have to be given to vignetting and image degradation effects.

The primary mirror substrate will be designed from low (zero) - expansion coefficient glass, such as Corning ULE or Schott Zerodur. Each mirror segment rests on 9 points of support and has three actuators for tip/tilt and piston alignment. The whole primary mirror is supported on a tensioned steel space frame truss. At the present time one of the main limiting factors in HET image quality is the distortion of this steel truss with temperature. In the HET baseline design, edge sensors were excluded for reasons of cost and it was planned to model the deformation of the truss with temperature. However, with experience this has proved difficult and as a consequence SALT will include edge sensors in its baseline design. Once the primary mirror is aligned by the centre of curvature alignment system the edge sensors take over and maintain the alignment.

The primary mirror truss is supported relative to the telescope structure by a 3-point kinematic mount. This ensures that stresses induced by the telescope structure are not transmitted to the mirror truss. Although the baseline design for the primary mirror support structure is a steel truss, other possible designs are being investigated. If the whole mirror and mirror support structure can be manufactured from low expansion material, then this could reduce by an order of magnitude the requirements on the maintenance of the segment alignment.

The initial alignment of the primary mirror on the HET was designed to be achieved using a lateral shearing interferometer located at the centre of curvature of the primary mirror. Although this system has the required resolution, difficulties have been experienced with its limited capture range. SALT plans to use a Shack-Hartmann sensor camera that will have the desired capture range and resolution.

Initially the primary mirror array will not be phased and consequently the diffraction limit of the telescope will be that of a single segment. Thus the piston of each segment has a loose tolerance (25 μm) but the tip/tilt of each segment is critical (0.0625 arcsecs) and the radius of curvature of each spherical segment is matched to a high degree (0.5 mm). In the longer term it is important to design SALT so that the possibility of phasing the mirror array is not excluded and thus achieve the diffraction limit of the complete primary mirror with adaptive optics.

2.3. Tracker and Spherical Aberration Corrector

Much of the complexity of the telescope is concentrated in the tracker beam assembly. The design of the spherical aberration corrector (SAC) for SALT represents a substantial improvement over that for the HET. The details of the SAC design are contained in the paper by O'Donoghue (2000). In brief, the SAC is a 4-element corrector comprising three conics and a general asphere, with the largest optic of diameter 0.5m (Fig. 3). The resulting image quality of less than 0.25 arcsec over an 8 arcmin field is superior to that produced by HET over its 4 arcmin field of view. The precision with which the SAC has to be moved in x,y,z in order to follow the focal sphere is demanding and at all times the SAC must be tilted in θ, ϕ to maintain its optical axis in alignment with the centre of curvature of the primary mirror. An addition to these motions, the SAC and the whole prime focus instrument platform can be rotated to compensate for the rotation of the field of view during an exposure. The performance of the tracker, and the quality of the primary mirror alignment will in the end determine the image quality over the duration of the observation.

The HET mirror coating is over-coated silver as this provides a superior reflectivity over all wavelengths (in comparison to aluminium) except for the bluest wavelengths (< 380 nm). This is especially important, as with the primary and the four-element reflective corrector, there are five surfaces before the light reaches the focal plane. Consequently any lack of reflectivity is raised to the fifth power. In practice it has been found that the HET coating has decayed more rapidly than expected, probably because the over-coating layer

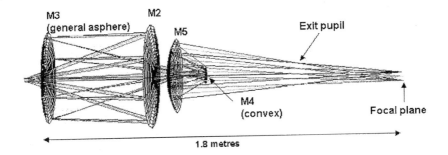

Figure 3. Layout of a design of spherical aberration corrector involving four mirrors. Light enters the system from the left. Three of the mirrors are conics (as indicated) and one is a general asphere (with up to 8th order coefficients). The overall length is less than 1800mm and the largest mirror is 500mm in diameter. The exit pupil of the system is marked.

was thinned too much in order to improve the ultraviolet performance. SALT is investigating using a multi-layer dielectric coating from Lawrence Livermore National Laboratory (Thomas, Wolfe & Farmer 1998) that has excellent ultraviolet response and a very durable over-coating but at the expense of a slightly reduced throughput at other wavelengths.

2.4. Prime Focus Instrument Platform

The Prime Focus Instrument Platform (PFIP) is a key component of the telescope. This system contains the spherical aberration corrector, the atmospheric dispersion corrector, the exit pupil baffle system, the acquisition CCD camera, the guide CCD cameras, and the calibration system (arcs, flatfields). It will also contain the fibre instrument feed that links the prime focus to the medium resolution and high resolution spectrographs in the environmentally controlled instrument room in the basement.

2.5. Image Quality and Environmental Control

The image size error budget from all contributions has a specification of encircled energy EE(80)=0.8 arcsec delivered at the prime focus over the 8 arcmin field of view, and a goal of EE(80)=0.6 arcsec. This specification is designed to avoid significantly degrading the median atmospheric seeing at the location of the telescope. At Sutherland DIMM seeing measurements have revealed that for 95% of the time the seeing (as measured 2m above ground level) is within a range of 0.5-2.0 arcsecs with a median seeing of 0.9 arcsec. Although the performance of the telescope will significantly degrade the best imaging (which only occurs $\sim$10% of the time), it will only degrade the median seeing of 0.9 arcsec to 1.1 arcsec.

To achieve this performance will require careful attention to minimising all sources of heat in the dome, removing unwanted heat effectively, and designing the telescope and dome to enable the system to equilibrate rapidly with the surroundings. In addition to cooling during the day to match the expected night time temperature, natural and/or forced ventilation will be used at night.

3. Science Drivers for SALT

The scientific case for SALT is based primarily on its spectroscopic capability in the optical and near-infrared (0.34 – 2.5 μm) wavelength region. The optical/infrared spectral region lies at the heart of modern observational astronomy. It is particularly rich in physical diagnostics that have developed from the accumulation of over a century of observation, associated laboratory experiments and theoretical work. Much of our understanding of the Universe (stars, galaxies, chemical composition, motions, expansion of the Universe) comes from spectroscopic analysis of celestial objects.

The SALT/HET design maximises the science achievable when the following criteria apply:

- uniformly distributed targets on the sky.

- targets have surface densities of a few degree^{-2} or $\sim$1 arcmin^{-2}. The two figures apply to targets accessible to the tracker (i.e. $\pm 6°$ of tracker motion) or accessible by the multi-object fibres/slitlets over the 8 arcmin field.

- variability studies confined to the time domain < 2 hours or > a day.

- high spatial resolution (i.e. substantially sub-arcsec) not required.

- observations limited to the wavelength domain $340nm < \lambda < 2500nm$ (possibly capable to 320 nm if coatings allow).

- high precision *absolute* photometry not required.

SALT instruments have to be chosen to match both these observing criteria and the science goals. The latter were the subject of a SALT/HET Workshop held in Cape Town (Buckley, 1998). As expected, there was considerable overlap with the science drivers for the HET (Còrdova 1990). Spectroscopy in the visible/near-IR (i.e. 350–1700 nm) will be the major observational mode delivering the best astrophysical returns. Clearly not all observing programmes will be feasible or competitive with other 8-m class telescopes. However, there are many projects that will be particularly well suited to SALT's queue-scheduling *modus operandi* (in fact many can *only* be done using this observing approach). If the science programs for SALT utilise its capabilities to the best possible advantage, then excellent scientific returns will be guaranteed for decades to come. Figure 4 shows the sensitivities for the HET suite of instruments, and similar figures are expected for SALT, at least for the fibre-fed instruments (MRS & HRS).

Key SALT programs discussed to date include: extra-Solar planet searches, stellar population studies in our Galaxy and nearby galaxies (i.e. the Magellanic Clouds and Local Group), extra-galactic surveys (e.g. in support of the *HST*

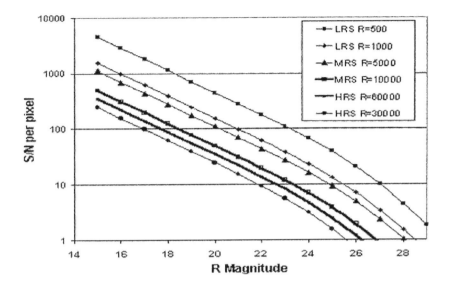

Figure 4. Expected performance of the HET spectrographs (Ramsey 1998). S/N ratio is calculated assuming a 3000-sec exposure in a dark sky with 1 arcsec seeing and assuming 3e$^-$ readout noise for the CCD.

Medium Deep Survey, *XMM* and *Chandra*, improvement in distance scales (e.g. from Cepheids, RR Lyrae stars and Supernovae), synoptic observations of stellar variability, gamma ray bursters, dynamical studies of stellar systems (clusters and galaxies) and characterising the dark matter content of the latter.

4. First-Light SALT Instrumentation

The choice of first-light instruments for SALT should aim to maximize the primary science goals. This choice is currently being discussed by the SALT Science Working Group (SSWG), which has representation from all of the SALT partners. Judging from the science proposals already presented, it seems a suite of efficient spectrographs, with some multi-object capability, and covering resolutions of a few hundred to ~100,000, would fulfill most of these requirements.

Current science drivers call for similar instrumentation to that at the HET, which comprises three spectrographs: two fibre-fed échelles and a prime focus instrument with some imaging capability. It is likely that any low-to-medium dispersion SALT spectrograph would be designed with some multi-object capability and dual visible/near-IR beams extending to at least the Hband (~1.7μm). The échelle approach for fibre-fed instruments is appealing, given the possibilities of near continuous wavelength coverage from the visible to near-IR (e.g. the Medium Resolution Spectrograph on HET will cover the entire spectral range ~400 1300 nm in one exposure), which maximises the information content for a given integration time. The latter is crucial for SALT, given its restricted viewing window.

Following discussion amongst various members of the SSWG, the possibility of developing a prime-focus imaging spectrograph (PFIS) with low-to-medium resolution, multi-object capability, plus Fabry-Perot and polarimetric modes, has been suggested. In comparison to the HET Low Resolution Spectrograph (LRS), the PFIS would have twice the field of view and reach higher spectral resolution. Recent optical technology advances will provide the highest possible efficiency single-order spectroscopy using grisms, volume phase holographic (VPH) gratings and Fabry-Perot etalons. VPH gratings offer high efficiency performance (~90%), and because of their nature, always working at the Bragg condition, can be "tuned" to any particular wavelength by rotating the grating and camera. They have to date been used in very few astronomical spectrographs (e.g. the LDSS++ instrument at the Anglo-Australian Observatory), but plans are underway for using them in several spectrographs currently under construction or in planning stages. With the addition of Fabry-Perot etalons and polarizing optics, the PFIS would be an extremely versatile instrument. Further details on the PFIS and other instrumentation options for SALT are contained in the paper by Buckley et al. (2000).

5. Implementation

5.1. Current Status

As of February 2000 the SALT partnership includes South Africa, HET Board, Poland, Rutgers University, Göttingen University, Wisconsin University and Carnegie Mellon University. The first partner to join SALT was the HET Board which agreed to contribute the HET plans, documentation, software, technical expertise, and HET equipment surplus to HET requirements (e.g. SSAC, UFOE) in return for observing time on SALT. In addition this provides two telescopes with similar capabilities covering the northern and southern hemispheres and enabling joint scientific collaborations between SALT and HET partners.

The present status of partners and committed funding is listed in Table 1.

Table 1. SALT Partnership – Committed Funding

Partner	Construction ($ million)	Operations ($ million)
South Africa	8.0	3.5
Poland	3.0	1.0
Rutgers University	1.2	1.0
Göttingen University	1.3	-
Wisconsin University	1.0	-
Total	14.5	5.5

In addition to the above partners, Carnegie Mellon University is also a founding partner having committed the first $50k down payment. As the cost of the telescope/facility is estimated at $16.5 million (see section 5.2) this means that the telescope/facility is 88% funded. It was on this basis and the fact that a number of other potential partners are actively fund raising that the Minister of Arts, Culture, Science and Technology in South Africa announced the "green light"' for SALT on 25 November 1999. This "green light" meant that the SALT Project Team could be recruited and the construction of SALT could commence.

5.2. Cost Estimate

The design of SALT provides large aperture capability at low cost. A feasibility study for the construction of SALT, maximising South African content, showed that the telescope/facility (excluding instrumentation) could be constructed for $16.474 million (at March 1999 price levels). Based on the HET experience three major instruments (prime focus imaging spectrograph, medium resolution spectrograph, high resolution spectrograph) are costed at $4.8 million. Allowing for a telescope/facility contingency cost of 10% and operations cost of $708k per year for 10 years leads to a total cost of $30 million.

Table 2. SALT Cost Estimate (March 1999)
$ million

Telescope / Facility	16.474
Contingency 10%	1.647
Instrumentation	4.800
Operations cost for 10 yrs	7.079
Total	30.000

These costs will be spread over the 5 year construction time scale of the project and the subsequent 10 years of science operations. Thus international agreements with partners are based on a 15 year time frame. The telescope time awarded to a given partner will depend on that partner's contribution relative to the total cost.

5.3. SALT Project Team and Schedule for SALT

The SALT Board is the autonomous governing body of the SALT Project. It has representatives from all partners in SALT. Two key appointments made by the SALT Board were (1) the SALT Project Scientist, Dr David Buckley, seconded from the South African Astronomical Observatory to the SALT Project from 1 December 1998 and (2) the SALT Project Manager, Mr Kobus Meiring, appointed to the SALT Project from 1 July 1999.

Since the declaration of the "green light" for SALT on 25 November 1999, rapid progress has been made in recruiting the SALT Project Team. The complete team together with their functions in the team are listed in Table 3.

Table 3. SALT Project Team

David Buckley	SALT Project Scientist
Willem Esterhuyse	Manager : Telescope Structure and Dome
Clifford Gumede	Control System Specialist
Nazli Hercules	Admin Assistant
Mariana de Kock	Manager : Facility
Mike Lomberg	SALT Business Manager
Kobus Meiring	SALT Project Manager
Leon Nel	Manager : Tracker and Payload
Gerhard Swart	SALT System Engineer
Arek Swat	Optical Specialist
Jian Swiegers	Manager : Mirror System

The proposed schedule for SALT and the main targets in each of the years are as follows:

- 1999/2000   concept development and site assessment
- 2000/2001   initial procurement and ground breaking
- 2001/2002   component development and installation
- 2002/2003   integration of mirror and tracker system
- 2003/2004   first light and commissioning
- 2004/2005   completion and science operations

References

Buckley, D. A. H. 1998, "Science with SALT", Proceedings of SALT/HET Workshop, 2-6 March 1998, Cape Town, South Africa (Cape Town:SAAO)

Buckley, D. A. H., O'Donoghue, D., Sessions, N. J., & Nordsieck, K. N. 2000, SPIE, 4008, in press

Còrdova, F. A. (ed.) 1990, The Spectroscopic Survey Telescope, Penn State University.

O'Donoghue, D. 2000, SPIE, 4003, in press

Ramsey, L. W., et al. 1998, SPIE, 3352, 34

Ramsey, L. W., Sebring, T. A., & Sneden, C. 1994, SPIE, 2199, 31

Sebring, T. A., Booth, J. A., Good, J. M., Krabbendam, V. L., Ray, F. B., & Ramsey, L. W. 1994, SPIE, 2199, 565

Sebring, T. A., & Ramsey, L. W. 1997, SPIE, 2871, 32

Thomas, N., Wolfe, J., Farmer, J. 1998, SPIE, 3352, 580

Microlensing 2000: A New Era of Microlensing Astrophysics
ASP Conference Series, Vol. 239, 2000
John Menzies and Penny D. Sackett, eds.

Galactic Exoplanet Survey Telescope (GEST): A Proposed Space-Based Microlensing Survey for Terrestrial Extra-Solar Planets

D. Bennett and S. H. Rhie

Physics Department, University of Notre Dame, 225 Nieuwland Science Hall, Notre Dame, IN 46556, USA

Abstract. We present a conceptual design for a space based gravitational microlensing planet search telescope which will be to detect extra solar planets with masses as low as that of Mars, at seperations of 1 AU and larger from their parent stars. This is the only proposed mission that would be sensitive to planets which have been ejected from the planetary systems that they formed in. The microlensing data would be collected by a diffraction limited, wide field imaging telescope of ~ 1.5 m aperture equipped with a large array of red-optimized CCD detectors. Such a system would be able to monitor $\sim 2 \times 10^8$ Galactic bulge stars, and it would discover $\gtrsim 10^4$ microlensing events due to normal stars plus a similar or possibly larger number of events due to planets that have been ejected from their stellar systems. The selected fields would be continuously observed at intervals of 20-30 minutes in order to detect the photometric microlensing signature of extra-solar planets which orbit the faint lensed stars. The number of detected planets will depend on the unknown abundance and the mass distribution of the extra-solar planets. If planetary systems are common but typically have planets of Neptune's mass or less, then GEST would detect ~ 1000 planets over a 3 year mission if we assume that most planetary systems are dominated by planets of about Neptune's' mass. We would also expect to detect about 100 planets of an Earth mass or smaller if such planets are common.

Some Closing Comments

Michael Feast
Department of Astronomy, University of Cape Town, Rondebosch, 7701, South Africa

There is something of a parallel between the way that microlensing has developed and the development of lensing – and I mean lensing with optical telescopes. Lippershey's invention was intially seen as having one major application, and this was essentially military. It was thus quickly taken up by the funding agencies of the day. It required the genius of Galileo to use this instrument to revolutionize astronomy (as well as profiting from the practical application). Microlensing burst on the scene, primarily through the suggestion of Paczyński and its application was seen, especially by physicists, as a way of detecting (or not detecting, as the case may be) dark matter. As such it, too, was taken up by funding agencies. But as with the telescope it has not taken astronomy long to take over. Not, of course, that astronomers are not interested in dark matter, but they are interested in much else as well.

This meeting has been important in taking stock of the present and future of microlensing in relation to astronomy generally, including the problem of dark matter. We have had a number of very clear and comprehensive reviews of topics relevant to microlensing and we have had reviews and other talks dealing with microlensing studies themselves. And these latter papers demonstrated rather clearly that we are still in the exciting "galilean" phase of microlensing studies with, it seemed, many uncertainties in the interpretation of results. This is, naturally, what makes the subject so interesting. Already on the first day we were faced with the question – how many of the events that are recorded are *really* due to microlensing? The problem of variable stars will not go away. Indeed microlensing surveys seem to be turning up novel types of variable stars (e.g., bumpers). Bumpers were only studied because they tend to mimic microlensing events. One wonders whether there are unrecognized types of variable stars in the microlensing survey data that are being passed over because they are obviously not microlensing events and are not being picked out as belonging to one of the known types of variable.

Even when we have convinced ourselves that an event is due to microlensing we are faced with the difficult task of deciding where the lens is located and what it is. In the case of microlensing towards the LMC, it has not been possible to decide to everyone's satisfaction whether the lenses are in our Galaxy or the LMC. And what are these lenses anyway? It is clear that microlensing events are trying to tell us something important – and possibly surprising – about the structure and composition of our Galaxy and/or the Magellanic Clouds. This is stimulating new work by other techniques on Galactic and Magellanic Cloud structure. In the circumstances we should not be too surprised if microlensing towards more distance galaxies threw up some quite unexpected results. The results presented at this meeting for microlensing towards M31, quasars and radio sources are especially impressive and have enormous potential.

As to the nature of most of the lenses, I got the impression that a majority of the participants hoped that they were white dwarfs although it is far from certain that this is what they are.

The complex light curves that are observed and computed for binary and limb-darkened events are quite remarkable. The information that can be obtained from these events is of great interest and importance. In certain areas the information to be obtained is unique. It is clear that this work requires the highest photometric accuracy, and combining observations from different sites requires a great deal of experience and skill. Nevertheless the potential results seem fully worth the effort. The number of events that had been followed intensively, with high accuracy and in several wavebands is still sufficiently small that one might anticipate major surprises if this work is expanded. Detailed spectroscopy of microlensing events remains a great opportunity for the future. One question that does not seem to have been satisfactorily answered yet is how unique a solution we are likely to get even from well sampled events.

The question of planetary detection was intensively debated both in the formal sessions and outside. We know so little about extrasolar planetary systems that it would be unwise to limit microlensing work simply because in certain areas planetary systems can be discovered by other means. For instance there seems no guarantee that the nature or frequency of planets round giant or other stars in the LMC or the Galactic Bulge can be safely inferred from observations of dwarf stars in the solar neighbourhood.

In all the areas discussed at this meeting: the detection of extrasolar planets, the study of stellar atmospheres, the nature of dark matter, the structure of our own and other galaxies (some at great distances) – one has the impression that we are just at the beginning. We are, as I said, in the galilean era of microlensing. The important thing is, like Galileo with his telescope, to use this new tool to the full, both directly and with follow-up observations. So little is known about many of the areas to which microlensing contributes that observation must lead the way. The prospects for surprising discoveries remains high and one can confidently predict that microlensing, or projects derived directly from microlensing studies, will figure largely in the programmes of the telescopes of the future about which we heard on the last day of the meeting.

Author Index

Albrow, M. D. 109, 135
Alcock, C. 244
Allard, F. 175
Allsman, R. A. 244
Alves, D. R. 244
An, J. H. 135
Arenou, F. 91
Aufdenberg, J. 175
Axelrod, T. S. 244

Baron, E. 175
Beaulieu, J. -P. 109, 135
Becker, A. C. 244
Bennett, D. P. 244, **270**, **393**
Binney, J. **231**
Bond, I. **33**, **153**
Bryce, H. M. **195**
Buckley, D. A. H. 382

Caldwell, J. A. R. 109, 135
Carr, B. J. **37**
Claeskens, J. C. **373**
Coleman, I. J. 204
Collaboration, P. 309
Cook, K. H. **27**, **63**, 244
Crotts, A. **318**

Dalal, N. 64
de Bruyn, A. G. 363
DePoy, D. L. 109, 135
Dominik, M. 109, 135
Drake, A. J. 244

Evans, N. W. **299**

Fassnacht, C. D. 363
Feast, M. **254**, **394**
Fischer, D. A. **116**
Freeman, K. C. 244

Gates, E. **271**
Gaudi, B. S. 109, **135**
Geha, M. 244
Gilmozzi, R. **372**
Giménez, A. **130**, **378**
Glicenstein, J.-F. **28**, **261**
Gould, A. **3**, 109, 135, 318
Graff, D. S. **73**

Gray, N. **204**
Grebel, E. K. **280**
Greenhill, J. 109, 135
Griest, K. 64, 244
Gyuk, G. **64**, 271, 318

Halbwachs, J. -L. 91
Han, C. **18**
Hansen, B. M. S. **82**
Hauschildt, P. H. **175**
Hendry, M. A. 195
Hill, K. 109, 135

Jean, C. 373

Kane, S. 109, 135
Kerins, E. **309**
Koopmans, L. V. E. **363**
Kuijken, K. 318

Lasserre, T. **54**
Lehner, M. J. 244

Marshall, S. L. 244
Martin, R. 109, 135
Mayor, M. 91
Meiring, K. 382
Menzies, J. **109**, 135, **164**, **341**
Minniti, D. 244

Naber, R. M. 109
Nelson, C. A. 244

O'Donoghue, D. 382

Peterson, B. A. 244
Pogge, R. W. 109, 135
Pollard, K. R. 109, 135
Popowski, P. **244**
Pratt, M. R. 244

Quinn, P. J. 244

Reid, I. N. **327**
Rhie, S. H. 393

Sackett, P. D. 109, 135, 164, **213**, 318, 341
Sahu, K. C. 109, 135
Schweitzer, T. B. A. 175
Stobie, R. S. **382**

Stubbs, C. W. 244
Surdej, J. 373
Sutherland, W. 244, 318

Tomaney, A. B. 244

Udry, S. **91**
Uglesich, R. 318

Valls-Gabaud, D. 195
Vandehei, T. 244
Vermaak, P. 109, 135, **144**

Wambsganss, J. **351**, 363
Watson, R. 109, 135
Welch, D. 244
Whitelock, P. 254
Widrow, L. 318
Wilkinson, M. I. 299
Williams, A. 109, 135

Yock, P. **160**

Zinnecker, H. **223**

Subject Index

– A –

AGB, 254
amplification, 144, 204
Andromeda
  galaxy, 299, 309
  Halo, 299
associations, 280, 327
astrometric
microlensing, 3, 18, 351
measurements, 91
atmosphere, 82

– B –

B1600+434, 351, 363
Baade's window, 231, 244
Backus-Gilbert, 204
baryonic dark matter, 37, 73
beige dwarfs, 73, 82
Big Bang nucleosynthesis, 37
binary
  eclipsing, 378
  frequency, 91, 164, 223
  lens, 3, 18, 109, 144, 213, 261
  rotation, 109
  spectroscopic, 91
  stars, 91
black hole, 37, 164, 270, 341
blending, 54, 144, 244
bright lens, 18
brown dwarfs, 37, 91, 164, 327, 341
bumper, 341

– C –

carbon stars, 73
caustic, 3, 144, 213, 351
  crossing, 3, 109, 213, 351, 363
  central, 144
centroid shift, 18
τ Cet, 116
Chang-Refsdal lens, 3
chemical enrichment, 73
cloud formation, 175
cold clouds, 37

compact objects, 37, 54
CORALIE, 91
CORAVEL, 91
COROT, 164
cosmological microlensing, 3, 351, 363
cusp, 213

– D –

dark lens, 18
dark matter, 73, 164, 271, 280, 363, 394
degeneracy, 3, 18, 204
detection
  efficiency, 54, 244, 261
  rate, 160
  threshold, 144
difference image analysis, 33, 244
direct opacity sampling, 175
Doppler technique, 116
drift scan, 373
dust formation, 175
dwarf galaxies, 280
Dyson spheres, 37, 116

– E –

eccentricity distribution, 91, 116
Einstein
  crossing time, 261, 351
  radius, 3, 351
  ring, 3, 135, 341
  timescale, 3, 309
ϵ CMa, 175
equation of state, 175
EROS, 28, 54, 73, 164, 261, 299, 309
EROS 99-BLG-1, 28
EROS trigger, 28
EROS1-LMC-1, 54
EROS1-LMC-2, 54
EROS2-LMC-3, 54
EROS2-LMC-4, 54
EROS2-SMC-97-1, 54
extended protodisk, 271

– F –

FAME, 91, 164
femtolensing, 3
finite source effects, 144, 195, 184, 204, 213
flat-spectrum sources, 363

– G –

GAIA, 91
Galactic
    bar, 231, 254
    Bulge, 28, 33, 231, 254, 394
    Centre, 28, 164, 231, 261
    Disk, 28, 261, 327
    halo, 54, 64, 73, 82, 271
    structure, 231, 254, 341
galilean phase, 394
Galileo, 394
γ Nor, 261
γ Sct, 261
gas cloud, 280, 341
gas content, 280
GEST, 393
Gl 229B, 327
Gl 644C (VB8), 327
GMAN, 270
gravitational imaging, 195

– H –

halo
    mass, 54, 73
    subdwarfs, 327
HD12661, 116
HET, 382
high magnification events, 135, 144, 153, 160
Hipparcos, 254
HST, 164, 244, 318, 327
Hubble Deep Field, 82, 271
Huchra's Lens, 3

– I –

ILMT, 373
image subtraction, 33, 153
IMF, 271

impact parameter, 244
infrared photometry, 164
INT, 309
INTEGRAL, 378
interferometry, 164
inverse problem, 204
irradiation, 175

– K –

Kepler, 164
kinematics, 254

– L –

L dwarfs, 175, 327, 341
lens mass, 37
lensing zone, 135, 144
Leo I, 299
LHS 3250, 82
life, 164
limb darkening, 109, 175, 195, 204, 213
limb polarization, 204
line blanketing, 175
liquid mirror telescope, 373
LMC, 28, 37, 54, 63, 64, 73, 164, 254, 271, 299, 309, 394
Local group, 280
long duration events, 244, 341

– M –

2MASS, 327
M dwarfs, 164, 175, 327
M31, 37, 164, 309, 318, 341, 394
MACHO, 28, 54, 63, 64, 73, 164, 244, 261, 270, 299, 309
MACHO 1998-BLG-35, 135
MACHO 95-BLG-30, 213
MACHO 96-BLG-3, 213
MACHO 96-BLG-5, 270
MACHO 97-BLG-28, 213
MACHO 97-BLG-41, 109, 213
MACHO 98-BLG-28, 109
MACHO 98-BLG-35, 153, 160
MACHO 98-BLG-6, 270
MACHO 98-SMC-1, 109, 164, 213, 299
MACHO 99-LMC-2, 153

MACHO LMC-9, 299
MACHOs, 37, 54, 64, 73, 82, 309, 341
MACHO-SMC-97-1, 54
macrolensing, 3, 37, 351
Magellanic Clouds, 33, 54, 231, 261
mass distribution, 116, 164
mass function, 91, 116, 341
mass ratio distribution, 91, 164
MEGA, 318
metallicity, 37, 73, 116, 130, 213, 254, 271, 280
microlensing alert, 28, 153
Milky Way, 37
Milky Way Halo, 299
Mira variables, 254, 318
Mkn 501, 73
MOA, 33, 153, 160
molecules, 175
MPS, 270

– N –

NGC5907, 37
NGST, 164
NLTE effects, 175

– O –

OGLE, 164, 244, 261, 351
OGLE 1998-BUL-14, 135
OGLE 1999-BUL-35, 135
OGLE 1999-BUL-36, 135
OGLE-7, 3
OMC, 378
open clusters, 327
optical depth, 64, 231, 244, 261, 249, 318, 341
OWL, 372

– P –

51 Peg, 116
parallax, 3, 18, 28, 54
period distribution, 91
photometric microlensing, 3
pixel lensing, 231, 299, 309
PL relation, 254
PLANET, 109, 135, 164
planets
 as companions, 135
 detection of, 144, 394
 extragalactic, 164
 extrasolar, 116, 130, 135, 160, 164, 393, 394
 free-floating, 223, 341
 frequency of, 164
 giant, 91, 223
 multiple, 144
 parent stars of, 91
 searches for, 91, 135, 160, 164, 271
 sensitivity to planets, 164
 systems of, 130
 terrestrial, 160
POINT-AGAPE, 299, 309
Population III, 37
profile fitting photometry, 153
proper motion, 3, 18, 73, 109, 327

– Q –

QSO0957+561, 341, 351
QSO1009+2956, 37
QSO2047-1009, 37
QSO2237+0305, 351
quasars, 37, 341, 351

– R –

radial oscillations, 195
radial-velocity surveys, 91, 116
radio microlensing, 341, 363
red clump giants, 28, 231, 244
red dwarfs, 37
rotation curves, 37, 64, 271, 280, 299

– S –

Sagittarius Dwarf, 73, 280
SALT, 164, 382
scale height, 64
scintillation, 363
self-lensing, 28, 64, 73, 341
shroud, 73
SIM, 3, 91, 164
Sloan survey, 82, 327
SMC, 28, 54, 64, 73, 299, 309
Solar Circle, 254

Solar Neighbourhood, 327
spectral energy distribution, 175
spirals, 280
star
    formation histories, 280
    spots, 175, 195
    variable, 394
stellar
    atmospheres, 73, 164, 175, 213
    mass function, 327
    populations, 280
    radius, 195
    remnants, 73
    rotation, 195
    tomography, 213
superluminal motion, 341, 363
supernovae, 37
synthetic spectra, 175

– T –

T dwarfs, 175, 327
tidal effects, 280
transit, 116, 164, 195, 341

– V –

VATT/Columbia survey, 318
VB 10, 175
velocity distribution, 64
VISTA, 223
VLA, 363
vote, 341

– W –

white dwarfs, 37, 73, 82, 164, 271
WIMPs, 37, 73

This book was prepared with the help of editor's tools written by Chris Biemesderfer and Jeannette Barnes (NOAO).

ASTRONOMICAL SOCIETY OF THE PACIFIC
CONFERENCE SERIES

and

INTERNATIONAL ASTRONOMICAL UNION
VOLUMES

Published
by

The Astronomical Society of the Pacific
(ASP)

ASP CONFERENCE SERIES VOLUMES
Published by the Astronomical Society of the Pacific

PUBLISHED: 1988 (* asterisk means OUT OF STOCK)

Vol. CS-1 PROGRESS AND OPPORTUNITIES IN SOUTHERN HEMISPHERE
OPTICAL ASTRONOMY: CTIO 25TH Anniversary Symposium
eds. V. M. Blanco and M. M. Phillips
ISBN 0-937707-18-X

Vol. CS-2 PROCEEDINGS OF A WORKSHOP ON OPTICAL SURVEYS FOR QUASARS
eds. Patrick S. Osmer, Alain C. Porter, Richard F. Green, and Craig B. Foltz
ISBN 0-937707-19-8

Vol. CS-3 FIBER OPTICS IN ASTRONOMY
ed. Samuel C. Barden
ISBN 0-937707-20-1

Vol. CS-4 THE EXTRAGALACTIC DISTANCE SCALE:
Proceedings of the ASP 100th Anniversary Symposium
eds. Sidney van den Bergh and Christopher J. Pritchet
ISBN 0-937707-21-X

Vol. CS-5 THE MINNESOTA LECTURES ON CLUSTERS OF GALAXIES
AND LARGE-SCALE STRUCTURE
ed. John M. Dickey
ISBN 0-937707-22-8

PUBLISHED: 1989

Vol. CS-6 SYNTHESIS IMAGING IN RADIO ASTRONOMY: A Collection of Lectures
from the Third NRAO Synthesis Imaging Summer School
eds. Richard A. Perley, Frederic R. Schwab, and Alan H. Bridle
ISBN 0-937707-23-6

PUBLISHED: 1990

Vol. CS-7 PROPERTIES OF HOT LUMINOUS STARS: Boulder-Munich Workshop
ed. Catharine D. Garmany
ISBN 0-937707-24-4

Vol. CS-8* CCDs IN ASTRONOMY
ed. George H. Jacoby
ISBN 0-937707-25-2

Vol. CS-9 COOL STARS, STELLAR SYSTEMS, AND THE SUN: Sixth Cambridge Workshop
ed. George Wallerstein
ISBN 0-937707-27-9

Vol. CS-10* EVOLUTION OF THE UNIVERSE OF GALAXIES:
Edwin Hubble Centennial Symposium
ed. Richard G. Kron
ISBN 0-937707-28-7

Vol. CS-11 CONFRONTATION BETWEEN STELLAR PULSATION AND EVOLUTION
eds. Carla Cacciari and Gisella Clementini
ISBN 0-937707-30-9

Vol. CS-12 THE EVOLUTION OF THE INTERSTELLAR MEDIUM
ed. Leo Blitz
ISBN 0-937707-31-7

PUBLISHED: 1991

Vol. CS-13 THE FORMATION AND EVOLUTION OF STAR CLUSTERS
ed. Kenneth Janes
ISBN 0-937707-32-5

ASP CONFERENCE SERIES VOLUMES
Published by the Astronomical Society of the Pacific

PUBLISHED: 1991 (* asterisk means OUT OF STOCK)

Vol. CS-14   ASTROPHYSICS WITH INFRARED ARRAYS
ed. Richard Elston
ISBN 0-937707-33-3

Vol. CS-15   LARGE-SCALE STRUCTURES AND PECULIAR MOTIONS IN THE UNIVERSE
eds. David W. Latham and L. A. Nicolaci da Costa
ISBN 0-937707-34-1

Vol. CS-16   Proceedings of the 3rd Haystack Observatory Conference on ATOMS, IONS, AND MOLECULES: NEW RESULTS IN SPECTRAL LINE ASTROPHYSICS
eds. Aubrey D. Haschick and Paul T. P. Ho
ISBN 0-937707-35-X

Vol. CS-17   LIGHT POLLUTION, RADIO INTERFERENCE, AND SPACE DEBRIS
ed. David L. Crawford
ISBN 0-937707-36-8

Vol. CS-18   THE INTERPRETATION OF MODERN SYNTHESIS OBSERVATIONS OF SPIRAL GALAXIES
eds. Nebojsa Duric and Patrick C. Crane
ISBN 0-937707-37-6

Vol. CS-19   RADIO INTERFEROMETRY: THEORY, TECHNIQUES, AND APPLICATIONS, IAU Colloquium 131
eds. T. J. Cornwell and R. A. Perley
ISBN 0-937707-38-4

Vol. CS-20   FRONTIERS OF STELLAR EVOLUTION:
50th Anniversary McDonald Observatory (1939-1989)
ed. David L. Lambert
ISBN 0-937707-39-2

Vol. CS-21   THE SPACE DISTRIBUTION OF QUASARS
ed . David Crampton
ISBN 0-937707-40-6

PUBLISHED: 1992

Vol. CS-22   NONISOTROPIC AND VARIABLE OUTFLOWS FROM STARS
eds. Laurent Drissen, Claus Leitherer, and Antonella Nota
ISBN 0-937707-41-4

Vol CS-23   ASTRONOMICAL CCD OBSERVING AND REDUCTION TECHNIQUES
ed. Steve B. Howell
ISBN 0-937707-42-4

Vol. CS-24   COSMOLOGY AND LARGE-SCALE STRUCTURE IN THE UNIVERSE
ed. Reinaldo R. de Carvalho
ISBN 0-937707-43-0

Vol. CS-25   ASTRONOMICAL DATA ANALYSIS, SOFTWARE AND SYSTEMS I - (ADASS I)
eds. Diana M. Worrall, Chris Biemesderfer, and Jeannette Barnes
ISBN 0-937707-44-9

Vol. CS-26   COOL STARS, STELLAR SYSTEMS, AND THE SUN:
Seventh Cambridge Workshop
eds. Mark S. Giampapa and Jay A. Bookbinder
ISBN 0-937707-45-7

Vol. CS-27   THE SOLAR CYCLE: Proceedings of the
National Solar Observatory/Sacramento Peak 12th Summer Workshop
ed. Karen L. Harvey
ISBN 0-937707-46-5

ASP CONFERENCE SERIES VOLUMES
Published by the Astronomical Society of the Pacific

PUBLISHED: 1992 (asterisk means OUT OF STOCK)

Vol. CS-28   AUTOMATED TELESCOPES FOR PHOTOMETRY AND IMAGING
eds. Saul J. Adelman, Robert J. Dukes, Jr., and Carol J. Adelman
ISBN 0-937707-47-3

Vol. CS-29   Viña del Mar Workshop on CATACLYSMIC VARIABLE STARS
ed. Nikolaus Vogt
ISBN 0-937707-48-1

Vol. CS-30   VARIABLE STARS AND GALAXIES
ed. Brian Warner
ISBN 0-937707-49-X

Vol. CS-31   RELATIONSHIPS BETWEEN ACTIVE GALACTIC NUCLEI
AND STARBURST GALAXIES
ed. Alexei V. Filippenko
ISBN 0-937707-50-3

Vol. CS-32   COMPLEMENTARY APPROACHES TO DOUBLE
AND MULTIPLE STAR RESEARCH, IAU Colloquium 135
eds. Harold A. McAlister and William I. Hartkopf
ISBN 0-937707-51-1

Vol. CS-33   RESEARCH AMATEUR ASTRONOMY
ed. Stephen J. Edberg
ISBN 0-937707-52-X

Vol. CS-34   ROBOTIC TELESCOPES IN THE 1990's
ed. Alexei V. Filippenko
ISBN 0-937707-53-8

PUBLISHED: 1993

Vol. CS-35* MASSIVE STARS: THEIR LIVES IN THE INTERSTELLAR MEDIUM
eds. Joseph P. Cassinelli and Edward B. Churchwell
ISBN 0-937707-54-6

Vol. CS-36   PLANETS AROUND PULSARS
ed. J. A. Phillips, S. E. Thorsett, and S. R. Kulkarni
ISBN 0-937707-55-4

Vol. CS-37   FIBER OPTICS IN ASTRONOMY II
ed. Peter M. Gray
ISBN 0-937707-56-2

Vol. CS-38   NEW FRONTIERS IN BINARY STAR RESEARCH: Pacific Rim Colloquium
eds. K. C. Leung and I.-S. Nha
ISBN 0-937707-57-0

Vol. CS-39   THE MINNESOTA LECTURES ON THE STRUCTURE
AND DYNAMICS OF THE MILKY WAY
ed. Roberta M. Humphreys
ISBN 0-937707-58-9

Vol. CS-40   INSIDE THE STARS, IAU Colloquium 137
eds. Werner W. Weiss and Annie Baglin
ISBN 0-937707-59-7

Vol. CS-41   ASTRONOMICAL INFRARED SPECTROSCOPY:
FUTURE OBSERVATIONAL DIRECTIONS
ed. Sun Kwok
ISBN 0-937707-60-0

ASP CONFERENCE SERIES VOLUMES
Published by the Astronomical Society of the Pacific

PUBLISHED: 1993 (* asterisk means OUT OF STOCK)

Vol. CS-42	GONG 1992: SEISMIC INVESTIGATION OF THE SUN AND STARS ed. Timothy M. Brown ISBN 0-937707-61-9
Vol. CS-43	SKY SURVEYS: PROTOSTARS TO PROTOGALAXIES ed. B. T. Soifer ISBN 0-937707-62-7
Vol. CS-44	PECULIAR VERSUS NORMAL PHENOMENA IN A-TYPE AND RELATED STARS, IAU Colloquium 138 eds. M. M. Dworetsky, F. Castelli, and R. Faraggiana ISBN 0-937707-63-5
Vol. CS-45	LUMINOUS HIGH-LATITUDE STARS ed. Dimitar D. Sasselov ISBN 0-937707-64-3
Vol. CS-46	THE MAGNETIC AND VELOCITY FIELDS OF SOLAR ACTIVE REGIONS, IAU Colloquium 141 eds. Harold Zirin, Guoxiang Ai, and Haimin Wang ISBN 0-937707-65-1
Vol. CS-47	THIRD DECENNIAL US-USSR CONFERENCE ON SETI -- Santa Cruz, California, USA ed. G. Seth Shostak ISBN 0-937707-66-X
Vol. CS-48	THE GLOBULAR CLUSTER-GALAXY CONNECTION eds. Graeme H. Smith and Jean P. Brodie ISBN 0-937707-67-8
Vol. CS-49	GALAXY EVOLUTION: THE MILKY WAY PERSPECTIVE ed. Steven R. Majewski ISBN 0-937707-68-6
Vol. CS-50	STRUCTURE AND DYNAMICS OF GLOBULAR CLUSTERS eds. S. G. Djorgovski and G. Meylan ISBN 0-937707-69-4
Vol. CS-51	OBSERVATIONAL COSMOLOGY eds. Guido Chincarini, Angela Iovino, Tommaso Maccacaro, and Dario Maccagni ISBN 0-937707-70-8
Vol. CS-52	ASTRONOMICAL DATA ANALYSIS SOFTWARE AND SYSTEMS II - (ADASS II) eds. R. J. Hanisch, R. J. V. Brissenden, and Jeannette Barnes ISBN 0-937707-71-6
Vol. CS-53	BLUE STRAGGLERS ed. Rex A. Saffer ISBN 0-937707-72-4

PUBLISHED: 1994

Vol. CS-54*	THE FIRST STROMLO SYMPOSIUM: THE PHYSICS OF ACTIVE GALAXIES eds. Geoffrey V. Bicknell, Michael A. Dopita, and Peter J. Quinn ISBN 0-937707-73-2
Vol. CS-55	OPTICAL ASTRONOMY FROM THE EARTH AND MOON eds. Diane M. Pyper and Ronald J. Angione ISBN 0-937707-74-0
Vol. CS-56	INTERACTING BINARY STARS ed. Allen W. Shafter ISBN 0-937707-75-9

ASP CONFERENCE SERIES VOLUMES
Published by the Astronomical Society of the Pacific

PUBLISHED: 1994 (* asterisk means OUT OF STOCK)

Vol. CS-57   STELLAR AND CIRCUMSTELLAR ASTROPHYSICS
eds. George Wallerstein and Alberto Noriega-Crespo
ISBN 0-937707-76-7

Vol. CS-58*   THE FIRST SYMPOSIUM ON THE INFRARED CIRRUS
AND DIFFUSE INTERSTELLAR CLOUDS
eds. Roc M. Cutri and William B. Latter
ISBN 0-937707-77-5

Vol. CS-59   ASTRONOMY WITH MILLIMETER AND SUBMILLIMETER WAVE
INTERFEROMETRY,
IAU Colloquium 140
eds. M. Ishiguro and Wm. J. Welch
ISBN 0-937707-78-3

Vol. CS-60   THE MK PROCESS AT 50 YEARS: A POWERFUL TOOL FOR ASTROPHYSICAL
INSIGHT, A Workshop of the Vatican Observatory --Tucson, Arizona, USA
eds. C. J. Corbally, R. O. Gray, and R. F. Garrison
ISBN 0-937707-79-1

Vol. CS-61   ASTRONOMICAL DATA ANALYSIS SOFTWARE AND SYSTEMS III - (ADASS III)
eds. Dennis R. Crabtree, R. J. Hanisch, and Jeannette Barnes
ISBN 0-937707-80-5

Vol. CS-62   THE NATURE AND EVOLUTIONARY STATUS OF HERBIG Ae/Be STARS
eds. Pik Sin Thé, Mario R. Pérez, and Ed P. J. van den Heuvel
ISBN 0-9837707-81-3

Vol. CS-63   SEVENTY-FIVE YEARS OF HIRAYAMA ASTEROID FAMILIES:
THE ROLE OF COLLISIONS IN THE SOLAR SYSTEM HISTORY
eds. Yoshihide Kozai, Richard P. Binzel, and Tomohiro Hirayama
ISBN 0-937707-82-1

Vol. CS-64*   COOL STARS, STELLAR SYSTEMS, AND THE SUN:
Eighth Cambridge Workshop
ed. Jean-Pierre Caillault
ISBN 0-937707-83-X

Vol. CS-65*   CLOUDS, CORES, AND LOW MASS STARS:
The Fourth Haystack Observatory Conference
eds. Dan P. Clemens and Richard Barvainis
ISBN 0-937707-84-8

Vol. CS-66*   PHYSICS OF THE GASEOUS AND STELLAR DISKS OF THE GALAXY
ed. Ivan R. King
ISBN 0-937707-85-6

Vol. CS-67   UNVEILING LARGE-SCALE STRUCTURES BEHIND THE MILKY WAY
eds. C. Balkowski and R. C. Kraan-Korteweg
ISBN 0-937707-86-4

Vol. CS-68*   SOLAR ACTIVE REGION EVOLUTION:
COMPARING MODELS WITH OBSERVATIONS
eds. K. S. Balasubramaniam and George W. Simon
ISBN 0-937707-87-2

Vol. CS-69   REVERBERATION MAPPING OF THE BROAD-LINE REGION
IN ACTIVE GALACTIC NUCLEI
eds. P. M. Gondhalekar, K. Horne, and B. M. Peterson
ISBN 0-937707-88-0

Vol. CS-70*   GROUPS OF GALAXIES
eds. Otto-G. Richter and Kirk Borne
ISBN 0-937707-89-9

ASP CONFERENCE SERIES VOLUMES
Published by the Astronomical Society of the Pacific

PUBLISHED: 1995 (* asterisk means OUT OF STOCK)

Vol. CS-71   TRIDIMENSIONAL OPTICAL SPECTROSCOPIC METHODS IN ASTROPHYSICS, IAU Colloquium 149
eds. Georges Comte and Michel Marcelin
ISBN 0-937707-90-2

Vol. CS-72   MILLISECOND PULSARS: A DECADE OF SURPRISE
eds. A. S Fruchter, M. Tavani, and D. C. Backer
ISBN 0-937707-91-0

Vol. CS-73   AIRBORNE ASTRONOMY SYMPOSIUM ON THE GALACTIC ECOSYSTEM: FROM GAS TO STARS TO DUST
eds. Michael R. Haas, Jacqueline A. Davidson, and Edwin F. Erickson
ISBN 0-937707-92-9

Vol. CS-74   PROGRESS IN THE SEARCH FOR EXTRATERRESTRIAL LIFE:
1993 Bioastronomy Symposium
ed. G. Seth Shostak
ISBN 0-937707-93-7

Vol. CS-75   MULTI-FEED SYSTEMS FOR RADIO TELESCOPES
eds. Darrel T. Emerson and John M. Payne
ISBN 0-937707-94-5

Vol. CS-76   GONG '94: HELIO- AND ASTERO-SEISMOLOGY FROM THE EARTH AND SPACE
eds. Roger K. Ulrich, Edward J. Rhodes, Jr., and Werner Däppen
ISBN 0-937707-95-3

Vol. CS-77   ASTRONOMICAL DATA ANALYSIS SOFTWARE AND SYSTEMS IV - (ADASS IV)
eds. R. A. Shaw, H. E. Payne, and J. J. E. Hayes
ISBN 0-937707-96-1

Vol. CS-78   ASTROPHYSICAL APPLICATIONS OF POWERFUL NEW DATABASES:
Joint Discussion No. 16 of the 22nd General Assembly of the IAU
eds. S. J. Adelman and W. L. Wiese
ISBN 0-937707-97-X

Vol. CS-79*   ROBOTIC TELESCOPES: CURRENT CAPABILITIES, PRESENT DEVELOPMENTS, AND FUTURE PROSPECTS FOR AUTOMATED ASTRONOMY
eds. Gregory W. Henry and Joel A. Eaton
ISBN 0-937707-98-8

Vol. CS-80*   THE PHYSICS OF THE INTERSTELLAR MEDIUM AND INTERGALACTIC MEDIUM
eds. A. Ferrara, C. F. McKee, C. Heiles, and P. R. Shapiro
ISBN 0-937707-99-6

Vol. CS-81   LABORATORY AND ASTRONOMICAL HIGH RESOLUTION SPECTRA
eds. A. J. Sauval, R. Blomme, and N. Grevesse
ISBN 1-886733-01-5

Vol. CS-82*   VERY LONG BASELINE INTERFEROMETRY AND THE VLBA
eds. J. A. Zensus, P. J. Diamond, and P. J. Napier
ISBN 1-886733-02-3

Vol. CS-83*   ASTROPHYSICAL APPLICATIONS OF STELLAR PULSATION,
IAU Colloquium 155
eds. R. S. Stobie and P. A. Whitelock
ISBN 1-886733-03-1

ATLAS   INFRARED ATLAS OF THE ARCTURUS SPECTRUM, 0.9 - 5.3 μm
eds. Kenneth Hinkle, Lloyd Wallace, and William Livingston
ISBN: 1-886733-04-X

ASP CONFERENCE SERIES VOLUMES
Published by the Astronomical Society of the Pacific

PUBLISHED: 1995 (* asterisk means OUT OF STOCK)

Vol. CS-84   THE FUTURE UTILIZATION OF SCHMIDT TELESCOPES, IAU Colloquium 148
eds. Jessica Chapman, Russell Cannon, Sandra Harrison, and Bambang Hidayat
ISBN 1-886733-05-8

Vol. CS-85* CAPE WORKSHOP ON MAGNETIC CATACLYSMIC VARIABLES
eds. D. A. H. Buckley and B. Warner
ISBN 1-886733-06-6

Vol. CS-86   FRESH VIEWS OF ELLIPTICAL GALAXIES
eds. Alberto Buzzoni, Alvio Renzini, and Alfonso Serrano
ISBN 1-886733-07-4

PUBLISHED: 1996

Vol. CS-87   NEW OBSERVING MODES FOR THE NEXT CENTURY
eds. Todd Boroson, John Davies, and Ian Robson
ISBN 1-886733-08-2

Vol. CS-88* CLUSTERS, LENSING, AND THE FUTURE OF THE UNIVERSE
eds. Virginia Trimble and Andreas Reisenegger
ISBN 1-886733-09-0

Vol. CS-89   ASTRONOMY EDUCATION: CURRENT DEVELOPMENTS,
FUTURE COORDINATION
ed. John R. Percy
ISBN 1-886733-10-4

Vol. CS-90   THE ORIGINS, EVOLUTION, AND DESTINIES OF BINARY STARS
IN CLUSTERS
eds. E. F. Milone and J. -C. Mermilliod
ISBN 1-886733-11-2

Vol. CS-91   BARRED GALAXIES, IAU Colloquium 157
eds. R. Buta, D. A. Crocker, and B. G. Elmegreen
ISBN 1-886733-12-0

Vol. CS-92* FORMATION OF THE GALACTIC HALO .... INSIDE AND OUT
eds. Heather L. Morrison and Ata Sarajedini
ISBN 1-886733-13-9

Vol. CS-93   RADIO EMISSION FROM THE STARS AND THE SUN
eds. A. R. Taylor and J. M. Paredes
ISBN 1-886733-14-7

Vol. CS-94   MAPPING, MEASURING, AND MODELING THE UNIVERSE
eds. Peter Coles, Vicent J. Martinez, and Maria-Jesus Pons-Borderia
ISBN 1-886733-15-5

Vol. CS-95   SOLAR DRIVERS OF INTERPLANETARY AND TERRESTRIAL DISTURBANCES:
Proceedings of 16th International Workshop National Solar
Observatory/Sacramento Peak
eds. K. S. Balasubramaniam, Stephen L. Keil, and Raymond N. Smartt
ISBN 1-886733-16-3

Vol. CS-96   HYDROGEN-DEFICIENT STARS
eds. C. S. Jeffery and U. Heber
ISBN 1-886733-17-1

Vol. CS-97   POLARIMETRY OF THE INTERSTELLAR MEDIUM
eds. W. G. Roberge and D. C. B. Whittet
ISBN 1-886733-18-X

ASP CONFERENCE SERIES VOLUMES
Published by the Astronomical Society of the Pacific

PUBLISHED: 1996 (* asterisk means OUT OF STOCK)

Vol. CS-98   FROM STARS TO GALAXIES: THE IMPACT OF STELLAR PHYSICS ON GALAXY EVOLUTION
eds. Claus Leitherer, Uta Fritze-von Alvensleben, and John Huchra
ISBN 1-886733-19-8

Vol. CS-99   COSMIC ABUNDANCES:
Proceedings of the 6th Annual October Astrophysics Conference
eds. Stephen S. Holt and George Sonneborn
ISBN 1-886733-20-1

Vol. CS-100   ENERGY TRANSPORT IN RADIO GALAXIES AND QUASARS
eds. P. E. Hardee, A. H. Bridle, and J. A. Zensus
ISBN 1-886733-21-X

Vol. CS-101   ASTRONOMICAL DATA ANALYSIS SOFTWARE AND SYSTEMS V – (ADASS V)
eds. George H. Jacoby and Jeannette Barnes
ISBN 1080-7926

Vol. CS-102   THE GALACTIC CENTER, 4th ESO/CTIO Workshop
ed. Roland Gredel
ISBN 1-886733-22-8

Vol. CS-103   THE PHYSICS OF LINERS IN VIEW OF RECENT OBSERVATIONS
eds. M. Eracleous, A. Koratkar, C. Leitherer, and L. Ho
ISBN 1-886733-23-6

Vol. CS-104   PHYSICS, CHEMISTRY, AND DYNAMICS OF INTERPLANETARY DUST, IAU Colloquium 150
eds. Bo Å. S. Gustafson and Martha S. Hanner
ISBN 1-886733-24-4

Vol. CS-105   PULSARS: PROBLEMS AND PROGRESS, IAU Colloquium 160
ed. S. Johnston, M. A. Walker, and M. Bailes
ISBN 1-886733-25-2

Vol. CS-106   THE MINNESOTA LECTURES ON EXTRAGALACTIC NEUTRAL HYDROGEN
ed. Evan D. Skillman
ISBN 1-886733-26-0

Vol. CS-107   COMPLETING THE INVENTORY OF THE SOLAR SYSTEM:
A Symposium held in conjunction with the 106th Annual Meeting of the ASP
eds. Terrence W. Rettig and Joseph M. Hahn
ISBN 1-886733-27-9

Vol. CS-108   M.A.S.S. -- MODEL ATMOSPHERES AND SPECTRUM SYNTHESIS:
5th Vienna - Workshop
eds. Saul J. Adelman, Friedrich Kupka, and Werner W. Weiss
ISBN 1-886733-28-7

Vol. CS-109   COOL STARS, STELLAR SYSTEMS, AND THE SUN: Ninth Cambridge Workshop
eds. Roberto Pallavicini and Andrea K. Dupree
ISBN 1-886733-29-5

Vol. CS-110   BLAZAR CONTINUUM VARIABILITY
eds. H. R. Miller, J. R. Webb, and J. C. Noble
ISBN 1-886733-30-9

Vol. CS-111   MAGNETIC RECONNECTION IN THE SOLAR ATMOSPHERE:
Proceedings of a Yohkoh Conference
eds. R. D. Bentley and J. T. Mariska
ISBN 1-886733-31-7

ASP CONFERENCE SERIES VOLUMES
Published by the Astronomical Society of the Pacific

PUBLISHED: 1996 (* asterisk means OUT OF STOCK)

Vol. CS-112 THE HISTORY OF THE MILKY WAY AND ITS SATELLITE SYSTEM
eds. Andreas Burkert, Dieter H. Hartmann, and Steven R. Majewski
ISBN 1-886733-32-5

PUBLISHED: 1997

Vol. CS-113 EMISSION LINES IN ACTIVE GALAXIES: NEW METHODS AND TECHNIQUES,
IAU Colloquium 159
eds. B. M. Peterson, F.-Z. Cheng, and A. S. Wilson
ISBN 1-886733-33-3

Vol. CS-114 YOUNG GALAXIES AND QSO ABSORPTION-LINE SYSTEMS
eds. Sueli M. Viegas, Ruth Gruenwald, and Reinaldo R. de Carvalho
ISBN 1-886733-34-1

Vol. CS-115 GALACTIC CLUSTER COOLING FLOWS
ed. Noam Soker
ISBN 1-886733-35-X

Vol. CS-116 THE SECOND STROMLO SYMPOSIUM:
THE NATURE OF ELLIPTICAL GALAXIES
eds. M. Arnaboldi, G. S. Da Costa, and P. Saha
ISBN 1-886733-36-8

Vol. CS-117 DARK AND VISIBLE MATTER IN GALAXIES
eds. Massimo Persic and Paolo Salucci
ISBN-1-886733-37-6

Vol. CS-118 FIRST ADVANCES IN SOLAR PHYSICS EUROCONFERENCE:
ADVANCES IN THE PHYSICS OF SUNSPOTS
eds. B. Schmieder. J. C. del Toro Iniesta, and M. Vázquez
ISBN 1-886733-38-4

Vol. CS-119 PLANETS BEYOND THE SOLAR SYSTEM
AND THE NEXT GENERATION OF SPACE MISSIONS
ed. David R. Soderblom
ISBN 1-886733-39-2

Vol. CS-120 LUMINOUS BLUE VARIABLES: MASSIVE STARS IN TRANSITION
eds. Antonella Nota and Henny J. G. L. M. Lamers
ISBN 1-886733-40-6

Vol. CS-121 ACCRETION PHENOMENA AND RELATED OUTFLOWS, IAU Colloquium 163
eds. D. T. Wickramasinghe, G. V. Bicknell, and L. Ferrario
ISBN 1-886733-41-4

Vol. CS-122 FROM STARDUST TO PLANETESIMALS:
Symposium held as part of the 108th Annual Meeting of the ASP
eds. Yvonne J. Pendleton and A. G. G. M. Tielens
ISBN 1-886733-42-2

Vol. CS-123 THE 12th 'KINGSTON MEETING': COMPUTATIONAL ASTROPHYSICS
eds. David A. Clarke and Michael J. West
ISBN 1-886733-43-0

Vol. CS-124 DIFFUSE INFRARED RADIATION AND THE IRTS
eds. Haruyuki Okuda, Toshio Matsumoto, and Thomas Roellig
ISBN 1-886733-44-9

Vol. CS-125 ASTRONOMICAL DATA ANALYSIS SOFTWARE AND SYSTEMS VI
eds. Gareth Hunt and H. E. Payne
ISBN 1-886733-45-7

ASP CONFERENCE SERIES VOLUMES
Published by the Astronomical Society of the Pacific

PUBLISHED: 1997 (* asterisk means OUT OF STOCK)

Vol. CS-126   FROM QUANTUM FLUCTUATIONS TO COSMOLOGICAL STRUCTURES
eds. David Valls-Gabaud, Martin A. Hendry, Paolo Molaro, and Khalil Chamcham
ISBN 1-886733-46-5

Vol. CS-127   PROPER MOTIONS AND GALACTIC ASTRONOMY
ed. Roberta M. Humphreys
ISBN 1-886733-47-3

Vol. CS-128   MASS EJECTION FROM AGN (Active Galactic Nuclei)
eds. N. Arav, I. Shlosman, and R. J. Weymann
ISBN 1-886733-48-1

Vol. CS-129   THE GEORGE GAMOW SYMPOSIUM
eds. E. Harper, W. C. Parke, and G. D. Anderson
ISBN 1-886733-49-X

Vol. CS-130   THE THIRD PACIFIC RIM CONFERENCE ON
RECENT DEVELOPMENT ON BINARY STAR RESEARCH
eds. Kam-Ching Leung
ISBN 1-886733-50-3

PUBLISHED: 1998

Vol. CS-131   BOULDER-MUNICH II: PROPERTIES OF HOT, LUMINOUS STARS
ed. Ian D. Howarth
ISBN 1-886733-51-1

Vol. CS-132   STAR FORMATION WITH THE INFRARED SPACE OBSERVATORY (ISO)
eds. João L. Yun and René Liseau
ISBN 1-886733-52-X

Vol. CS-133   SCIENCE WITH THE NGST (Next Generation Space Telescope)
eds. Eric P. Smith and Anuradha Koratkar
ISBN 1-886733-53-8

Vol. CS-134   BROWN DWARFS AND EXTRASOLAR PLANETS
eds. Rafael Rebolo, Eduardo L. Martin, and Maria Rosa Zapatero Osorio
ISBN 1-886733-54-6

Vol. CS-135   A HALF CENTURY OF STELLAR PULSATION INTERPRETATIONS:
A TRIBUTE TO ARTHUR N. COX
eds. P. A. Bradley and J. A. Guzik
ISBN 1-886733-55-4

Vol. CS-136   GALACTIC HALOS: A UC SANTA CRUZ WORKSHOP
ed. Dennis Zaritsky
ISBN 1-886733-56-2

Vol. CS-137   WILD STARS IN THE OLD WEST: PROCEEDINGS OF THE 13th NORTH
AMERICAN WORKSHOP ON CATACLYSMIC VARIABLES
AND RELATED OBJECTS
eds. S. Howell, E. Kuulkers, and C. Woodward
ISBN 1-886733-57-0

Vol. CS-138   1997 PACIFIC RIM CONFERENCE ON STELLAR ASTROPHYSICS
eds. Kwing Lam Chan, K. S. Cheng, and H. P. Singh
ISBN 1-886733-58-9

Vol. CS-139   PRESERVING THE ASTRONOMICAL WINDOWS:
Proceedings of Joint Discussion No. 5 of the 23rd General Assembly of the IAU
eds. Syuzo Isobe and Tomohiro Hirayama
ISBN 1-886733-59-7

ASP CONFERENCE SERIES VOLUMES
Published by the Astronomical Society of the Pacific

PUBLISHED: 1998 (* asterisk means OUT OF STOCK)

Vol. CS-140   SYNOPTIC SOLAR PHYSICS --18th NSO/Sacramento Peak Summer Workshop
eds. K. S. Balasubramaniam, J. W. Harvey, and D. M. Rabin
ISBN 1-886733-60-0

Vol. CS-141   ASTROPHYSICS FROM ANTARCTICA:
A Symposium held as a part of the 109th Annual Meeting of the ASP
eds. Giles Novak and Randall H. Landsberg
ISBN 1-886733-61-9

Vol. CS-142   THE STELLAR INITIAL MASS FUNCTION: 38th Herstmonceux Conference
eds. Gerry Gilmore and Debbie Howell
ISBN 1-886733-62-7

Vol. CS-143*   THE SCIENTIFIC IMPACT OF THE GODDARD HIGH RESOLUTION SPECTROGRAPH (GHRS)
eds. John C. Brandt, Thomas B. Ake III, and Carolyn Collins Petersen
ISBN 1-886733-63-5

Vol. CS-144   RADIO EMISSION FROM GALACTIC AND EXTRAGALACTIC COMPACT SOURCES, IAU Colloquium 164
eds. J. Anton Zensus, G. B. Taylor, and J. M. Wrobel
ISBN 1-886733-64-3

Vol. CS-145   ASTRONOMICAL DATA ANALYSIS SOFTWARE AND SYSTEMS VII – (ADASS VII)
eds. Rudolf Albrecht, Richard N. Hook, and Howard A. Bushouse
ISBN 1-886733-65-1

Vol. CS-146   THE YOUNG UNIVERSE GALAXY FORMATION AND EVOLUTION AT INTERMEDIATE AND HIGH REDSHIFT
eds. S. D'Odorico, A. Fontana, and E. Giallongo
ISBN 1-886733-66-X

Vol. CS-147   ABUNDANCE PROFILES: DIAGNOSTIC TOOLS FOR GALAXY HISTORY
eds. Daniel Friedli, Mike Edmunds, Carmelle Robert, and Laurent Drissen
ISBN 1-886733-67-8

Vol. CS-148   ORIGINS
eds. Charles E. Woodward, J. Michael Shull, and Harley A. Thronson, Jr.
ISBN 1-886733-68-6

Vol. CS-149   SOLAR SYSTEM FORMATION AND EVOLUTION
eds. D. Lazzaro, R. Vieira Martins, S. Ferraz-Mello, J. Fernández, and C. Beaugé
ISBN 1-886733-69-4

Vol. CS-150   NEW PERSPECTIVES ON SOLAR PROMINENCES, IAU Colloquium 167
eds. David Webb, David Rust, and Brigitte Schmieder
ISBN 1-886733-70-8

Vol. CS-151   COSMIC MICROWAVE BACKGROUND AND LARGE SCALE STRUCTURES OF THE UNIVERSE
eds. Yong-Ik Byun and Kin-Wang Ng
ISBN 1-886733-71-6

Vol. CS-152   FIBER OPTICS IN ASTRONOMY III
eds. S. Arribas, E. Mediavilla, and F. Watson
ISBN 1-886733-72-4

Vol. CS-153   LIBRARY AND INFORMATION SERVICES IN ASTRONOMY III -- (LISA III)
eds. Uta Grothkopf, Heinz Andernach, Sarah Stevens-Rayburn, and Monique Gomez
ISBN 1-886733-73-2

ASP CONFERENCE SERIES VOLUMES
Published by the Astronomical Society of the Pacific

PUBLISHED: 1998 (* asterisk means OUT OF STOCK)

Vol. CS-154   COOL STARS, STELLAR SYSTEMS AND THE SUN: Tenth Cambridge Workshop
eds. Robert A. Donahue and Jay A. Bookbinder
ISBN 1-886733-74-0

Vol. CS-155   SECOND ADVANCES IN SOLAR PHYSICS EUROCONFERENCE:
THREE-DIMENSIONAL STRUCTURE OF SOLAR ACTIVE REGIONS
eds. Costas E. Alissandrakis and Brigitte Schmieder
ISBN 1-886733-75-9

PUBLISHED: 1999

Vol. CS-156   HIGHLY REDSHIFTED RADIO LINES
eds. C. L. Carilli, S. J. E. Radford, K. M. Menten, and G. I. Langston
ISBN 1-886733-76-7

Vol. CS-157   ANNAPOLIS WORKSHOP ON MAGNETIC CATACLYSMIC VARIABLES
eds. Coel Hellier and Koji Mukai
ISBN 1-886733-77-5

Vol. CS-158   SOLAR AND STELLAR ACTIVITY: SIMILARITIES AND DIFFERENCES
eds. C. J. Butler and J. G. Doyle
ISBN 1-886733-78-3

Vol. CS-159   BL LAC PHENOMENON
eds. Leo O. Takalo and Aimo Sillanpää
ISBN 1-886733-79-1

Vol. CS-160   ASTROPHYSICAL DISCS: An EC Summer School
eds. J. A. Sellwood and Jeremy Goodman
ISBN 1-886733-80-5

Vol. CS-161   HIGH ENERGY PROCESSES IN ACCRETING BLACK HOLES
eds. Juri Poutanen and Roland Svensson
ISBN 1-886733-81-3

Vol. CS-162   QUASARS AND COSMOLOGY
eds. Gary Ferland and Jack Baldwin
ISBN 1-886733-83-X

Vol. CS-163   STAR FORMATION IN EARLY-TYPE GALAXIES
eds. Jordi Cepa and Patricia Carral
ISBN 1-886733-84-8

Vol. CS-164   ULTRAVIOLET–OPTICAL SPACE ASTRONOMY BEYOND HST
eds. Jon A. Morse, J. Michael Shull, and Anne L. Kinney
ISBN 1-886733-85-6

Vol. CS-165   THE THIRD STROMLO SYMPOSIUM: THE GALACTIC HALO
eds. Brad K. Gibson, Tim S. Axelrod, and Mary E. Putman
ISBN 1-886733-86-4

Vol. CS-166   STROMLO WORKSHOP ON HIGH-VELOCITY CLOUDS
eds. Brad K. Gibson and Mary E. Putman
ISBN 1-886733-87-2

Vol. CS-167   HARMONIZING COSMIC DISTANCE SCALES IN A POST-HIPPARCOS ERA
eds. Daniel Egret and André Heck
ISBN 1-886733-88-0

Vol. CS-168   NEW PERSPECTIVES ON THE INTERSTELLAR MEDIUM
eds. A. R. Taylor, T. L. Landecker, and G. Joncas
ISBN 1-886733-89-9

ASP CONFERENCE SERIES VOLUMES
Published by the Astronomical Society of the Pacific

PUBLISHED: 1999 (* asterisk means OUT OF STOCK)

Vol. CS-169   11th EUROPEAN WORKSHOP ON WHITE DWARFS
eds. J.-E. Solheim and E. G. Meištas
ISBN 1-886733-91-0

Vol. CS-170   THE LOW SURFACE BRIGHTNESS UNIVERSE, IAU Colloquium 171
eds. J. I. Davies, C. Impey, and S. Phillipps
ISBN 1-886733-92-9

Vol. CS-171   LiBeB, COSMIC RAYS, AND RELATED X- AND GAMMA-RAYS
eds. Reuven Ramaty, Elisabeth Vangioni-Flam, Michel Cassé, and Keith Olive
ISBN 1-886733-93-7

Vol. CS-172   ASTRONOMICAL DATA ANALYSIS SOFTWARE AND SYSTEMS VIII
eds. David M. Mehringer, Raymond L. Plante, and Douglas A. Roberts
ISBN 1-886733-94-5

Vol. CS-173   THEORY AND TESTS OF CONVECTION IN STELLAR STRUCTURE:
First Granada Workshop
ed. Álvaro Giménez, Edward F. Guinan, and Benjamín Montesinos
ISBN 1-886733-95-3

Vol. CS-174   CATCHING THE PERFECT WAVE: ADAPTIVE OPTICS AND
INTERFEROMETRY IN THE 21st CENTURY,
A Symposium held as a part of the 110th Annual Meeting of the ASP
eds. Sergio R. Restaino, William Junor, and Nebojsa Duric
ISBN 1-886733-96-1

Vol. CS-175   STRUCTURE AND KINEMATICS OF QUASAR BROAD LINE REGIONS
eds. C. M. Gaskell, W. N. Brandt, M. Dietrich, D. Dultzin-Hacyan,
and M. Eracleous
ISBN 1-886733-97-X

Vol. CS-176   OBSERVATIONAL COSMOLOGY: THE DEVELOPMENT OF GALAXY SYSTEMS
eds. Giuliano Giuricin, Marino Mezzetti, and Paolo Salucci
ISBN 1-58381-000-5

Vol. CS-177   ASTROPHYSICS WITH INFRARED SURVEYS: A Prelude to SIRTF
eds. Michael D. Bicay, Chas A. Beichman, Roc M. Cutri, and Barry F. Madore
ISBN 1-58381-001-3

Vol. CS-178   STELLAR DYNAMOS: NONLINEARITY AND CHAOTIC FLOWS
eds. Manuel Núñez and Antonio Ferriz-Mas
ISBN 1-58381-002-1

Vol. CS-179   ETA CARINAE AT THE MILLENNIUM
eds. Jon A. Morse, Roberta M. Humphreys, and Augusto Damineli
ISBN 1-58381-003-X

Vol. CS-180   SYNTHESIS IMAGING IN RADIO ASTRONOMY II
eds. G. B. Taylor, C. L. Carilli, and R. A. Perley
ISBN 1-58381-005-6

Vol. CS-181   MICROWAVE FOREGROUNDS
eds. Angelica de Oliveira-Costa and Max Tegmark
ISBN 1-58381-006-4

Vol. CS-182   GALAXY DYNAMICS: A Rutgers Symposium
eds. David Merritt, J. A. Sellwood, and Monica Valluri
ISBN 1-58381-007-2

Vol. CS-183   HIGH RESOLUTION SOLAR PHYSICS: THEORY, OBSERVATIONS,
AND TECHNIQUES
eds. T. R. Rimmele, K. S. Balasubramaniam, and R. R. Radick
ISBN 1-58381-009-9

ASP CONFERENCE SERIES VOLUMES
Published by the Astronomical Society of the Pacific

PUBLISHED: 1999 (* asterisk means OUT OF STOCK)

Vol. CS-184 THIRD ADVANCES IN SOLAR PHYSICS EUROCONFERENCE:
MAGNETIC FIELDS AND OSCILLATIONS
eds. B. Schmieder, A. Hofmann, and J. Staude
ISBN 1-58381-010-2

Vol. CS-185 PRECISE STELLAR RADIAL VELOCITIES, IAU Colloquium 170
eds. J. B. Hearnshaw and C. D. Scarfe
ISBN 1-58381-011-0

Vol. CS-186 THE CENTRAL PARSECS OF THE GALAXY
eds. Heino Falcke, Angela Cotera, Wolfgang J. Duschl, Fulvio Melia,
and Marcia J. Rieke
ISBN 1-58381-012-9

Vol. CS-187 THE EVOLUTION OF GALAXIES ON COSMOLOGICAL TIMESCALES
eds. J. E. Beckman and T. J. Mahoney
ISBN 1-58381-013-7

Vol. CS-188 OPTICAL AND INFRARED SPECTROSCOPY OF CIRCUMSTELLAR MATTER
eds. Eike W. Guenther, Bringfried Stecklum, and Sylvio Klose
ISBN 1-58381-014-5

Vol. CS-189 CCD PRECISION PHOTOMETRY WORKSHOP
eds. Eric R. Craine, Roy A. Tucker, and Jeannette Barnes
ISBN 1-58381-015-3

Vol. CS-190 GAMMA-RAY BURSTS: THE FIRST THREE MINUTES
eds. Juri Poutanen and Roland Svensson
ISBN 1-58381-016-1

Vol. CS-191 PHOTOMETRIC REDSHIFTS AND HIGH REDSHIFT GALAXIES
eds. Ray J. Weymann, Lisa J. Storrie-Lombardi, Marcin Sawicki,
and Robert J. Brunner
ISBN 1-58381-017-X

Vol. CS-192 SPECTROPHOTOMETRIC DATING OF STARS AND GALAXIES
ed. I. Hubeny, S. R. Heap, and R. H. Cornett
ISBN 1-58381-018-8

Vol. CS-193 THE HY-REDSHIFT UNIVERSE:
GALAXY FORMATION AND EVOLUTION AT HIGH REDSHIFT
eds. Andrew J. Bunker and Wil J. M. van Breugel
ISBN 1-58381-019-6

Vol. CS-194 WORKING ON THE FRINGE:
OPTICAL AND IR INTERFEROMETRY FROM GROUND AND SPACE
eds. Stephen Unwin and Robert Stachnik
ISBN 1-58381-020-X

PUBLISHED: 2000

Vol. CS-195 IMAGING THE UNIVERSE IN THREE DIMENSIONS:
Astrophysics with Advanced Multi-Wavelength Imaging Devices
eds. W. van Breugel and J. Bland-Hawthorn
ISBN 1-58381-022-6

Vol. CS-196 THERMAL EMISSION SPECTROSCOPY AND ANALYSIS OF DUST,
DISKS, AND REGOLITHS
eds. Michael L. Sitko, Ann L. Sprague, and David K. Lynch
ISBN: 1-58381-023-4

Vol. CS-197 XVth IAP MEETING DYNAMICS OF GALAXIES:
FROM THE EARLY UNIVERSE TO THE PRESENT
eds. F. Combes, G. A. Mamon, and V. Charmandaris
ISBN: 1-58381-24-2

ASP CONFERENCE SERIES VOLUMES
Published by the Astronomical Society of the Pacific

PUBLISHED: 2000 (* asterisk means OUT OF STOCK)

Vol. CS-198 EUROCONFERENCE ON "STELLAR CLUSTERS AND ASSOCIATIONS: CONVECTION, ROTATION, AND DYNAMOS"
eds. R. Pallavicini, G. Micela, and S. Sciortino
ISBN: 1-58381-25-0

Vol. CS-199 ASYMMETRICAL PLANETARY NEBULAE II: FROM ORIGINS TO MICROSTRUCTURES
eds. J. H. Kastner, N. Soker, and S. Rappaport
ISBN: 1-58381-026-9

Vol. CS-200 CLUSTERING AT HIGH REDSHIFT
eds. A. Mazure, O. Le Fèvre, and V. Le Brun
ISBN: 1-58381-027-7

Vol. CS-201 COSMIC FLOWS 1999: TOWARDS AN UNDERSTANDING OF LARGE-SCALE STRUCTURES
eds. Stéphane Courteau, Michael A. Strauss, and Jeffrey A. Willick
ISBN: 1-58381-028-5

Vol. CS-202 PULSAR ASTRONOMY – 2000 AND BEYOND, IAU Colloquium 177
eds. M. Kramer, N. Wex, and R. Wielebinski
ISBN: 1-58381-029-3

Vol. CS-203 THE IMPACT OF LARGE-SCALE SURVEYS ON PULSATING STAR RESEARCH, IAU Colloquium 176
eds. L. Szabados and D. W. Kurtz
ISBN: 1-58381-030-7

Vol. CS-204 THERMAL AND IONIZATION ASPECTS OF FLOWS FROM HOT STARS: OBSERVATIONS AND THEORY
eds. Henny J. G. L. M. Lamers and Arved Sapar
ISBN: 1-58381-031-5

Vol. CS-205 THE LAST TOTAL SOLAR ECLIPSE OF THE MILLENNIUM IN TURKEY
eds. W. C. Livingston and A. Özgüç
ISBN: 1-58381-032-3

Vol. CS-206 HIGH ENERGY SOLAR PHYSICS – *ANTICIPATING HESSI*
eds. Reuven Ramaty and Natalie Mandzhavidze
ISBN: 1-58381-033-1

Vol. CS-207 NGST SCIENCE AND TECHNOLOGY EXPOSITION
eds. Eric P. Smith and Knox S. Long
ISBN: 1-58381-036-6

ATLAS VISIBLE AND NEAR INFRARED ATLAS OF THE ARCTURUS SPECTRUM 3727-9300 Å
eds. Kenneth Hinkle, Lloyd Wallace, Jeff Valenti, and Dianne Harmer
ISBN: 1-58381-037-4

Vol. CS-208 POLAR MOTION: HISTORICAL AND SCIENTIFIC PROBLEMS, IAU Colloquium 178
eds. Steven Dick, Dennis McCarthy, and Brian Luzum
ISBN: 1-58381-039-0

Vol. CS-209 SMALL GALAXY GROUPS, IAU Colloquium 174
eds. Mauri J. Valtonen and Chris Flynn
ISBN: 1-58381-040-4

Vol. CS-210 DELTA SCUTI AND RELATED STARS: Reference Handbook and Proceedings of the 6th Vienna Workshop in Astrophysics
eds. Michel Breger and Michael Houston Montgomery
ISBN: 1-58381-043-9

ASP CONFERENCE SERIES VOLUMES
Published by the Astronomical Society of the Pacific

PUBLISHED: 2000 (* asterisk means OUT OF STOCK)

Vol. CS-211 MASSIVE STELLAR CLUSTERS
eds. Ariane Lançon and Christian M. Boily
ISBN: 1-58381-042-0

Vol. CS-212 FROM GIANT PLANETS TO COOL STARS
eds. Caitlin A. Griffith and Mark S. Marley
ISBN: 1-58381-041-2

Vol. CS-213 BIOASTRONOMY `99: A NEW ERA IN BIOASTRONOMY
eds. Guillermo A. Lemarchand and Karen J. Meech
ISBN: 1-58381-044-7

Vol. CS-214 THE Be PHENOMENON IN EARLY-TYPE STARS, IAU Colloquium 175
eds. Myron A. Smith, Huib F. Henrichs and Juan Fabregat
ISBN: 1-58381-045-5

Vol. CS-215 COSMIC EVOLUTION AND GALAXY FORMATION:
STRUCTURE, INTERACTIONS AND FEEDBACK
The 3rd Guillermo Haro Astrophysics Conference
eds. José Franco, Elena Terlevich, Omar López-Cruz, and Itziar Aretxaga
ISBN: 1-58381-046-3

Vol. CS-216 ASTRONOMICAL DATA ANALYSIS SOFTWARE AND SYSTEMS IX
eds. Nadine Manset, Christian Veillet, and Dennis Crabtree
ISBN: 1-58381-047-1   ISSN: 1080-7926

Vol. CS-217 IMAGING AT RADIO THROUGH SUBMILLIMETER WAVELENGTHS
eds. Jeffrey G. Mangum and Simon J. E. Radford
ISBN: 1-58381-049-8

Vol. CS-218 MAPPING THE HIDDEN UNIVERSE: THE UNIVERSE BEHIND THE MILKY WAY
THE UNIVERSE IN HI
eds. Renée C. Kraan-Korteweg, Patricia A. Henning, and Heinz Andernach
ISBN: 1-58381-050-1

Vol. CS-219 DISKS, PLANETESIMALS, AND PLANETS
eds. F. Garzón, C. Eiroa, D. de Winter, and T. J. Mahoney
ISBN: 1-58381-051-X

Vol. CS-220 AMATEUR - PROFESSIONAL PARTNERSHIPS IN ASTRONOMY:
The 111th Annual Meeting of the ASP
eds. John R. Percy and Joseph B. Wilson
ISBN: 1-58381-052-8

Vol. CS-221 STARS, GAS AND DUST IN GALAXIES: EXPLORING THE LINKS
eds. Danielle Alloin, Knut Olsen, and Gaspar Galaz
ISBN: 1-58381-053-6

PUBLISHED: 2001

Vol. CS-222 THE PHYSICS OF GALAXY FORMATION
eds. M. Umemura and H. Susa
ISBN: 1-58381-054-4

Vol. CS-223 COOL STARS, STELLAR SYSTEMS AND THE SUN:
Eleventh Cambridge Workshop
eds. Ramón J. García López, Rafael Rebolo, and María Zapatero Osorio
ISBN: 1-58381-056-0

Vol. CS-224 PROBING THE PHYSICS OF ACTIVE GALACTIC NUCLEI
BY MULTIWAVELENGTH MONITORING
eds. Bradley M. Peterson, Ronald S. Polidan, and Richard W. Pogge
ISBN: 1-58381-055-2

ASP CONFERENCE SERIES VOLUMES
Published by the Astronomical Society of the Pacific

PUBLISHED: 2001 (* asterisk means OUT OF STOCK)

Vol. CS-225   VIRTUAL OBSERVATORIES OF THE FUTURE
eds. Robert J. Brunner, S. George Djorgovski, and Alex S. Szalay
ISBN: 1-58381-057-9

Vol. CS-226   12th EUROPEAN CONFERENCE ON WHITE DWARFS
eds. J. L. Provencal, H. L. Shipman, J. MacDonald, and S. Goodchild
ISBN: 1-58381-058-7

Vol. CS-227   BLAZAR DEMOGRAPHICS AND PHYSICS
eds. Paolo Padovani and C. Megan Urry
ISBN: 1-58381-059-5

Vol. CS-228   DYNAMICS OF STAR CLUSTERS AND THE MILKY WAY
eds. S. Deiters, B. Fuchs, A. Just, R. Spurzem, and R. Wielen
ISBN: 1-58381-060-9

Vol. CS-229   EVOLUTION OF BINARY AND MULTIPLE STAR SYSTEMS
A Meeting in Celebration of Peter Eggleton's 60th Birthday
eds. Ph. Podsiadlowski, S. Rappaport, A. R. King, F. D'Antona, and L. Burderi
IBSN: 1-58381-061-7

Vol. CS-230   GALAXY DISKS AND DISK GALAXIES
eds. Jose G. Funes, S. J. and Enrico Maria Corsini
ISBN: 1-58381-063-3

Vol. CS-231   TETONS 4: GALACTIC STRUCTURE, STARS, AND
THE INTERSTELLAR MEDIUM
eds. Charles E. Woodward, Michael D. Bicay, and J. Michael Shull
ISBN: 1-58381-064-1

Vol. CS-232   THE NEW ERA OF WIDE FIELD ASTRONOMY
eds. Roger Clowes, Andrew Adamson, and Gordon Bromage
ISBN: 1-58381-065-X

Vol. CS-233   P CYGNI 2000: 400 YEARS OF PROGRESS
eds. Mart de Groot and Christiaan Sterken
ISBN: 1-58381-070-6

Vol. CS-234   X-RAY ASTRONOMY 2000
eds. R. Giacconi, S. Serio, and L. Stella
ISBN: 1-58381-071-4

Vol. CS-235   SCIENCE WITH THE ATACAMA LARGE MILLIMETER ARRAY (ALMA)
ed. Alwyn Wootten
ISBN: 1-58381-072-2

Vol. CS-236   ADVANCED SOLAR POLARIMETRY: THEORY, OBSERVATION, AND
INSTRUMENTATION, The 20th Sacramento Peak Summer Workshop
ed. M. Sigwarth
ISBN: 1-58381-073-0

Vol. CS-237   GRAVITATIONAL LENSING: RECENT PROGRESS AND FUTURE GOALS
eds. Tereasa G. Brainerd and Christopher S. Kochanek
ISBN: 1-58381-074-9

Vol. CS-238   ASTRONOMICAL DATA ANALYSIS SOFTWARE AND SYSTEMS X
eds. F. R. Harnden, Jr., Francis A. Primini, and Harry E. Payne
ISBN: 1-58381-075-7

Vol. CS-239   MICROLENSING 2000: A NEW ERA OF MICROLENSING ASTROPHYSICS
ed. John Menzies and Penny D. Sackett
ISBN: 1-58381-076-5

INTERNATIONAL ASTRONOMICAL UNION (IAU) VOLUMES
Published by the Astronomical Society of the Pacific

PUBLISHED: 1999

Vol. No. 190   NEW VIEWS OF THE MAGELLANIC CLOUDS
eds. You-Hua Chu, Nicholas B. Suntzeff, James E. Hesser, and David A. Bohlender
ISBN: 1-58381-021-8

Vol. No. 191   ASYMPTOTIC GIANT BRANCH STARS
eds. T. Le Bertre, A. Lèbre, and C. Waelkens
ISBN: 1-886733-90-2

Vol. No. 192   THE STELLAR CONTENT OF LOCAL GROUP GALAXIES
eds. Patricia Whitelock and Russell Cannon
ISBN: 1-886733-82-1

Vol. No. 193   WOLF-RAYET PHENOMENA IN MASSIVE STARS AND STARBURST GALAXIES
eds. Karel A. van der Hucht, Gloria Koenigsberger, and Philippe R. J. Eenens
ISBN: 1-58381-004-8

Vol. No. 194   ACTIVE GALACTIC NUCLEI AND RELATED PHENOMENA
eds. Yervant Terzian, Daniel Weedman, and Edward Khachikian
ISBN: 1-58381-008-0

PUBLISHED: 2000

Vol. XXIVA   TRANSACTIONS OF THE INTERNATIONAL ASTRONOMICAL UNION
REPORTS ON ASTRONOMY 1996-1999
ed. Johannes Andersen
ISBN: 1-58381-035-8

Vol. No. 195   HIGHLY ENERGETIC PHYSICAL PROCESSES AND MECHANISMS FOR EMISSION FROM ASTROPHYSICAL PLASMAS
eds. P. C. H. Martens, S. Tsuruta, and M. A. Weber
ISBN: 1-58381-038-2

Vol. No. 197   ASTROCHEMISTRY: FROM MOLECULAR CLOUDS TO PLANETARY SYSTEMS
eds. Y. C. Minh and E. F. van Dishoeck
ISBN: 1-58381-034-X

Vol. No. 198   THE LIGHT ELEMENTS AND THEIR EVOLUTION
eds. L. da Silva, M. Spite, and J. R. de Medeiros
ISBN: 1-58381-048-X

PUBLISHED: 2001

IAU SPS   ASTRONOMY FOR DEVELOPING COUNTRIES
Special Session of the XXIV General Assembly of the IAU
ed. Alan H. Batten
ISBN: 1-58381-067-6

Vol. No. 196   PRESERVING THE ASTRONOMICAL SKY
eds. R. J. Cohen and W. T. Sullivan, III
ISBN: 1-58381-078-1

Vol. No. 200   THE FORMATION OF BINARY STARS
eds. Hans Zinnecker and Robert D. Mathieu
ISBN: 1-58381-068-4

Vol. No. 203   RECENT INSIGHTS INTO THE PHYSICS OF THE SUN AND HELIOSPHERE: HIGHLIGHTS FROM SOHO AND OTHER SPACE MISSIONS
eds. Pål Brekke, Bernhard Fleck, and Joseph B. Gurman
ISBN: 1-58381-069-2

INTERNATIONAL ASTRONOMICAL UNION (IAU) VOLUMES
Published by the Astronomical Society of the Pacific

PUBLISHED: 2001

Vol. No. 204   THE EXTRAGALACTIC INFRARED BACKGROUND AND ITS COSMOLOGICAL
IMPLICATIONS
eds. Martin Harwit and Michael G. Hauser
ISBN: 1-58381-062-5

Vol. No. 205   GALAXIES AND THEIR CONSTITUENTS
AT THE HIGHEST ANGULAR RESOLUTIONS
eds. Richard T. Schilizzi, Stuart N. Vogel, Francesco Paresce, and Martin S. Elvis
ISBN: 1-58381-066-8

Vol. XXIVB   TRANSACTIONS OF THE INTERNATIONAL ASTRONOMICAL UNION
REPORTS ON ASTRONOMY
ed. Hans Rickman
ISBN: 1-58381-087-0

Complete lists of proceedings of past IAU Meetings are maintained at the
IAU Web site at the URL: http://www.iau.org/publicat.html

Volumes 32 - 189 in the IAU Symposia Series may be ordered from:

Kluwer Academic Publishers
P. O. Box 117
NL 3300 AA Dordrecht
The Netherlands

Kluwer@wKap.com

All book orders or inquiries concerning ASP or IAU volumes listed should be directed to the:

The Astronomical Society of the Pacific Conference Series
390 Ashton Avenue
San Francisco CA 94112-1722 USA

Phone: 415-337-2126
Fax: 415-337-5205

E-mail: catalog@astrosociety.org
Web Site: http://www.astrosociety.org